POVERTY AND DEVELOPMENT IN THE 1990s

Edited by
Tim Allen and Alan Thomas
for an Open University Course Team

OXFORD UNIVERSITY PRESS

in association with

The Open
University

1992

This book has been printed on paper produced from pulps bleached without use of chlorine gases and produced in Sweden from wood from continuously farmed forests. The paper mill concerned, Papyrus Nymölla AB, is producing bleached pulp in which dioxin contaminants do not occur.

Published in the United Kingdom by Oxford University Press, Oxford
in association with
The Open University, Milton Keynes

Oxford University Press, Walton Street, Oxford OX2 6DP

Oxford New York Toronto
Delhi Bombay Calcutta Madras Karachi
Petaling Jaya Singapore Hong Kong Tokyo
Nairobi Dar es Salaam Cape Town
Melbourne Auckland

and associated companies in
Berlin Ibadan

Oxford is a trade mark of Oxford University Press

The Open University, Walton Hall, Milton Keynes MK7 6AA

First published in the United Kingdom 1992

British Library Cataloguing in Publication Data

Data available

ISBN 0–19–877330–7
ISBN 0–19–877331–5 (Pbk)

Library of Congress Cataloging in Publication Data

Data available

ISBN 0–19–877330–7
ISBN 0–19–877331–5 (Pbk)

Edited, designed and typeset by The Open University

Printed in Great Britain by the Alden Press, Oxford

CONTENTS

THE OPEN UNIVERSITY
U208 *THIRD WORLD DEVELOPMENT*
COURSE TEAM AND AUTHORS

Tom Hewitt, Lecturer in Development Studies, The Open University (Course Team Chair)

Ben Crow, Lecturer in Development Studies, The Open University (Course Team Chair)

Tim Allen, Lecturer in Development Studies, The Open University

Carolyn Baxter, Course Manager, The Open University

Henry Bernstein, Senior Lecturer in Agricultural and Rural Development, Institute of Development Policy and Management, University of Manchester

Krishna Bharadwaj, Professor, Centre for Economic Studies and Development, Jawaharlal Nehru University, New Delhi, India

Suzanne Brown, Course Manager, The Open University

Janet Bujra, Lecturer in Sociology, Department of Social and Economic Studies, University of Bradford

Angus Calder, Staff Tutor in Arts, The Open University, Edinburgh

Kate Crehan, Associate Professor, New School for Social Research, New York, USA

Sue Dobson, Graphic Artist, The Open University

Harry Dodd, Print Production Controller, The Open University

Kath Doggett, Project Control, The Open University

Joshua Doriye, Professor, Institute of Finance and Management, Dar es Salaam, Tanzania

Chris Edwards, Senior Lecturer in Economics, School of Development Studies, University of East Anglia

Diane Elson, Lecturer in Development Economics, University of Manchester (Part Assessor)

Sheila Farrant, Tutor Counsellor and Assistant Staff Tutor, The Open University, Cambridge (Course Reader)

Jo Field, Project Control, The Open University

Jayati Ghosh, Associate Professor, Centre for Economic Studies and Development, Jawaharlal Nehru University, New Delhi, India

Heather Gibson, Lecturer in Economics, University of Kent at Canterbury

Angela Grunsell, Oxfam Primary Education, London (Course Reader)

Liz Gunner, Lecturer in Literature, Languages and Cultures of Africa, School of Oriental and African Studies, London (Course Reader)

Garry Hammond, Senior Editor, The Open University

Barbara Harriss, Lecturer in Agricultural Economics and Governing Body Fellow, Wolfson College, University of Oxford

John Harriss, Director, Centre for Development Studies, London School of Economics (Part Assessor)

Pamela Higgins, Graphic Designer, The Open University

Caryl Hunter-Brown, Liaison Librarian, The Open University

Gillian Iossif, Lecturer in Statistics, The Open University

Rhys Jenkins, Reader in Economics, School of Development Studies, University of East Anglia

Hazel Johnson, Lecturer in Development Studies, The Open University

Sabrina Kassam, Research Assistant, The Open University

Andrew Kilmister, Lecturer in Economics, School of Business Studies, Oxford Polytechnic

Patti Langton, Producer, BBC

Christina Lay, Course Manager, The Open University

Anthony McGrew, Lecturer in Government, Social Science Faculty, The Open University

Maureen Mackintosh, Reader in Economics, Kingston Polytechnic

Mahmood Mamdani, Professor, Centre for Basic Research, Kampala, Uganda

Charlotte Martin, Teacher and Open University Tutor (Course Reader)

Mahmood Messkoub, Lecturer in Economics, University of Leeds

Richard Middleton, Staff Tutor in Arts, The Open University, Newcastle

Alistair Morgan, Lecturer in Institute of Educational Technology, The Open University

Eleanor Morris, Producer, BBC

Ray Munns, Cartographer, The Open University

Kathy Newman, Secretary, The Open University

Debbie Payne, Secretary, The Open University

Ruth Pearson, Lecturer in Economics, School of Development Studies, University of East Anglia

Richard Pindar, Training consultant, Sheffield (Course Reader)

David Potter, Professor of Government, The Open University

Janice Robertson, Editor, The Open University

Carol Russell, Editor, The Open University

Vivian von Schelling, Lecturer in Development Studies, Polytechnic of East London

Gita Sen, Fellow (Professor), Centre for Development Studies, Kerala, India

Meg Sheffield, Senior Producer, BBC

Paul Smith, Lecturer in Environmental Studies, The Open University

Ines Smyth, Senior Lecturer, Institute of Social Studies, The Hague, Netherlands; Research Associate, Department of Applied Social Studies and Social Research, University of Oxford

Hilary Standing, Lecturer in Social Anthropology, School of African and Asian Studies, University of Sussex (Part Assessor)

John Taylor, Head of Centre for Chinese Studies, South Bank Polytechnic (Course Reader)

Alan Thomas, Senior Lecturer in Systems, The Open University

Steven Treagust, Research Assistant, The Open University

Euclid Tsakalotos, Lecturer in Economics, University of Kent at Canterbury

Gordon White, Professorial Fellow, Institute of Development Studies, University of Sussex

David Wield, Senior Lecturer in Technology Policy, The Open University

Gordon Wilson, Staff Tutor in Technology, The Open University, Leeds

Philip Woodhouse, Lecturer in Agricultural and Rural Development, Institute of Development Policy and Management, University of Manchester

Peter Worsley, Emeritus Professor, University of Manchester (External Assessor)

Marc Wuyts, Professor of Applied Quantitative Economics, Institute of Social Studies, The Hague, Netherlands

The Course Team would like to acknowledge the financial support of Oxfam and the European Community in the preparation of U208 *Third World Development*.

INTRODUCTION

ALAN THOMAS

Figure 1   Ugandan children helping to build their own school.

This book is about development in what has become known as the 'Third World', although the term 'Third World' is not used in the title of the book; this is because it may soon, as will be explained, have outlived its usefulness. In practice, 'Third World' has come to be a synonym for the poorer, southern, less developed countries of the world as a group. Poverty is an age-old concern, but the term 'development', like 'Third World', was hardly in use before the last 40 years. Alternative meanings of 'development'

are hotly contested and indeed the whole terrain of debate around development is changing extremely fast and in very uncertain ways, to an extent not foreseen even a few years ago.

So a book introducing poverty and development *in the 1990s* is a great challenge. In many ways, extrapolating from the experience of the past 40 years is likely to be a very poor guide to the next ten. But the past is all we have to go on. We have to find modes of analysis that will help in understanding and assessing future experience in different parts of the world, however new and unexpected they may be.

During the decade of the eighties there were a great many changes in how development issues were perceived and how world poverty was tackled. New issues came to the forefront, notably the environment, international debt, and the question of gender relations. Neo-liberalism, with its emphasis on market mechanisms, became, at least for the time being, the dominant way of thinking about development in terms of development theory and in terms of influence on policy.

Major changes in the world at large also influence the position of the Third World and thinking about development. At the beginning of the 1990s the pace of world change seemed to be greater than ever. The 'end of the Cold War' brought so-called democratization in Eastern Europe, including the reunification of Germany in October 1990, and continued changes in the Soviet Union. The Gulf War which followed the crisis precipitated by the Iraqi invasion of Kuwait in August 1990 underlined the fact that the world now apparently had the USA as its sole superpower. We have hardly had time to begin to analyse the implications of these changes for development, and there is every reason to suppose that further changes will occur that affect our field of study throughout the 1990s.

At the same time, the questions of the appalling poverty of large numbers of the world's people, with continuing enormous inequities between rich and poor, and the apparent inability of national governments and international agencies to mount a concerted and successful development effort to remedy the situation, remains as potent as ever. Given the increasing inequality in wealth between different parts of the Third World over the past decade, and the new importance of questions such as environment, debt and gender, I believe it is even more urgent to address such issues.

You may well be motivated to read this book by a wish to answer the very practical question: 'What can be done?' There are of course no easy

prescriptions, not least because of the wide variety of interests (often conflicting) at stake, and the great range of agencies that might potentially be involved in 'doing something' about any one of the varied issues that make up the field of development.

By whom is development being done? To whom? These are also potent questions. They should particularly be asked when 'solutions' are put forward that start 'We should…' without making it clear who 'We' are and what interests 'We' represent. In fact, rather than pursue the impossible aim of giving 'solutions' to development 'problems', the overall aim of this book is to provide you with means to begin *analysing* and *assessing answers* to the question 'What can be done about world poverty and development?'

By the end of the book, the question will have been raised as to how far development in the 1990s is likely to involve new and different issues from those thought of previously as most important. To put it another way, is the Third World and its development about to be transformed by the impact of recent events and world changes? Or, is the current highly turbulent state of affairs simply a transition to a new phase of development in which more or less the same concerns, the same theoretical frameworks, the same analytical tools, will continue?

Q What is the 'Third World' and does it still mean something in the 1990s?

Q Does 'development' mean something new in the 1990s?

In this Introduction we take a first closer look at these questions before I explain how the rest of the book is arranged and give some guidance on how to read it.

1 The 'Third World' from the 1950s to the 1990s

The term 'Third World' came into use in the postwar period of the Cold War, at the same time as the growth of international institutions surrounding the United Nations. It was originally a political or ideological concept, roughly denoting the search for a different approach from either capitalist ('First World') or communist ('Second World') – not necessarily a *middle* way, but certainly a distinctive, positive force.

This idea emerged first in France. There and elsewhere the independent Left were seeking a new form of democratic participation; neither the capitalist parties nor the organized communist parties on the Eastern European model provided for the sort of direct democracy they were after. In 1956 the Suez crisis provoked great disillusionment with the capitalist parties, but the Soviet invasion of Hungary in the same year generated a similar disillusion with Eastern European state socialism. This double crisis also gave Nasser, of Egypt, and Tito, of Yugoslavia, independent stature, and with India's Nehru

they became the driving force behind the non-aligned movement (Box 1).

In the words of Peter Worsley, author of *The Third World* (first published in 1964):

> "What the Third World originally was, then, is clear: it was the non-aligned world. It was also a world of poor countries. Their poverty was the outcome of a more fundamental identity: that they had all been colonized."
>
> (Worsley, 1979, p.102)

The 'fundamental identity' of having all (or almost all) been colonized allowed for solidarity in anti-colonialism and backing national independence movements, at a time when independence was recent for some and still to be achieved for others.

Box 1   The non-aligned movement

Non-aligned conferences have been held approximately every four years since 1961. The members are simply those states and national independence movements who are invited, and accept an invitation, to the conferences.

A forerunner of the non-aligned conferences was the 'Bandung' conference held in Indonesia in 1955 as an *Afro-Asian* solidarity conference. It included virtually all the independent Asian and African countries of the time, including those like Pakistan and Thailand with clear military and economic allegiance to the West. The non-aligned movement (NAM) came about as they were joined by increasing numbers of newly independent Black African countries, starting with Ghana in 1957, for whom the achievement of independence from colonial rule translated into a vehement desire for freedom from foreign military domination. The 'non-aligned philosophy' which arose emphasized certain positive principles, notably *peaceful co-existence* and *anti-colonialism*.

The main criteria for invitation to the first non-aligned conference, held in Belgrade,

Yugoslavia, in 1961, were given by Willetts (1978, pp.18–19) as:

1   an independent policy based on the coexistence of states with different political and social systems and non-alignment or a trend in favour of such a policy;

2   consistent support to movements for national independence;

3   non-membership of a multilateral military alliance concluded in the context of Great Power conflicts.

Since 1961 the NAM had changed a great deal even before the collapse of Eastern European state socialism and the apparent end of the Cold War in 1989 and 1990. Most notably, at the Third Non-Aligned Conference in Lusaka in 1970, a third 'basic aim' of non-alignment was added to those of peaceful coexistence and anti-colonialism: namely the struggle for *economic independence*. In recent years the NAM has functioned more as an economic pressure group over common problems such as debt and less as a political bloc.

There were also movements in the cultural field towards creating a new, positive identity to counter what was seen as the divisive and alienative character of Westernized industrial culture. Frantz Fanon, for example, who was born in Martinique but spent many years helping in the fight for Algerian independence from France, published several books aimed at countering 'the colonization of the personality'. Fanon's writing epitomizes the hope that something positive would come from the new force represented by the many Black African and other states gaining independence in the 1950s and 1960s. The extract in Box 2 is from *The Wretched of the Earth*, published in 1961 just before Algerian independence.

Worsley, however, in the same 1979 article, went on to point out how, as time went on, divisions tended to break up the rather fragile unity among Third World countries. By 1979, both economic differentiation and political polarization were destroying the original idea of the Third World as the non-aligned world.

In economic terms there were enormous and growing differences between the 'newly industrializing countries' (NICs), including the four East Asian 'dragons' (Taiwan, South Korea, Singapore, Hong Kong) and some others such as Brazil and Mexico, and the continuing poor agrarian countries such as most of those of sub-Saharan Africa. East Asia and East Africa are completely different places – far more so than, say, the United States and parts of Latin America. In political terms, Worsley himself points out that from the beginning 'Third World countries were overwhelmingly a sub-set of capitalist countries'. Nevertheless, by 1979: 'The choice has become, increasingly, polarized between capitalism and some variant of communism. In the wake of the US defeat [in Vietnam], a few countries have crossed into the other camp. But the choice has been between the two camps, not in some 'Third' direction' (Worsley, 1979, p.108).

Throughout the 1980s, these divisions if anything became even more pronounced. Then at

---

Box 2   **Extract from *The Wretched of the Earth***

The Third World today faces Europe like a colossal mass whose aim should be to try to resolve the problems to which Europe has not been able to find the answers.

But let us be clear: what matters is to stop talking about output, and intensification, and the rhythm of work.

No, there is no question of a return to Nature. It is simply a very concrete question of not dragging men to mutilation, of not imposing upon the brain rhythms which very quickly obliterate it and wreck it. The pretext of catching up must not be used to push a man around, to tear him away from himself or his privacy, to break and kill him...

So, Comrades, let us not pay tribute to Europe by creating states, institutions and societies which draw their inspiration from her.

Humanity is waiting for something from us other than such an imitation, which would be almost an obscene caricature.

If we want to turn Africa into a new Europe, and America into a new Europe, then let us leave the destiny of our countries to the Europeans. They will know how to do it better than the most gifted among us.

But if we want humanity to advance a step further, if we want to bring it up to a different level than that which Europe has shown it, then we must invent and we must make discoveries...

For Europe, for ourselves and for humanity, comrades, we must turn over a new leaf, we must work out new concepts, we must set foot a new man.

(Fanon, 1961, pp.253–55)

the end of that decade came the sudden changes in Eastern Europe linked to the slogan of 'democratization'. It looked briefly as though there was a possibility of the new form of democratic participation envisaged by those whose ideas of a third way gave rise to the 'Third World' label back in the 1950s. However, it soon became clear that what was happening amounted to the collapse of the Soviet model of state socialism in Europe. One of Worsley's two 'camps' that previously polarized the world was enormously, perhaps fatally, weakened.

The United States in particular saw an outbreak of triumphalism. In the words of Henry Kissinger: 'There has been a war between capitalism and socialism and capitalism has won!' The outcome of the Gulf War seemed to underline American world dominance. American commentators referred to 'the new world order', using phrases such as 'Pax Americana' and 'the coming American century'.

The essay 'The End of History', by US State Department analyst Francis Fukuyama (1989), had already come to epitomize this popular feeling. In it Fukuyama put forward a grand view of history as the working out of struggles between great ideological principles. He caught the mood of the moment in the West by arguing that the fusion of liberal democracy and industrial capitalism now represents the only viable basis for modern human society.

Perhaps the term 'Third World' has lost any useful meaning. Indeed, a book was published in 1986 called *The End of the Third World* (Harris, 1986), thus foreshadowing Fukuyama's phraseology. There are three reasons behind this argument:

1 As pointed out above, the 'Third World' seems more than ever disparate rather than representing any kind of unity.

2 The logic of trying to find a third alternative to two dominant models seems to have disappeared, since one of the two models has demonstrably failed, at least in Europe.

3 If its political meaning is lost, the term 'Third World' implies mainly a world of poverty, lack of industrialization and so on; the idea of an 'end' to this Third World would express the hope or intention to do away once and for all with such problems.

But there are difficulties with each of these arguments.

1 The first point seems to make the Third World unimportant, when there is already a problem of Western attention being drawn away from the Third World to Eastern Europe. However, even if one discounts the East Asian NICs (which probably do not represent a way forward for the rest of the Third World) those remaining poor countries still contain the majority of the world's population.

2 The second point ignores the fact that China and some other smaller Third World states such as Cuba continue with 'socialist' development models. It also implies that distinctive Third World concerns have gone away, whereas in fact they are very much still there.

3 Looking at the problems of the Third World shows the continuing need to ask 'Is there an alternative?' As Galbraith (1990) has argued in a reply to Fukuyama, a concerted attack on poverty cannot be mounted within a pure capitalist framework. In a free market system, public resources are not likely to be mobilized in support of policies aimed directly at improving health, providing employment or protecting the environments which enable people to make their living. There are strong grounds, which come from the Third World itself and its concerns, for continuing to seek a new alternative (or at least a strong modification) to capitalist industrialization.

All this implies either dropping the term 'Third World' or looking for a different kind of definition. The *Third World Guide*, published annually by a group of independent journalists in Latin America, uses the following definition, by the Egyptian economist Ismail-Sabri Abdalla:

> "All those nations which, during the process of formation of the existing world order, did

not become rich and industrialized...A historical perspective is essential to understand what is the Third World, because by definition it is the periphery of the system produced by the expansion of world capital."

(quoted in *Guia del Tercer Mundo*, 1981, p.6)

This definition combines recognition of these human issues of poverty with an emphasis on common historical explanations. It relates the Third World to the formation of the international capitalist system (a process that occurred over several centuries of European colonialist domination of the world – see Chapters 8–10 of this book) rather than to the polarized post-war world political order of the 'Cold War'. Thus it is as valid as ever with the end of that Cold War.

Another possibility is to continue to emphasize the concerns of the poor countries but to use different language. One can talk in terms of 'less developed countries' (LDCs) and 'advanced industrial countries' (AICs). However, this fails to emphasize the common ground and the scope for political and economic solidarity that arises from being in a common position in relation to global capital. The idea of 'the South' is one that has gained ground recently, and with it the notion of the 'North–South divide' placing industrialized countries, both Western capitalist countries and Eastern bloc state socialist countries, on one side and the non-industrialized countries (along with the NICs) on the other. Among several new groups formed has been the South Commission, which published *The Challenge to the South* (South Commission, 1990).

In this book, several authors continue to use 'Third World' while others use different language to refer to the poorer countries of the world as a group. Note, however, that whatever language is chosen and whatever definition is used it must contain an implicit view of the world. Abdalla's view, for example, is strongly critical about the effects of the world capitalist system on Third World countries and negative about those countries' future prospects under the same system.

2 Changing views on development

'Development' is a positive word that is almost synonymous with 'progress'. Although it may entail disruption of established patterns of living, over the long term it implies increased living standards, improved health and well-being for all, and the achievement of whatever is regarded as a general good for society at large.

In this book we are primarily concerned with development at the level of societies rather than individuals or even localities. What people learn from their experiences may be regarded as 'personal development' and particular building projects may be called 'developments', but these are not the focus for considering the development (or lack of development) of the Third World. This is not to say that such aspects are unimportant. One of the most influential attempts at defining what is meant by development is based on the idea of creating the conditions for 'the realization of the potential of human personality' (Seers, 1969; see Chapter 6 of this book). And development as building is an important idea when considering *how* development occurs or may be brought about. The relationship between development at local levels and at national or societal levels also brings in the idea of equity between various localities or between different social groups or classes.

Already we can see that the very word 'development', seemingly an idea of which everyone must approve, hides a number of debates. For example, what aspects should be included when considering development, and how should it be measured? Is it primarily an economic concept? or should social aspects be of equal or even of greater importance? Should ideals such as equity, political participation and so on be included in a *definition* of development – or regarded as additional desirable elements which may actually be in conflict with the achievement of development itself?

In trying to answer the question whether development means the same thing as it has done and whether the meaning will remain the same throughout the 1990s, we have to realize that there have always been such debates – the question is whether what is debated is changing. We have already noted that certain issues are being given increased importance, and this is reflected in the way the meaning of development is debated. Thus, debates over the environment, for example, have given rise to the concept of 'sustainable development' (see Chapter 5), though there is as yet no consensus on how to relate this to conventional definitions of development. There have also been moves to include specific mention of the impact of changes on women in definitions of development, and more generally to analyse development in terms of gender relations (see Chapter 15).

The particular form taken by gender relations in any given social context is one example of the importance of culture. Worsley, in a later book, refers to culture (in the sense of a shared set of values) as 'the missing factor' in many studies of development (Worsley, 1984, p.41). In the present book there is also an attempt to include culture, with a view, as Worsley suggests, to 'examining the interplay between economic and political institutions and the rest of social life' (ibid., p.59). However, although development clearly always has cultural implications, as it has implications for gender relations and for the environment, it is not possible to discuss every aspect at once. Development is intrinsically interdisciplinary. The various chapters of this book are written by different authors with different primary foci; although all are writing outside conventional disciplinary boundaries, you will still find more emphasis on the cultural in some, and less (or none) in others.

However, there is no evidence of any major new change in how development is likely to be viewed in the 1990s. Indeed, despite all the debates, there is some agreement on two points. First, tackling poverty is of basic importance. The World Bank's *World Development Report 1990* comes close to a definition of development in such terms with its view that 'Reducing poverty is the fundamental objective of economic development' (p.24). Second, development is a multi-faceted process with political and social as well as economic aspects.

So far we have looked at what development is in terms of how to recognize whether development has taken place. The question of *how* development occurs is equally important. Development can be seen in two rather different ways: (1) as an *historical process of social change* in which societies are transformed over long periods; and (2) as consisting of *deliberate efforts aimed at progress* on the part of various agencies, including governments, all kinds of organizations and social movements. However, as an historical process development is certainly not necessarily positive; for their part, development efforts do not all succeed.

The two ways of looking at how development takes place are of course related. The idea of development as historical social change does not negate the importance of 'doing development'. Historical processes incorporate millions of deliberate actions. Conversely, one's view of what efforts are likely to work is bound to be coloured by one's view of history. For example, Abdalla's definition of the Third World, quoted above, embodied the view that the historical process which resulted in the development of the industrialized world was the same process in which the Third World did not become developed.

We already noted that this definition of Abdalla's implies a particular view of the world – one which has in fact informed much thinking on how to achieve development. Put simply, in this kind of view, Western capitalist industrialization created structures in which Third World economies were dependent and which tended to lead to and maintain underdevelopment.

This is one version of a 'dependency view', which is in turn one of a number of views which may be grouped under the heading of *structuralism*. In general, such views are concerned with underlying social and economic structures and see development as involving changes in these structures.

An essentially different view is that of *neo-liberalism*, which has its emphasis on the importance of market relations, and which has been in the ascendancy, not only with the main international agencies such as the World Bank but also with increasing numbers of Third World governments, and others, for the past decade. These two competing approaches entail different views of history, of what is meant by the Third World, of how to achieve development and who should be the agents of it.

Although *neo-liberalism* and *structuralism* both incorporate a range of variations, there are other approaches to development which fall outside these two labels. In Chapter 6 both these terms will be explained in more detail, as well as other approaches, notably *populism* and *interventionism*. The point to realize for the moment is that in an area of debate such as development, definitions and explanations are not cut and dried. They carry implications about one's view of the world that can lead into wide-ranging political, moral and theoretical disputes.

3 The structure of this book and how to read it

The book is arranged in three parts. The first part presents the idea of the 'Third World' as 'a world of problems'. Chapters 1 to 5 deal in turn with hunger and famine, diseases, unemployment, population, and environmental degradation – each of which relates to an aspect of poverty and may be seen as one of the 'problems' endemic in the 'Third World'. Then Chapter 6 discusses what is meant by 'development' in relation to such concerns and introduces alternative views on how development occurs in the context of global capitalism and a world of nation states.

The second part, on the making of the Third World, analyses the historical context from which current concerns arose. Chapter 7 looks at pre-capitalist diversity; Chapters 8 to 10 consider European colonialism in relation to the development of capitalism on a world scale; and Chapters 11 and 12 discuss post-colonial influences that have also shaped today's Third World.

The third part explains further some current issues and concepts useful for understanding development in the 1990s. Chapter 13 sets the scene by relating the Third World to its position in the 'new' global order and Chapter 14 relates development to the current trend towards the 'democratization' of Third World states. Chapters 15, 16 and 17 underline the necessity of including consideration, respectively, of technology, gender relations, and wider aspects of culture in any assessment of development. Particular cases are used to emphasize the importance of class and ethnicity (Chapter 18) and the politics of cultural expression (Chapter 19). The concluding chapter returns to the discussion of what general options are feasible for development in the 1990s, and in particular whether any such options are at all viable unless based, like previously dominant models of development, on large-scale industrialization.

By the end of the book you should be in a better position to begin to analyse or at least to ask the right kind of questions about any given example of 'development' in the coming years.

Throughout the book various devices are used to help you in your reading and study, most of which have been introduced already. Within the first introductory paragraphs of each chapter there is an emphasized Question or Questions, to which the rest of the chapter should provide some sort of answer. Each chapter ends with a Summary or Conclusion which you should be able to link back to the chapter Question(s). Key concepts are emphasized in bold lettering and have a boxed summary discussion of their meaning nearby. Numbered Boxes are used for long examples, cases, illustrations or specific explanations that can be taken separately from the main flow of the argument. Photographs and cartoons may be used not just to illustrate the text but to give additional examples, so you should look carefully at the captions. Tables are used to provide data to back up arguments in the text; note that different authors may have used different sources so that not all the tables are precisely compatible with each other.

Concepts may be mentioned without definition that are discussed more fully later. (This has happened quite frequently in this Introduction, for instance.) You should find that the Index emphasizes the place in the text where a concept is explained or where the boxed summary discussion of a concept occurs.

Apart from the Index, other points of reference are the map at the end of this Introduction and the List of Acronyms, Abbreviations and Organizations at the end of the book.

Sometimes the text asks you to make some calculations on a table, look carefully for certain features of an argument in an extract, consider your own views on a subject before going on, etc. Of course, you can simply read on and treat such questions as rhetorical – but in general you will gain more understanding through stopping briefly to try to answer the question posed.

Finally, please note that although this is a textbook rather than a collection of academic articles, and the editors have endeavoured to arrange it so that conceptual material is explained and built on throughout the book, it is written by a variety of authors with differing expertise as well as differing views. Not all the disagreements, points of overlap, possible cross-references, etc. have been pointed out, though I hope there are not too many places where one author actually contradicts another! You should find it useful to try to make as many links for yourself as you can between the arguments in different chapters, so that you get as rounded a picture as possible of the complexity of development in the 1990s.

Countries and major cities of the world

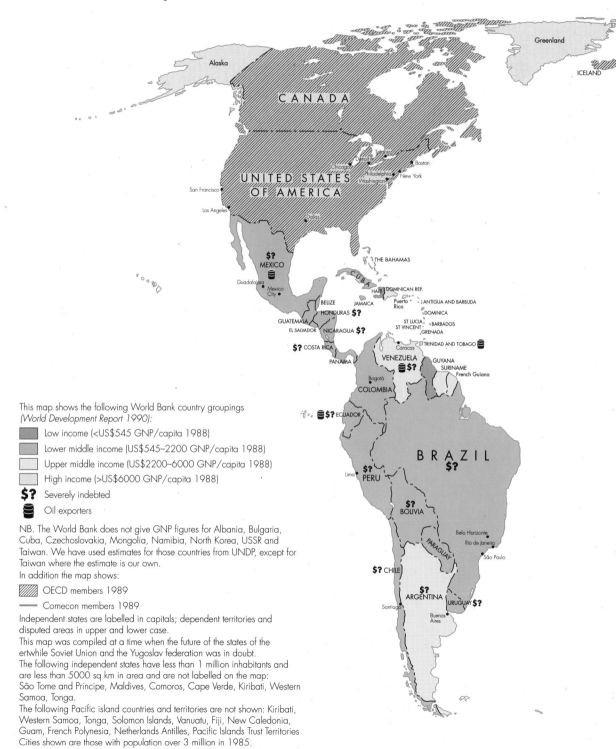

This map shows the following World Bank country groupings
(World Development Report 1990):

	Low income (<US$545 GNP/capita 1988)
	Lower middle income (US$545–2200 GNP/capita 1988)
	Upper middle income (US$2200–6000 GNP/capita 1988)
	High income (>US$6000 GNP/capita 1988)

$? Severely indebted

🛢 Oil exporters

NB. The World Bank does not give GNP figures for Albania, Bulgaria,
Cuba, Czechoslovakia, Mongolia, Namibia, North Korea, USSR and
Taiwan. We have used estimates for those countries from UNDP, except for
Taiwan where the estimate is our own.

In addition the map shows:

▨ OECD members 1989

— Comecon members 1989

Independent states are labelled in capitals; dependent territories and
disputed areas in upper and lower case.

This map was compiled at a time when the future of the states of the
erstwhile Soviet Union and the Yugoslav federation was in doubt.

The following independent states have less than 1 million inhabitants and
are less than 5000 sq km in area and are not labelled on the map:
São Tome and Principe, Maldives, Comoros, Cape Verde, Kiribati, Western
Samoa, Tonga.

The following Pacific island countries and territories are not shown: Kiribati,
Western Samoa, Tonga, Solomon Islands, Vanuatu, Fiji, New Caledonia,
Guam, French Polynesia, Netherlands Antilles, Pacific Islands Trust Territories

Cities shown are those with population over 3 million in 1985.

A WORLD OF PROBLEMS?

1

UNDERSTANDING FAMINE AND HUNGER

BEN CROW

"Nourishment is fundamental. The story of human history, reduced to essentials, revolves around the basic requirements for life."

(Rotberg, 1983)

The question of hunger is one of the most pressing problems that must be solved in the developing world. This is why the first chapter of this book aims to offer some understanding of how hunger arises. It begins by introducing some analytical approaches to the problem. Some of the ideas are difficult, and many of the issues raised are controversial, with disagreements between alternative schools of thought.

Q What are the processes that can lead to famine and hunger?

There are few simple and general answers to this question which are not misleading. However, it is satisfying to report that there is the beginning of consensus on some approaches to understanding the problem. These suggest how hunger could be reduced and how famine could be prevented. The work of economist Amartya Sen has been influential, and this chapter is indebted to two of his books (Sen, 1981; Drèze & Sen, 1989). The policy implications may not be straightforward, socially, historically and politically, and discussion of implementation is beyond the scope of this chapter. The conclusion of the chapter is, nevertheless, one of hope.

1.1 Famine and hunger – some starting points

An important distinction

We need to start by distinguishing between famine and chronic undernutrition. During the 1970s and 1980s, people in the industrialized world were made aware of famine, particularly in Africa, through television and newspaper reports of starvation. These were accounts of the most acute manifestation of hunger. **Famine** is a crisis in which starvation from insufficient intake of food is associated with sharply increased death rates.

Famine is, however, not the only form of hunger or undernutrition. In many parts of the world where famine has not occurred in recent years, sustained nutritional deprivation is, nevertheless, experienced by a significant proportion of the population. This long-term condition of **chronic hunger** is rarely given international focus but it may kill more people globally than the acute crisis of famine does.

> **Famine:** Acute starvation associated with a sharp increase in mortality.
>
> **Chronic hunger:** Sustained nutritional deprivation.

The terms famine and hunger have broader meanings in other contexts. An early English dictionary defines famine as 'a scarcity of food' (Johnson, 1810), and the term is sometimes also used as a general term for a shortage. The same dictionary explains hunger as 'desire of food [and] the pain felt from fasting'. These general usages are inadequate for analysis because they imply connections (food shortage in the case of famine, and desire or pain in the definition of hunger) which may not further our understanding of the causes of famine and hunger. Later in this chapter, I will explain why a general notion of food shortage is unhelpful as a way of understanding famine.

You don't have to promise them anything, sir. This is not what has been declared famine area — it is further up

Figure 1.1

Famine is an emotive word with powerful connotations. Thus it is important to use specific or technical definitions, such as those summarized in the keyword boxes, as a basis for analysis. In particular, the distinction between famine and chronic hunger is useful because the causes of the two appear to be different, and because responses to the crisis of famine and to the sustained deprivation of chronic hunger may also be different.

In fact, famine and chronic undernutrition are concentrated in different areas. It has been noted that in Africa famines tend to occur in drier areas and chronic undernutrition tends to be concentrated in areas with more rainfall. Before we draw any rapid conclusion that there is a direct relation between climate and the form of hunger, however, we should examine the cases of China and India. These two countries, encompassing almost half the world's population and wide ranges of climatic conditions, demonstrate contrasting experiences with famine and chronic undernutrition (Box 1.1). We will return later in the chapter to the lessons of success and failure which can be drawn.

The global scale of hunger

To put these events in context we can examine some estimates of the global scale of hunger. One recent estimate (*The Hunger Report*, Chen, 1990) suggests that 31% of the world's population is undernourished. Another estimate (World Bank, 1986) indicates that this figure may overstate the global level of hunger by a factor of four. In order to evaluate these estimates it is

---

Box 1.1   The India–China contrast

In India, famine appears to have been avoided in the last three decades but chronic hunger continues at a high level. Social changes introduced in China after the revolution in 1949 significantly reduced chronic hunger, but the largest famine of the twentieth century took place in China in the years of 'The Great Leap Forward', 1958–61. It is estimated that 15–25 million people may have died in this famine.

Chronic hunger in India may nevertheless raise death rates to the extent that two to three million more die each year than would have died if nutritional levels were adequate. Thus, despite Indian success in preventing famine, failure to resolve chronic hunger in India leads to increased deaths equivalent to the great Chinese famine every seven to eight years (Drèze & Sen, 1989).

necessary to understand something about how they have been made.

There are difficulties in finding out how many people in the world suffer from famine or hunger. These difficulties are of three kinds.

1   There are considerable difficulties in identifying undernutrition in one individual. This is primarily because there is no single medical test which will unequivocally indicate the level of nutrition or undernutrition.

2   This difficulty is compounded by uncertainty about the minimum dietary energy supply below which life and activity may be jeopardized. Box 1.2 describes the main approaches to measurement and current estimates of minimum required nutrition.

3   Because there is no one test of undernutrition, many indirect measures are used to estimate the level of nutrition for large groups of people and whole national populations. All are subject to reservations, particularly in the case of global estimates of hunger.

Overall estimates of global hunger are generally made in one of two ways.

1   *Income measures* of hunger estimate the average income of poor households and compare that income with the cost of the food necessary to provide a minimum level of nutrition.

2   Measures of *national food availability* compare average availability of food per head with a minimum necessary supply. An estimate is made of the overall availability of food in a country (production plus imports and minus exports). This total quantity is divided by the population of the country to obtain a figure of food availability per head. This figure, generally calculated in weight of grain equivalent, is converted into kilocalories of dietary energy supply and compared with a minimum level of kilocalories (see Box 1.2).

Box 1.2   Measuring undernutrition

It is surprisingly difficult to measure undernutrition. There are two general approaches at the individual level, focusing on nutritional or medical indications.

1   Measures of *nutritional intake* estimate how much food a person is eating and assess the adequacy of that quantity of dietary energy supply. In practice, however, it is difficult and expensive to measure what people eat. As a result it is only feasible to measure the nutritional intake of small numbers of people. In addition, doubts arise because the measurement and constant observation required to weigh each item of food throughout the day is intrusive and likely to influence eating behaviour. One nutritionist writes that measuring nutritional intake is 'the most difficult, expensive and probably the least satisfactory way of identifying malnourished people' (Payne, 1990, p.16).

2   Measures of *nutritional status* use some physiological characteristic to determine whether an individual appears to be undernourished. Undernourished children tend to be small for their age so a comparison of the height of one individual against the average height for a well-nourished child of that age will provide an indication of undernutrition. Similarly, comparison of upper-arm circumference with typical measurements for well-fed children will provide another indication of undernutrition. Some body measurements, including the overall size of an adult, may provide indications of past levels of nutrition (undernutrition in early childhood is associated with smaller body size), whereas other measurements (such as arm circumference) may provide a better guide to current nutrition.

In order to establish undernutrition, levels of nutritional intake are compared with minimum dietary energy requirements, measured in kilocalories of energy. Accepted standards have been revised downwards in recent decades from 2830 kcal/day in 1957, and 2450 kcal/day in 1985, to latest figures of 2200 kcal/day for someone undertaking little activity and a 'survival requirement' of 1550 kcal/day (Payne, 1990, p.15). One reason given for the reduction in accepted standards of dietary energy requirements has been recognition of the small average body size prevailing in many developing countries, even though small body size may itself be a consequence of previous undernutrition.

The estimate contained in *The Hunger Report* says that 1570 million people (31% of world population) had a 'dietary energy supply less than nutritional requirements' in the years 1984–86 (Chen, 1990). Although this estimate uses units related to nutrition it is actually a measure of national food availability. The authors identified 49 countries 'where the total dietary energy supply [that is, overall food availability divided by population] was less than that required for health, growth and productive work'. The figure of 1570 million is the total number of people living in those 49 countries. This is, therefore, an indirect measure of hunger. It takes no account of differences in command over food between rich and poor within one country.

A World Bank report, *Poverty and Hunger*, used an income measure to estimate the level of global hunger in 1985. It produced two estimates, both of them lower than *The Hunger Report* estimate above, but one estimate was double the other. The World Bank estimated that between 340 million and 730 million people in the Third World 'did not have enough income to obtain enough energy from their diet' (World Bank, 1986, p.1). The estimate of 340 million includes those with insufficient income to purchase adequate food to avoid serious health risks and stunted growth in children. The estimate of 730 million is based on the minimum nutrition required to allow an active working life.

The World Bank figures were derived from estimates of household income levels. From aggregate national incomes an estimate was made of the shares going to rich and poor. Then the Bank's researchers estimated the proportion of poor households' incomes spent on food. From this they calculated the level of nutritional energy which such a budget could buy and compared that with a minimum nutritional norm.

The World Bank's 1986 report considered that economic growth had reduced the proportion of undernourished people in the developing world between 1970 and 1980. But, because of population growth, the absolute number of people with inadequate nutrition had increased. The regional concentration of hunger had increased

during this period. Two-thirds of the world's hungry live in South Asia and a fifth in sub-Saharan Africa. In South Asia and sub-Saharan Africa, both the proportion and the absolute number of undernourished people had increased markedly.

Thus, the best estimates of world hunger vary considerably, but even the smallest is a significant proportion of world population. At the least, if we take the World Bank's lower estimate of income requirement not allowing for 'an active working life', there were 340 million people undernourished. Since most people, particularly in the Third World, are required to have an active working life, then the higher figure of 730 million people is a better estimate. The estimate based on overall food availability doubles that figure. As Drèze & Sen note in relation to debates about nutritional requirements: 'The complexity of the relationship between nutritional intakes and nutritional status should not make us lose sight of the fact that the magnitude of uncontroversial deprivation in this hungry world is enormously large' (Drèze & Sen, 1989, pp.41–2).

In order to understand what can lead to famine and hunger we need an approach which transcends the simple assumptions of these global estimates. Because they deal with national averages, they tend to ignore the nutritional circumstances of particular groups in the population, the poor and the vulnerable. The income-based calculations of the World Bank make a stab at overcoming this limitation but still deal in overall national figures. Another limitation of these measures, for an analysis of hunger or famine, is that they do not distinguish how people get access to food. In order to describe how people actually obtain command over the food they need to survive, we need to examine the diversity of livelihoods.

Diversity of livelihoods

One of the striking characteristics of Third World economies is the wide diversity of economic activity and the forms of livelihood arising from them. Advanced, high-technology factories may coexist with agriculture for direct consumption that uses very simple tools, and with the direct use of natural products through hunting and gathering. A generalized or abstracted concept of each person having a cash income is removed from this reality of diverse livelihoods and incomes. Some people do receive a cash income in the form of a regular wage from an employer; others catch fish or tend livestock and either consume or sell the products; many seek whatever employment is available on a daily or a seasonal basis and are paid in cash or food; others engage in the small-scale trade of goods they have made; and many combine some of these occupations with production on the land for consumption and sale.

The simplest distinction among livelihoods is that between people who are *self-employed* and those who are wage labourers. People who are self-employed have some control over the tools (and land in the case of agriculture) required for their production. These people include peasant farmers, pastoralists (living from livestock), fishing communities, small craft producers and petty traders. A third category of livelihoods includes those who are able to be *employers of wage labour*. These differences of livelihood are related to the nature and organization of **production** and **exchange** and are the basis of **social class** distinctions.

Production: A process in which human energy is expended to transform natural products into goods for consumption. It involves interaction between people and nature. The simplest production process has three elements: (1) the work done by the people, i.e. the human labour; (2) the subject of that work, the raw materials of nature and of previous production; and (3) the tools and skills used in the work. Thus, agriculture, or production on the land, requires (a) people to provide labour power; (b) at a minimum, adequately fertile land, rainfall and seeds of the crop to be grown; and (c) a hoe or a plough for tilling the soil and the skills and practices of agriculture.

Exchange: A process that is required (as is specialization) for production to go beyond the simplest levels. Most familiarly, exchange is the process of buying and selling of goods. It also includes the paying of wages and rents or interest. More generally, exchange is the process and the social circuits through which goods are distributed among a population.

Social class: The distinction between different social groups according to the ways in which they make their living, particularly between those who own means of production (land, factories, machines) and those who do not (who sell their labour power to cultivate the land or work in the factories). We will be concerned particularly to distinguish between peasants (those who produce on the land, using family labour, partly for their own consumption and partly for sale), wage labourers (those who sell their labour power to make a living as agricultural and industrial workers), pastoralists (who tend livestock, often nomadically, to produce meat and milk for consumption and exchange), and others.

Different occupation groups, or different social classes, have different kinds of 'income'. For example:

(a) Fishing communities do not get a regular wage. Each person's 'wage' depends on the size of the catch from the river or the sea, the frequency of that catch, how it is divided among the community, how much of the catch is exchanged and at what price.

(b) Agricultural labourers may get a daily wage but the amount of money they have to spend will depend first on whether there is work (which may only be available in some seasons, and not reliably even then) and second on the prevailing wage rate (which may go down if there are many people seeking few jobs). Then, the usefulness of the wage in combating hunger will depend on the amount of food it can buy.

The different characteristics of these livelihoods arise from the organization of production. The process of generalizing from these diverse livelihoods to the idea of an average income is useful when trying to estimate the scale of global hunger, but an understanding of the characteristics of each economic activity, and the reliability of the livelihood it supports, is required for the analysis of hunger or famine.

1.2 Famine vulnerability

"Famine is the closing scene of a drama whose most important and decisive acts have been played out behind closed doors."

(Dessalegn Rahmato, 1987)

The causes of famine are complex. The final crisis frequently has a long and untold history of deterioration behind it, with interacting causes which may be difficult to unravel. There are directions of inquiry which assist our understanding, but one conclusion I hope you will take from this chapter is to beware of simple explanations.

Amartya Sen (1981), in a study of several recent famines, started by finding out which occupation groups or livelihoods had been most affected. This study suggested that there are differences between famines and that an examination of the *command* over food provided by different livelihoods may give some understanding of the vulnerabilities leading to those different famines.

To illustrate the factors associated with famine we can examine one of the worst and most publicized famines of the 1980s, the famine in Ethiopia in 1984 and 1985. We will be looking specifically at events affecting peasants or agriculturalists in Wollo province because there is a careful, scholarly account of the circumstances of famine in that region by Dessalegn Rahmato (1987). I will supplement that account with

factors affecting famine among pastoralists or livestock herders in the same province ten years earlier, and provide a brief contrast with a description of some elements influencing famine in Bangladesh in 1974.

Famine in Wollo 1984–85

Wollo was one of the provinces of Ethiopia hardest hit in 1984–85. Some of the most harrowing images of the 1984–85 Ethiopian famine came from the town of Korem in Wollo, where large numbers of peasants and pastoralists gathered to seek relief in the final stages of the crisis (Figure 1.2).

At this stage, when television pictures portrayed an open plain filled with huddled figures waiting for relief food and medicine, famine victims expressed an image of helplessness. In earlier stages of the crisis, this helplessness had been preceded by feverish activity:

"Neighbours and friends decide to pool their resources the better to withstand the hardship; agreements are reached between relatives or friends to dispose of assets in turns, and to support each other in the meantime; measures are taken to remove livestock to...other areas less exposed to the crisis; arrangements are made to sell livestock to peasants in one's own community or a neighbouring one with the understanding that in the end the sellers will rent the animals for farming purposes; markets both in neighbouring communities and in distant ones, especially those reported to be free from social or ecological stress, are frequently monitored, and the information disseminated widely; distress signals are sent out to relatives living in urban areas or in other provinces..."

(Dessalegn Rahmato, 1987, p.23)

The television pictures from the plains around Korem depicted the last stage of a four-stage crisis: (1) austerity; (2) temporary migration (of which the feverish activity described above was a part); (3) divestment and asset disposal (the selling of possessions); and (4) exhaustion and

Figure 1.2 Korem relief camp, Wollo, 1984.

Figure 1.3 Families leaving home: collecting water to go to Korem relief camp, Wollo, 1984.

dispersal. The final stage of mass migration is a 'collective articulation of grievances' and an attempt to escape from intolerable privation (Figure 1.3).

Dessalegn Rahmato notes that 'the Wollo peasantry is diligent, frugal and highly skilled, and yet this same peasantry has been the victim of all the major famines that have occurred in this country in the last one hundred years.' Why should skilled, hard-working farmers be subject to crisis so often?

The explanation that was widely accepted for this famine is one that is commonly given for many famines: it was a famine caused by drought. This explanation of causation through climatic abnormality, be it flood, drought, frost or cyclone, is probably the explanation of famine most widely accepted. In its simplest version this sort of explanation is sometimes extended to make a distinction between 'man-made' and 'natural' famines. Man-made famines would include those connected with war or displaced populations.

From this distinction an implication is often drawn that 'natural' famines are unavoidable but those caused by human intervention arise from some human failing. However, this distinction can be deeply misleading because:

> "Famine is, by its very nature, a social phenomenon (it involves the inability of large groups of people to establish command over food in the society in which they live), but the forces influencing such occurrences may well include, *inter alia*, developments in physical nature in addition to social processes... it has to be recognized that even when the prime mover in a famine is a natural occurrence such as a flood or a drought, what its impact will be on the population will depend on how society is organized."
>
> (Drèze & Sen, 1989, p.46)

In general, the natural disaster explanation of famine posits a direct sequence of causes starting with climatic abnormality, leading to failure of harvests and a reduction in food available and the amount that people have to eat. However, drought, flood, frost and cyclones occur frequently and persistently all over the world and are only associated with famine on rare occasions. On those occasions the weather is generally only one precipitating factor among several, such as war, social disruption and disturbance of exchange, and these precipitating factors act upon a deeper vulnerability of livelihoods. Almost all recent famines in Africa have been associated with the social disruption of war as well as climatic abnormality (Figure 1.4).

In Wollo, persistent drought (combined with unusual frosts) was one of the factors precipitating the crisis among the peasantry, but the origins of the crisis are to be found in social conditions and economic and political changes unrelated to the drought.

Dessalegn Rahmato writes of 'famine hiding behind the mountains', alluding to the hidden factors contributing to famine vulnerability. He does not claim to understand the full story of the famine among the peasantry in Wollo but he

Figure 1.4   Drought by itself does not necessarily cause famine. The Shawata Project in central Tigray, in the middle of a drought area, harnesses irregular and light rainfall so that the irrigated land is able to yield two good harvests a year. The project was, however, bombed by Ethiopian government forces during the civil war.

identifies three critical factors behind the longstanding and deepening vulnerability of agricultural livelihoods there.

1 *A stagnant form of production*, discouraging innovation and unresponsive to growing population and deteriorating soil conditions. Agriculture in Wollo is the cultivation of cereals and pulses and the keeping of livestock, primarily for direct consumption (subsistence). There is little connection between agriculture and urban economic activity, and little that is produced in agriculture is exchanged. There are few economic activities other than agriculture and few opportunities for employment. Because agriculture is not integrated with other economic activity, and focused on subsistence, there is little opportunity to incorporate new technologies. A low standard of living is socially accepted and there is little support for innovation. 'Peasants in the Northeast region as a whole are, even in normal conditions, and in times free of environmental and social stress, not too far removed from the precipice of starvation and death' (Dessalegn Rahmato, 1987, p.102).

2 *The rise in the external obligations,* in food, money and labour, of the peasantry over the preceding ten years. Like governments in many Third World countries, the government of Ethiopia obtained food for the cities partly by purchasing it at fixed prices from the peasants. Each peasant household had to deliver a predetermined quota of grain. The price given by the government for this quota was substantially below, and for some crops only a third of, the price available in local markets. These government quotas were not lifted during the famine (but were reduced after it). Some peasants were forced to sell livestock in order to buy food to deliver to the government. Taxes were also exacted throughout the famine period. In fact, with cruel irony, the victims of the famine were required to pay a special famine levy. This levy was established to help victims of the famine after the crisis was recognized in October 1984, and it was collected in Wollo as in other parts of Ethiopia. One further burden on the peasantry, contributing to the vulnerability of their economy and their poverty, was the requirement that they contribute labour to communal and state projects. This work included large-scale soil conservation projects, work on the land of those assigned to fight in the civil war, and work on state plantations. This unpaid labour took up at least one day per week. All these obligations had arisen since the change of government in 1974, and they subtracted from the basic necessities of peasant consumption.

3 The third critical factor contributing to the profound vulnerability of peasant livelihoods arose from the preceding two factors. With a dead-end form of production and increasing external obligations, much of the Wollo peasantry was reduced to *acute poverty*. Households had little reserve with which to resist periodic crisis.

These three factors provide a brief summary account of the main causes of famine vulnerability among peasants in one province of Ethiopia in 1984–85. We will turn later to related but distinct factors associated with famine among pastoralists. What should already be clear is that the explanation of famine involves several complex and interacting factors, with the organization of production and exchange at the centre of the picture.

The vulnerability of peasant livelihoods in Wollo has to do with all three elements of production identified earlier (the work of the people, their interaction with nature, and the tools and skills they use). Dessalegn Rahmato's account particularly focuses on the stagnation, or lack of development, in the form of production, its inability to respond to deteriorating natural conditions (environmental stress) and growing population. There is not space in this chapter to explore why this form of subsistence production, and the technical skills and social relationships which sustain it, is not generating change. Chapters 4 and 5 begin to examine the social factors contributing to ecological deterioration and to population change.

If we now turn from production to exchange, much of the feverish activity leading up to the famine was focused on a new involvement with market exchange. As subsistence activity fails, and as a wide range of survival strategies are exhausted, peasants attempt to sell livestock, house materials, tools and cooking utensils, stores of wealth such as jewellery and, by no means least, their ability to work for others, their labour power:

> "Crisis survival brings sudden and dramatic changes in peasants' economic thinking for a brief period. In this phase of the crisis we see the peasant desperately trying to break out of the subsistence system, and plunge into the cash economy and the exchange system. The market now assumes greater importance than before as the peasant becomes aware that survival depends on acquiring cash to purchase food... However, the food deficit peasant is at a disadvantage in these circumstances because what he has to offer [labour power and saleable assets] is not in demand, while what he wishes to acquire, i.e. food, is both relatively scarce and in high demand. On the other hand, the peasant with surplus food and the rural trader stand to gain in the exchange process."

(Dessalegn Rahmato, 1987, pp.167–9)

It is time now to stand back from the particular features of famine in Wollo.

Approaches to analysing famine

There have been two general categories of approach to the understanding of famine in recent years. The first looks at the relationship between population and overall food supply. It asks the question: will there be enough food for this number of people? Food availability in an economy is total production, plus imports and minus exports of food. The 'food availability decline' thesis lies behind ideas that a flood or a drought, or a harvest failure for some other reason, is the direct and simple cause of famine.

A second approach, pioneered by Amartya Sen (1981) is now tending to replace the 'food availability decline' approach because it seems to give greater understanding of what happens during a famine and why people may go hungry. It concentrates on the ways in which individuals and households gain command over food. This

entitlement approach identifies two basic characteristics, **endowments** and **entitlements**, which give access to livelihoods and food:

> "For example, a peasant has his land, labour power and a few other resources which together make up his endowment. Starting from this endowment, he can produce a bundle of food that will be his. Or by selling his labour power, he can get a wage and with that buy commodities including food. Or, he can grow some cash crops and sell them to buy food and other commodities. There are many other possibilities."

(Sen, 1981, p.46)

Endowment: The owned assets and personal capacities which an individual or household can use to establish entitlement to food.

Entitlement: The relationships, established by trade, direct production or sale of labour power, through which an individual or household gains access to food. *Direct entitlement* is access to food gained through own production and consumption. *Exchange entitlement* is that command over food which is achieved by selling labour power in order to buy food. *Trade entitlement* is the sale of produce to buy food.

Despite Sen's use of the pronoun 'he', his ideas apply equally to individual women and men and to households. Endowments are those capacities and owned assets which can be used to establish entitlement to food, and entitlements are forms of livelihood or social relationships which give command over food. Sen further distinguishes different forms of entitlement: direct entitlement through own production; trade entitlement through sale of commodities; exchange entitlement through sale of labour power.

This form of analysis thus rests on ideas about the relations between classes or occupation groups within an economy. Entitlement ideas encompass social relations both of exchange and of production, and they are not directly concerned with levels of food production.

In the case of the Wollo famine among peasant agricultural producers, we can identify the failure of two different kinds of entitlement to food.

There was a failure of direct entitlement, i.e. a failure of direct production for own consumption (subsistence). It is in this aspect of the crisis that drought played an important precipitating role. As we have seen, the explanation of this failure of direct entitlement requires an understanding of the organization of production and its ability to cope with changing conditions.

In the later stages of the crisis, the peasants were also engulfed by the failure of their attempt to establish exchange entitlements. One of the advances in the recent understanding of famine has been a recognition of the way that the operation of market exchange may exacerbate or initiate famine. We will look briefly at instances of this below. For the Wollo peasantry, their 'plunge into the cash economy' could not bring adequate command over food. Many people were seeking paid work from an economy which generated little employment at the best of times. Many households were also trying to sell livestock, tools, utensils, houses and jewellery, and prices were reduced to fractions of those expected in more normal times. Those able to sell assets or find employment found that the relative scarcity of food had pushed prices much higher than normal. These two price movements (prices of saleable assets down, prices of food up) interacted with one another. As a result, the proceeds of the sale of livestock, for example, could buy only a fraction of the food that could have been realized by a similar transaction previously.

If we identify the first of Dessalegn Rahmato's key factors, the stagnant form of production, as the underlying vulnerability leading to the collapse of the peasants' direct entitlement, it is clear that the rise in external obligations contributed to the peasants' 'plunge into the cash economy' and the failure of their exchange entitlement.

Famine among pastoralists in Wollo in 1973

As in the case of peasant agriculture, pastoral production is a form of livelihood which requires considerable skill and environmental knowledge. Pastoral production rests upon the movement of herds of livestock (goats, sheep, camel and cows) to find areas of adequate grazing. Skill is required to predict the availability of grazing, through changing seasons and climatic conditions. Planning and skill are also required to combine the different rates of reproduction of different animals so as to maintain herd size and output of milk, meat and other products.

In most societies where pastoralism still exists, the requirement for access to extensive, uncultivated land is being constrained by the expansion of agriculture and the restrictions of states (enforcing territorial boundaries, for example). So it was in the years preceding the famine in Wollo in 1973.

In the case of the nomadic pastoralists from the Afar community of Wollo, a simple explanation of how they came to be destitute would suggest that the drought killed their livestock, and without animal products they could not survive. This view of a single direct entitlement failure is too simple because several other factors exacerbated their vulnerability and contributed to the crisis.

One factor was the incorporation of key riverside grazing, previously used by the pastoralists during the dry season, into a commercial scheme for growing sugar and cotton for export. The pastoralists' dry weather refuge had been 'crucially curtailed by the alienation of land for commercial agriculture' (Sen, 1981, p.105, citing Flood, 1975). This magnified the impact of the drought.

Secondly, the trade entitlement of the pastoralists fell sharply. Pastoralists trade in animals and other animal products for food and the goods they need, notably grain. In this area they meet about half their nutritional requirements by eating agricultural products obtained through this exchange. Normally, pastoralists can acquire calories more cheaply by selling animals and buying grain than by eating animal products. During the famine the price of grain rose above normal levels and the value of animals and animal products fell below normal levels. For the pastoralist, these two price movements combined to make the net exchange rate (animals for grain) considerably worse. In a later famine, in Harerghe province in southern Ethiopia in 1974, these price changes caused a loss of grain entitlement of 62–72%. In other words, where the sale of a particular sheep, goat or camel had realized, say, 100 kg of maize during normal times, only 28–38 kg would have been received during the famine. Stasistics for this southern famine indicate that this trade entitlement loss was much more significant than the loss of entitlement due to the death of livestock (Figure 1.5).

Famine in Bangladesh in 1974

Famine occurred in Bangladesh in 1974 and chose its victims disproportionately, not from peasants or pastoralists but from among those primarily depending on sale of their labour power in the countryside. Economic activity in the Bangladesh countryside is diverse and highly integrated. Peasant production of rice, wheat, jute and pulses is both for sale and for own consumption. Most peasants, particularly those able to produce on a larger scale, employ others at some points in the agricultural cycle. There is also considerable small-scale economic activity in petty trading, fishing, food processing and other production. However, at least half of all rural households live primarily by selling their labour power to others.

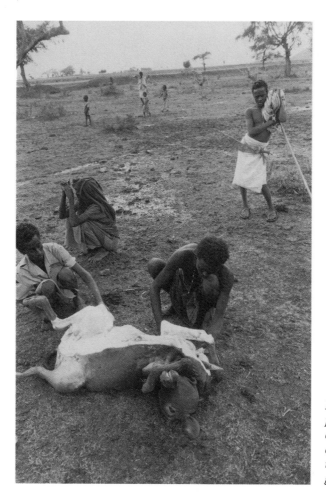

Figure 1.5 Afar pastoralists skin a calf during the drought. They will try to sell the skin on a glutted market.

Retrospective analysis of the events of 1974 suggests that famine arose primarily from the loss of exchange entitlement among rural workers. In this case (as in others) there appears to have been adequate food available in the country. One counterintuitive feature of response to climatic abnormality in Bangladesh is that excess rainfall leading to floods and the destruction of agricultural crops in low-lying areas also increases production on higher ground. The increased production in flood years tends to exceed lost production. So it was in 1974.

What caused starvation was a rapid rise in food prices (notably rice) combining with a drop in employment opportunities (partly resulting from the flood) and probably a fall in wages. The cause of the food price inflation is not fully understood. It was influenced by social and political upheaval, expectations among traders that prices would rise further, and a breakdown of negotiations between the government and foreign aid donors over food aid supply which led to the government losing influence over market stability. While the interaction of these (and other) causes of the food price rise are not well explained, this famine can be clearly distinguished as one arising from a failure of exchange entitlement.

Comparing famines

Finally, before leaving this brief introduction to recent understanding of famine, look at Table 1.1, which reproduces in summary form Amartya Sen's analysis of four famines. (Note that the discussion above on Wollo in 1973 focused on a smaller group of pastoralists who were affected as well as the farmers highlighted in the table.) The table indicates some of the diversity of livelihood collapses which can lead to acute hunger. It also confirms that by no means all famines are preceded by a decline in food available within the economy as a whole. In fact, the distribution of food may be largely independent of overall levels of food availability. Famines do occur when harvests have increased. Table 1.1 does not give a comprehensive or simple answer to what causes famine. It does provide questions, lines of investigation, which can guide empirical investigation of famine. Think how you would add a line to the table for the 1984 Wollo famine. What further information would you need beyond that in the section on Wollo above?

Table 1.1   Comparative analysis of four famines

Which famine?	Was there a food availability collapse?	Which occupation group provided the largest number of famine victims?	Did that group suffer substantial endowment loss?	Did that group suffer exchange entitlement shifts?	Did that group suffer direct entitlement failure?	Did that group suffer trade entitlement failure?
Bengal 1943	No	Rural labour	No	Yes	No	Yes
Ethiopia (Wollo) 1973	No	Farmer	A little, yes	Yes	Yes	No
Ethiopia (Harerghe) 1974	Yes	Pastoralist	Yes	Yes	Yes	Yes
Bangladesh 1974	No	Rural labour	Earlier, yes	Yes	No	Yes

Source: Sen, A. (1981) *Poverty and Famines: an essay on entitlement and deprivation,* p.163, Oxford University Press, Oxford.

1.3 Hunger: chronic undernutrition and poverty

'The phenomenon of endemic hunger is much more pervasive, it affects many times the number of people who are threatened by famines,' note Drèze & Sen (1989, p.267). The estimates of global hunger we examined earlier were attempts to quantify the total number of people suffering from chronic undernutrition. The number suffering from famine is much smaller.

Whereas the catastrophe of famine occurs through the widespread *breakdown* of livelihoods for specific social classes or occupational groups, chronic hunger arises through a *continuous failure to generate* sufficient livelihoods.

As such, chronic hunger is one aspect, probably the most fundamental, of a wider set of deprivations understood as **poverty**. Chronic hunger and poverty are closely related. It is rare for wealthy households to go hungry, even in a famine. Warnock notes that there is 'widespread agreement with the World Bank that undernutrition is largely a reflection of poverty' (Warnock, 1988, p.10).

Poverty is one of the underlying characteristics contributing to the issues raised in Chapters 1 to 5 of this book, and will be considered again in Chapter 6. It is necessary here only to introduce some of the ways in which poverty has been defined and measured.

The most common is an income- or consumption-centred approach: the poor are identified through having an income (including own production) insufficient to provide a minimum standard of living. This is similar to the income-based estimate of global hunger discussed earlier. It is a quantitatively defined standard of poverty and is often the most useful for making broad comparisons between standards of living in different regions of the world.

Two further possibilities need to be borne in mind. One is that poverty may be understood as relative to prevailing social standards, rather than defined at an absolute global level. Thus,

Poverty: Alternative understandings include:

Poverty line A simple definition where poverty is identified as an income below a minimal standard. The difficulties of obtaining cross-society comparability and of dealing with direct entitlements make this definition of limited use.

Consumption-based poverty line 'The expenditure necessary to buy a minimum standard of nutrition and other basic necessities and a further amount that varies from country to country, reflecting the cost of participating in the life of society' (World Bank, 1990, p.26).

Poverty as failure of capabilities Rather than focusing on a command over goods, as entitlement ideas do, this definition highlights basic human capabilities (to take part in society, to obtain health care, to achieve an adequate standard of living) and the failure to achieve them.

standards of living are higher within the advanced, industrialized world, and poverty is recognized at a higher level of living or income. The second is to define poverty in terms of capabilities rather than command over goods. Thus a basic freedom from poverty is understood in terms of what people can do and what they can be. An adequate life requires the capability to get enough food, adequate health care, access to clean water and sanitation, and to be a functioning member of society.

We can understand a little more about chronic hunger and how it has been reduced by examining the success story of China after the revolution in 1949.

Over a 30-year period, the Chinese revolution made great strides in the reduction of poverty and hunger through (1) effective provision of jobs within production brigades; (2) provision of health services throughout the countryside; (3) effective distribution of food to both cities and countryside; and (4) in total establishing a level of social security.

The key indicators of China's success in hunger reduction compared with India's are the level of life expectancy (the number of years newborn children would live if subject to prevailing mortality risks; see Chapter 2) and the level of mortality among children (Box 1.3). These measures of social well-being do not measure chronic undernutrition alone. An increase in life expectancy and a reduction of the level of childhood deaths may be caused both by nutritional improvements and by improvements in health and sanitation (as is discussed in more detail in the next chapter). In the case of China, analysis of nutritional status and mortality patterns confirms that 'China achieved a remarkable transition in health and nutrition' (Drèze & Sen, 1989, p.204).

Starting from similarly low levels at independence for India (in 1947) and the revolution in China (1949), China had begun, by 1979, to approach the levels of life expectancy and under-five mortality achieved in industrialized countries. Thus, in China, life expectancy at birth is in the upper 60s, compared with the mid- to upper 50s in India. The under-five mortality in China is 47 per thousand (i.e. 47 children out of every thousand die before they reach the age of five) compared with 154 in India.

We arrive at the paradox, identified by Drèze & Sen, of Chinese and Indian experience of famine and hunger. In China there has been some success in the reduction of chronic hunger but the great famine of 1958–61 was not prevented. By contrast, there has been some success in preventing famine in India, but little progress in the reduction of chronic hunger. Drèze & Sen attribute Indian success in famine prevention to relatively effective popular representation which enforces government response to distress. Chinese success in reducing hunger they identify with a large-scale transformation of social organization, providing the capacity for livelihood generation and effective public action to ensure access to nutrition, health facilities and social support.

1.4 Women, men and hunger

Thus far, we have focused on the social relations apparently giving whole households command over food. There has been concern in recent decades about the distribution of food *within* the household. This concern arises for two reasons. First, there is evidence, from some parts of the world and among some social classes, that there are higher levels of undernutrition among women and girls than among men and boys. Secondly, unequal food distribution within households may be the cause of higher death rates observed among women and girls in some parts of the world, although differential access to health care may also be an important factor (Harriss, 1989; Drèze & Sen, 1989, chapter 4).

As with individual nutritional intake, food distribution within households is difficult to measure systematically. However, in the case of India, at least, there is evidence of systematic discrimination in the allocation of food within households in some regions. Women eat last and least. Men take a disproportionate share of household food resources at the expense of other household members, and women and children may get less than their nutritional requirements. Certain high-status foods, such as meat or fish, may be reserved for men and rarely eaten by women. Male babies may be breast fed longer than female babies (Harriss, 1989).

Attempts to understand the reason for this sort of discrimination raise difficult questions of social organization. There are no fully convincing explanations, but there are two links which may provide the beginnings of understanding.

1 Higher levels of female mortality occur in areas where female participation in production is less pronounced. It is suggested that in these areas female children may be valued less highly than male children, and female children may be fed and looked after less carefully (Bardhan, 1974).

2 An important factor may be female autonomy, i.e. the degree of control that women have over their lives, over wages, the sale of

Box 1.3   The relationship between hunger and mortality

The relation between hunger and death is not straightforward. Post-mortem medical analysis rarely attributes death to starvation, even when famine conditions prevail. The immediate cause of death is frequently identified as an infectious illness, such as measles or diarrhoeal disease. There is a two-way relation between illness and undernutrition. Undernourished people get more severely ill and stay ill for longer than well-fed people in the same conditions. Undernutrition reduces immune response. Also, illness can contribute to undernutrition by reducing the body's capacity to deal with food (Martorell, 1988). These interactions have particular significance in circumstances of multiple deprivation characteristic of poverty, in which inadequate command over food is combined with poor access to health care, absence of clean water and inadequate sanitation.

The figure indicates how poverty (i.e. socio-economic status) can be related to ill-health, hunger and inadequate growth in children (nutritional status).

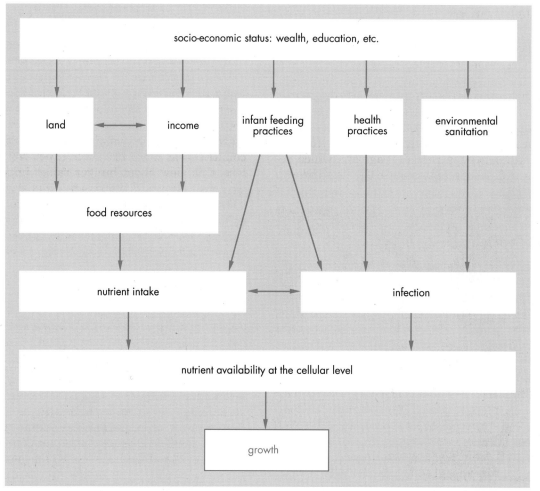

Examples of mechanisms through which poverty influences growth in children.

output, property and food distribution. Where women have greater control (i.e. more autonomy) gender differentials of mortality and undernutrition are less (Dyson & Moore, 1983; G. Sen in *Wuyts, Mackintosh & Hewitt, 1992*).

There is a lot still to be learned about how this gender discrimination operates in practice, but both of these points reinforce the need to know more about the detailed roles of men and women in production.

1.5 Hunger and social change – issues for the analysis of development

In most of this chapter, I have focused primarily on production and livelihoods as the central topic of analysis. I have mentioned only in passing some of those questions which are frequently raised when people in the industrialized world discuss hunger. For example, I reproduce as Figure 1.6 a recent advertisement for the *New Internationalist*, a British magazine which reports on development issues in the Third World.

Figure 1.6 Advertisement for the New Internationalist.

Even in this brief cartoon, which contains only a few sentences telling potential readers what they might expect to read about in the *New Internationalist,* some broad explanations for hunger and famine are suggested. Studies of the causes of hunger and poverty cover a very wide ground. However, a basic understanding of production and livelihoods is essential if the broader arguments about historical, political and international causation are to be evaluated and put to use.

We can take two of the points made in the cartoon to illustrate this.

1   The view that changes made by colonial rule laid the foundation for hunger is not confined to the *New Internationalist.* Much analysis of hunger (e.g. Warnock, 1988), including that of some historians writing from the Third World, has taken a similar view. However, the generality of the argument has to be restricted. It does not provide much purchase, for example, upon the cases of famine in either Ethiopia or China because the experience of colonial rule in those countries was either brief or indirect. Nevertheless, this view about hunger raises important questions. How did the expansion of European influence and rule change economic activity in different parts of the Third World? This question is addressed in subsequent chapters, particularly Chapter 9.

2   The idea that the introduction of cash crop production led to hunger is also not confined to the *New Internationalist.* There is a considerable body of writing (e.g. George, 1976; Buchanan, 1982) which suggests that the growing of crops for sale may displace food crop cultivation, resulting in a decline of food output and reduced household command over food. This view also raises important questions, though we need to be cautious about trying to apply it as a universal explanation of famine or hunger. The example of peasant production in Wollo highlights limitations of 'self-sufficient' or 'traditional' forms of livelihood. It is not necessarily the case that food self-sufficiency ensures stable and sufficient command over food.

The cash crop view of hunger causation calls attention to a wider set of social changes which play an important role in hunger and poverty, sometimes exacerbating vulnerabilities at the same time as they hold out the potential for overcoming those vulnerabilities. The cash crop view of hunger refers only to a part of a wider process of commercialization or *commoditization*. As peasant production becomes integrated with wider economic activity, it is not just the output of that agriculture which is bought and sold as a commodity. Land, tools and labour power (the work of people) are also increasingly bought and sold. These are fundamental changes in social organization and can lead to increased vulnerabilities. For example:

> "Poor people who possess no means of production excepting their own labour power, which they try to sell for a wage in order to buy enough food, are particularly vulnerable to labour market conditions. A decline in wages *vis-à-vis* food prices, or an increase in unemployment, can spell disaster for this class...The class of landless wage labourers has indeed recurrently produced famine victims in modern times... The importance of the vulnerability of wage labourers can be particularly acute in the intermediate phase in which the class of wage labourers has become large...but a system of social security has not yet developed."
>
> (Drèze & Sen, 1989, pp.5–6)

The general point about both the colonialism view and the cash crop view is that they raise important questions which cannot be adequately assessed without an understanding of employment and production.

The question with which we started this chapter is, in practice, a very broad one. I tried to narrow it down from even broader questions of 'what can be done about hunger?' or even 'what causes hunger?' so that I could take on a topic which might be tackled within the short space available. In reality, it has many implications. For instance, the two views about hunger briefly discussed above raise some of the main issues in development studies. What is the role of (colonial) government? How does economic integration change livelihoods?

If we were trying to understand all the diverse causes of poverty and hunger, and even more so if we attempted to make suggestions about what could be done, then we would ask about the nature of industrialization, what sorts of livelihoods it provides, and what sorts it destroys, and the rate at which it does so. We would also need to investigate the rate and pattern of economic growth and the nature of social support and social security (provided by local collective action as well as by government).

The causes and responses to hunger also take us into the great questions of social and political organization, the great struggles between capitalist forms of economic organization and its alternatives. These issues all come up again later in this book. In fact, sooner or later, the subject of hunger is raised in almost every debate in the study of development. To reverse the meaning of the quotation with which I started this chapter, the explanation of hunger raises questions about all aspects of human history.

1.6 Some principles for action

I promised to say something about recent ideas for action on famine and hunger. There are no panaceas in development; the great questions of development can rarely be resolved by simple technical inventions or unproblematic intervention by governments. Nevertheless, the analysis focused on livelihoods has made clear the principles on which response to famine and hunger can be effective. In some cases responses along such lines have proved to be effective.

One conclusion which can be drawn from the analysis of famine is that entitlements or livelihoods can be protected and replaced. Those states (including India and several African states) that have established institutional procedures for undertaking this task have been among those where famine has not occurred for many years. Such procedures often include the capacity to implement large-scale employment schemes at short notice.

A second conclusion is that effective popular representation in government (through elections, open political opposition and unfettered news coverage) may ensure that governments respond to famine.

The third conclusion is that the reduction of hunger entails the more difficult task of generating new livelihoods and entitlements. As such, this encompasses some of the broader aims of development, and constitutes the subject matter of development studies, rather than simply of this chapter.

Summary

1   There is a distinction between famine (a sharp increase in mortality associated with starvation) and chronic hunger (chronic undernourishment due to inadequate food consumption).

2   Examination of the social relationships which provide different classes or occupational groups with command over food (i.e. entitlements) provides a better explanation of hunger and famine than does the level of food availability.

3   Famines occur because entitlements are made vulnerable by social, economic and climatic change, whereas chronic hunger is related to poverty and continuous failure to generate sufficient entitlements.

4   A key element in the prevention of famine is the protection and creation of livelihoods, which can in principle be achieved through public action. Public action could also reduce chronic hunger through providing jobs, health care, education and elements of social security; this raises questions about development in general.

2

DISEASES OF POVERTY

GORDON WILSON

In 1989 *The Observer* printed an article under the front-page headline *The 'silent genocide' of millions of children*. It began 'Eleven million children are dying every year in the developing world because of the unwillingness of the rich countries to help them.'

In developing this argument, however, the article (most of which is reprinted below) begged several questions. You should think about these as you read it.

The 'silent genocide' of millions of children

"Eleven million children are dying every year in the developing world because of the unwillingness of the rich countries to help them. The deaths are unnecessary because the diseases from which the children suffer can all be prevented or successfully treated.

These are the shocking conclusions of a new, and virtually unnoticed, report by the World Health Organization, the authoritative watchdog body that monitors health trends throughout the world.

In one of the most comprehensive surveys ever undertaken into childhood mortality, the organization reveals that four million die from diarrhoea a year when a simple packet of sugar and salts, costing less than 7p a time, could save almost two-thirds of them from the lethal dehydration it causes.

'For $50m (£31m), or the cost of a modest commercial building in New York City, we could cut deaths from diarrhoeal diseases by two million children per year,' claims Dr Hiroshi Nakajima, Director-General of the organization.

In what Dr Nakajima describes as 'the silent genocide', the report reveals that another three million children die a year from six infectious diseases: polio, tetanus, measles, diphtheria, whooping cough and tuberculosis, when the cost of vaccinating them against these killers costs about £6 a head. In Britain, less than 20 children die a year from them because of widespread vaccination.

Dr Nakajima said last week: 'For less than $1bn (£600m), the cost of 20 modern military planes, the world could control these illnesses.'

In the developing world, about 2.8 million die from these diseases annually and at least another three million are disabled.

In a catalogue of suffering and handicap, the report reveals that another four million children die every year due to acute respiratory diseases, mostly pneumonia. In Britain, the figure is about 150…

Commenting on the report's implications, Dr Nakajima said 'This is a preventable tragedy because the developed world has the resources and technology to end common diseases worldwide.

'The missing ingredient is the will to help the developing countries. Simply, the rich countries must transfer technology, health, manpower and money, because the poorest countries can't help themselves.' "

(*The Observer*, 1 October 1989)

The argument of the article is clear. The poorer countries cannot help themselves, and the rich countries are unwilling to assist even though they could solve the problem – for example, by transferring technology (Figure 2.1).

But what has been left out of the picture?

1 The practical problems of distribution of vaccines and medicines on a worldwide scale are not mentioned.

2 Nothing is said about the causes of diarrhoea and infectious diseases.

3 Medical technology, in the form of vaccines and medicines, is the only approach considered. Do other valid approaches exist?

4 Are all inhabitants of Third World countries equally at risk, or are some countries more prone to disease than others? Within countries are some occupation groups or social classes (see Chapter 1) more at risk than others?

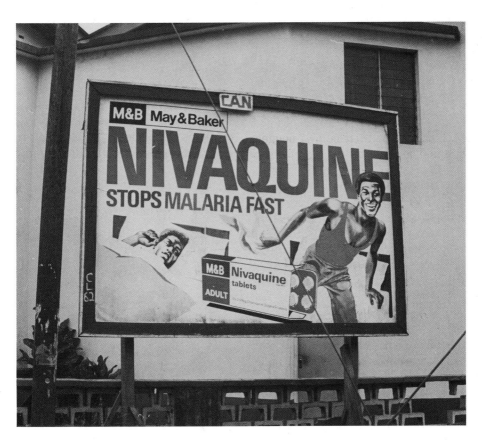

Figure 2.1 A cheap biomedical remedy for Nigeria. However, huge quantities of chloroquine-based drugs (like Nivaquine) have been distributed throughout Africa without wiping out malaria. It remains a major killer, and there are now resistant strains. The signs are that it is on the increase.

I shall return to this article several times during the chapter. It argues a particular line, but from the second and third of my 'omissions' especially, you should note there is an important debate here.

Q Are the health needs of the Third World best served by biomedical interventions; or will an attack on underlying causes of diseases be more effective?

I am about to engage in this debate, but first there is a need to look a little closer at some general information about diseases and their incidence in the Third World.

I use the word 'diseases' for a reason. Although *The Observer* article was written by the paper's *health* correspondent and refers to a report by the World *Health* Organization, what is actually discussed are kinds of infirmity. What is being promoted is not health, in the sense of a state of complete physical, social and mental well-being, but the eradication of diseases. In practice, health care policies are often talked about in this very limited sense, and most of the discussion here will focus specifically on diseases of poverty and what can be done about them.

One further point needs to be kept in mind from the outset. Disease statistics may not be reliable, not only because of difficulties of collection, but also because figures may be deliberately manipulated, perhaps to exaggerate a problem in order to obtain resources, or to 'prove' that a particular programme is working effectively. Occasionally, the data obtained seem to be little more than guesswork. For example, Table 2.4 later in this chapter includes an under-five mortality rate for Uganda in 1980. Data for that figure must have been collected soon after the Tanzanian invasion and the fall of Idi Amin. There was still fighting in parts of the country, and state services were incapacitated. Is it likely that the figure in the table accurately reflects the national situation? Who collected the data and how? We should therefore be careful to concentrate only on what seem particularly significant trends. Broad brush strokes are all that are possible.

2.1 Variations between countries in the prevalence of diseases

Look at Tables 2.1, 2.2 and 2.3 on the following pages. Note any immediate points that they highlight: for example, by comparing the highest and lowest figures in each column.

You may notice points like the following:

- **Life expectancy** in African countries, Central and South America, and Asia (excluding Japan) is usually much lower and the under-five mortality rate much higher than in Western Europe, Japan and the USA.

- In Malaysia, the state with the lowest incidence of poverty (Selangor) has the lowest infant **mortality rate**, whereas the state with the highest incidence of poverty (Kelantan) has the highest infant mortality rate.

- Diseases of the circulatory system account for the greatest number of deaths per 100 000 population in the USA, whereas in Guatemala the greatest number is due to infectious and parasitic diseases.

Mortality rates and **life expectancy**: The *infant mortality rate* is the number of deaths in the first year of life per 1000 live births. The *under-five mortality rate* is the number of children who die before the age of five for every 1000 live births. *Life expectancy* is the average length of life or the expectation of life at birth. Infant and under-five mortality rates are strongly correlated with adult mortality. If infant or under-five mortality is high, adult mortality is likely to be high and life expectancy low. They can, therefore, be useful indicators of susceptibility to diseases. Also, health care policies in Third World countries are often directed at children and changes in infant or under-five mortality rates are a way of assessing these policies.

Table 2.1 Estimates of under-five mortality rate (U5MR) and life expectancy for selected countries (1988 data)

Country	U5MR	Life expectancy at birth (years)
Africa		
Mozambique	298	47
Sierra Leone	266	41
Malawi	262	47
Uganda	169	51
Ghana	146	54
Egypt	125	61
America		
Peru	123	62
Brazil	85	65
Jamaica	22	74
Cuba	18	74
USA	13	75
Asia		
Bangladesh	188	51
Pakistan	166	57
India	149	58
Indonesia	119	56
China	43	70
Sri Lanka	43	70
Japan	8	78
Europe		
Yugoslavia	28	72
Poland	18	71
UK	11	75
Sweden	7	77

Source: UNICEF (1990) *State of the World's Children,* Oxford University Press, Oxford.

Table 2.2 Infant mortality and incidence of poverty in Malaysia

State	Incidence of poverty[a] (%)	Infant mortality rate per 1000 live births
Selangor	8.6	13.4
Johor	12.2	17.0
Negeri Sembilan	13.0	18.3
Penang	13.4	15.1
Pahang	15.7	19.1
Melaka	15.8	16.0
Perak	20.3	19.9
Terenganu	28.9	22.2
Sarawak	31.9	19.5
Sabah	33.1	22.7
Perlis	33.7	18.4
Kedah	36.6	21.0
Kelantan	39.2	23.1

[a]A measure of absolute poverty which refers to a proportion of the population unable to meet specified minimum needs, i.e. the group that cannot afford a minimum nutritionally adequate diet plus essential non-food requirements.

Source: UNICEF (1988) *Asian and Pacific Atlas of Children in National Development,* UNICEF East Asia and Pakistan Regional Office with the co-operation of the UN Economic and Social Commission for Asia and the Pacific.

Table 2.3   Age-standardized death rate by selected cause in Guatemala (1984) and the USA (1987)

Cause of death	Age-standardized death rate per 100 000 population[a]	
	Guatemala	USA
Suicide and self-inflicted injury	0.9	10.8
Motor vehicle traffic accidents	1.5	18.0
Diseases of the digestive system	36.8	20.6
Malignant neoplasms (cancers)	56.9	131.8
Injury and poisoning	62.3	53.6
Diseases of the circulatory system (including heart attack, stroke, etc.)	119.4	222.5
Diseases of the respiratory system	164.4	39.7
Other infectious and parasitic diseases	216.0	7.8

[a]The age-standardized death rate is the number of deaths that occur in a given population per year after adjustment for the different age distributions between countries. To take these examples: Guatemala has a higher proportion of younger people and a lower proportion of older people among its population compared with the USA. Unless allowance is made for this difference, meaningful comparison between the two countries is difficult. This death rate is often quoted per 1000 population, per 10 000 population, or per 100 000 population as here.

Source: WHO (1989) *World Health Statistics Annual*, World Health Organization, Geneva.

Let us now look at the tables in more detail to see whether there are any general implications we can draw from these initial impressions.

Table 2.1 suggests that people in Third World countries are more susceptible to diseases than those living in Western, industrialized societies. Under-five mortality rates are generally much higher and life expectancies are generally much lower in the Third World countries listed. It is worth while noting the variations, however. Twelve countries in the table, all of which are conventionally thought of as Third World, have under-five mortality rates between 85 (Brazil) and 298 (Mozambique). Then there is a large gap before we come to the countries with medium to low under-five mortality rates, extending from China and Sri Lanka (each 43) to Sweden (7). This second group is dominated by the industrialized West, includes the East European countries, but also includes four 'Third World' countries: China and Sri Lanka being joined by Jamaica (22) and Cuba (18). There is a similar pattern if we examine life expectancy, with China, Sri Lanka, Cuba and Jamaica almost comparable to the industrialized countries (Figure 2.2). We shall look again at some of these 'exceptions' later in the chapter, but for the moment note how they illustrate that generalizations about diseases in the Third World are dangerous.

Figure 2.2   A Third World exception: pensioners exercising at a street corner of Havana, Cuba.

Table 2.4 shows that the under-five mortality rate for virtually all countries has declined in the last 30 or so years. That is good news, but how does the position in Third World countries compare with that in Western countries, such as the UK? Is the Third World actually catching up with the West with regard to under-five mortality?

Figure 2.3 shows graphically the changes in the infant mortality rate for five countries compared with that of the UK over the period 1960–88.

Table 2.4 Changes in the under-five mortality rate since 1960 for selected countries

Country	Under-five mortality rate		
	1960	*1980*	*1988*
Africa			
Mozambique	330	258	298
Sierra Leone	386	300	266
Malawi	364	300	262
Uganda	224	187	169
Ghana	224	165	146
Egypt	300	164	125
America			
Peru	233	144	123
Brazil	160	103	85
Jamaica	88	29	22
Cuba	87	27	18
USA	30	16	13
Asia			
Bangladesh	262	211	188
Pakistan	277	192	166
India	282	180	149
Indonesia	235	145	119
China	202	56	43
Sri Lanka	113	58	43
Japan	40	12	8
Europe			
Yugoslavia	113	36	28
Poland	70	24	18
UK	27	16	11
Sweden	20	9	7

Source: UNICEF (1990) *State of the World's Children,* Oxford University Press, Oxford.

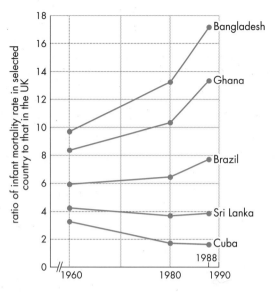

Figure 2.3 Changes in infant mortality relative to the UK in selected countries.

2.2 Variations within countries in the prevalence of diseases

We have seen that there are considerable variations between countries in the prevalence of diseases as measured by under-five mortality rate, but what is the situation *within* countries? Are all sectors of the population equally susceptible to disease, or are some sectors less so than others?

Age

It is difficult to look in detail at variations in mortality by age for many Third World countries because data are inaccurate or do not exist. Table 2.5, however, compares male deaths in Ecuador with those in the UK to illustrate the general points.

Table 2.5 Male deaths by age in Ecuador and the UK

Age	Age-specific death rate per 100 000 males		Ratio Ecuador:UK
	Ecuador (1987)	UK (1988)	
0	5214.5	1019.2	5.1
1–4	376.6	44.7	8.6
5–14	87.7	24.8	3.5
15–24	164.8	80.3	2.0
25–34	260.8	94.2	2.8
35–44	378.7	174.4	2.2
45–54	647.7	511.9	1.3
55–64	1300.8	1595.2	0.8
65–74	2856.2	4123.9	0.7
Over 75	11 009.5	11 121.2	1.0

Source: WHO (1989) *World Health Statistics Annual,* World Health Organization, Geneva.

Although, generally, Ecuador has a higher mortality rate than the UK, the difference, as measured by the ratio between the death rates for the two countries, is relatively small apart from the first years of life. This is usual in Third World countries. It points particularly to diseases that attack the young as being the major killers in the Third World. The mortality rate is actually slightly higher in the UK after the age of 55, reflecting the impact of cancer and heart disease as major killers in an industrialized country.

Gender

Throughout the world, life expectancy tends to be higher for women than for men. Look now at the first column of Table 2.6 which ranks the life expectancy of women relative to that of men for the countries listed in Table 2.1.

All but one of the 21 countries listed has female life expectancy greater than that of men. However, with one or two exceptions (which should make you heed my earlier warning about generalizations), the difference in life expectancy between men and women is consistently lower in the Third World countries listed, and the one country where women do not have a higher life expectancy than men (Bangladesh) is in the Third World.

The greater difference in favour of women in the industrial countries has been interpreted as women having a 'natural' biological advantage. If this is true, the smaller difference in Third World countries suggests that women have acquired some 'nurtured' disadvantage. This may be due to unequal distribution of food and access to medical care between men and women in a household (see Chapter 1); to the amount and nature of the work women do; or because they experience more pregnancies than do Western women, with associated higher risk.

In its 1990 *World Development Report*, the World Bank reports a similar pattern with respect to the under-five mortality rate, with females generally having a lower under-five mortality rate than males, but the gap being narrower in Third World countries. Also, in several low-income countries, the position is reversed, with females having the higher under-five mortality rate. The World Bank concludes: 'This suggests differential treatment of males and females with respect to food and medical care' (World Bank, 1990, p.259). Other writers have indicated an 'excess' female mortality among children in rural Bangladesh (Chen, Huq & D'Souza, 1980, p.9) and greater incidence of disease among female children in Calcutta, India (Sen, 1984).

Education is also strongly associated with good health. Indeed, a World Bank report on mortality decline in the developing world commented on 'the extremely powerful role of literacy in determining a population's level of mortality' and suggested that this factor carries far more weight than many others, such as income growth (World Bank, 1985, p.120). Another World Bank publication, on child mortality in urban Brazil in the 1970s, concluded: 'increased maternal education accounted for a larger share (34%) of the mortality decline between 1970 and 1976 than any other single factor, including access to piped water' (World Bank, 1983, Abstract).

The second and third columns of Table 2.6 suggest that, in general, women in Third World countries are not educated to the same level as males. We can, therefore, add this to the possible reasons why these women may be more susceptible to diseases than women in the industrial countries. A note of caution, however: the education argument is persuasive and accords with our 'common sense', but the precise nature of that education is important. Studies in Indonesia have shown that children of highly educated mothers are not always less susceptible to diseases than those of their uneducated counterparts. The reasons suggested are that highly educated mothers have shorter birth intervals between children (traditionally birth intervals are long in Indonesia) and they end breastfeeding earlier, two factors that mitigate against good health in their offspring (Heering, 1990).

Town and country

Collecting and interpreting statistics based on where people live presents numerous problems.

Table 2.6   Life expectancy and education of women relative to that of men in selected countries

Country	Life expectancy: females as a percentage of males (1987)	Adult literacy rate: females as a percentage of males (1989)	Secondary school enrolment: females per 100 males (1986–88)
Bangladesh	98.6	51	46
Pakistan	100.0	48	42
India	100.0	51	54
Jamaica	100.0	n/a	108
Malawi	103.1	60	60
China	104.4	68	74
Egypt	104.5	51	73
Cuba	105.0	100a	108
Indonesia	105.1	78	n/a
Sri Lanka	106.2	91	110
Peru	106.5	86	90
Uganda	106.7	64	56
Ghana	106.8	67	65
Mozambique	107.2	40	57
Japan	107.6	–	102
UK	107.9	–	104
Sierra Leone	108.0	55	48
Sweden	108.0	–	102
Yugoslavia	108.5	89	96
Brazil	108.5	96	128
USA	109.8	–	101
Poland	111.8	n/a	105

n/a, data not available; a1981 data, age 10 years and older;

–, in West European countries, the USA and Japan, it is assumed that adult literacy for both men and women is 100%.

Source: UNICEF (1990) *State of the World's Children*, Oxford University Press, Oxford.

For a start, how do you define such things as 'urban' and 'rural'? What size does a town have to be before its inhabitants are classified as urban? How are people classified who depend on cities for their livelihoods and are within easy reach of their services but live outside them? In particular, hospitals are more likely to be situated in the cities than in the countryside. If seriously ill rural dwellers are sent to these urban hospitals and subsequently die in them, whose death statistics – rural or urban – are they?

Secondly, death registration records are probably more complete for urban areas, which can make the rural populations appear less susceptible to diseases by comparison. Thirdly, the statistics may not be a true reflection of differing conditions in the countryside and towns. If, for example, healthy young adults from the country migrate to the towns to seek work, they leave behind a population that is comparatively more vulnerable to disease. This has the effect of improving the urban statistics and making worse the rural ones.

Table 2.7 gives data on rural and urban mortality for a sample of countries. In each country infant mortality is higher in rural areas, although the extent of the difference ranges from being marginal in Kenya to rural infant mortality being 2.7 times greater than urban infant mortality in Mexico.

Obviously, a multitude of factors combine and it is impossible to predict what the precise urban–rural health split will be in each instance. Factors which might make urban populations less susceptible to diseases are:

* greater access to medical services;

* greater access to clean water and sanitation, where this is not negated by other factors (see below);

* better educational opportunities;

* generally higher incomes, enhancing the above factors; higher incomes also mean better nutrition.

Factors which might make urban populations more susceptible to diseases include:

* overcrowding, enhancing the transmission of infectious diseases;

* insanitary conditions in slums and shanty towns.

In addition, urban populations in many countries are likely to experience greater incidence of accidents and violence.

Poverty

Poverty is known to have a strong influence on the prevalence of diseases worldwide, as you would expect from your reading of Chapter 1 which relates chronic hunger to poverty. Figure 2.4 is a graph of the data in Table 2.2, relating incidence of poverty to the infant mortality rate in the administrative states of Malaysia.

Looking at Figure 2.4, would you say it supports the argument that poverty is a strong influence on the prevalence of diseases?

Table 2.7 Infant mortality rate and urban/rural residence for selected countries

Country	Infant mortality rate		
	Rural	Urban	Ratio rural:urban
Sub-Saharan Africa			
Côte d'Ivoire	121	70	1.7
Ghana	87	67	1.3
Kenya	59	57	1.0
Asia			
India	105	57	1.8
Indonesia	74	57	1.3
Philippines	55	42	1.3
Thailand	43	28	1.5
Latin America			
Guatemala	85	65	1.3
Mexico	79	29	2.7
Panama	28	22	1.3
Peru	101	54	1.9

Source: World Bank (1990) World Development Report 1990, World Bank, Washington DC, p.31.

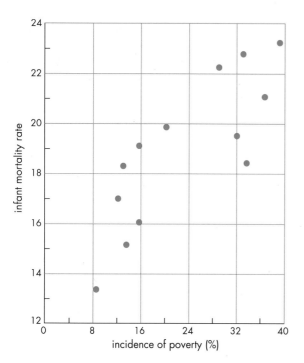

Figure 2.4 Infant mortality and incidence of poverty in Malaysia.

The answer is 'yes'. Figure 2.4 is known as a *scatter diagram*. Such diagrams are useful for showing visually whether or not there is any correlation between two sets of indicators, in this case poverty and health indicators. Although the points on Figure 2.4 are scattered about (hence the term 'scatter diagram'), it is possible to discern a definite overall pattern. In general, in the states of Malaysia the greater the incidence of poverty, the greater the infant mortality rate. This is called a *positive correlation* between these two indicators. This does not mean necessarily that poverty is itself a direct cause of disease, but it lies behind other factors that probably are: poor education and medical services, poor and insanitary living conditions, poor access to safe water and poor nutrition.

2.3 'Third World diseases' and their causes

Look again at *The Observer* article which opened this chapter. Notice how mortality in the Third

World is particularly associated with diarrhoeal diseases, acute respiratory diseases (e.g. pneumonia, bronchitis, influenza), and other infectious diseases (e.g. measles, tuberculosis, typhoid and malaria).

Now turn back to Table 2.3 and the causes of death in Guatemala. Infectious diseases of all varieties (including diseases of the respiratory and digestive system) are by far the most common cause of death. In contrast, these diseases are relatively unimportant in the USA (although diseases of the respiratory system and the digestive system are significant causes of death among older people). In the USA, diseases of the circulatory system (heart attack, stroke, etc.), followed by cancers, are the major killers.

The United Nations Children's Fund (UNICEF) and the World Health Organization (WHO) have also classified the major children's diseases in the developing world and their results are shown in Figure 2.5. The prevalence of diarrhoeal

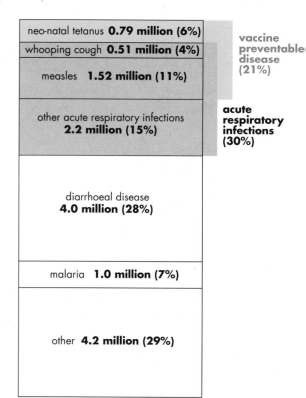

Figure 2.5 Annual deaths of children under five by main causes.

infections is the most striking feature. Infectious diseases feature strongly, as do respiratory diseases. But it should be stressed that many diseases are also interrelated. Diarrhoea is strongly related to malnutrition, for example, and UNICEF and the WHO state that malnutrition is a contributory cause in approximately one-third of all child deaths. Measles makes a child highly susceptible to respiratory infections, and pneumonia may therefore be the ascribed cause of a death for which measles is primarily responsible.

The argument of *The Observer* article, that infectious diseases are the cause of thousands of unnecessary deaths in the Third World, does indeed seem to be confirmed by the WHO and UNICEF statistics. But is this an adequate explanation? What are the causes of the causes?

Diarrhoeas are generally transmitted via contaminated water or food. Malaria is transmitted by mosquitoes which thrive in stagnant water. Worm infestations such as roundworm and hookworm are also highly prevalent in many Third World countries, another impact of poor sanitation and lack of clean water (Figure 2.6). Respiratory infections are not transmitted by water or food, but their incidence and severity are exacerbated by inadequate air pollution controls and by overcrowding. Moreover, all these diseases have to be set against a background of poor nutrition which weakens resistance to getting them in the first place and to fighting them off once they occur.

Does this mean, therefore, that rather than define the diseases themselves as the causes of the problem, we should view them as *symptoms* of more underlying causes: dirty water, poor sanitation, polluted air, overcrowding and poor nutrition?

Well, yes and no. Let us step back further. We have already noted the differences in the prevalence of diseases by age, gender, place of residence, and poverty. Do these differences reflect different degrees of access to good sanitation,

Figure 2.6 Collecting water in Lagos, Nigeria. The rapid growth of the city has made it difficult for the authorities to provide sufficient clean water.

clean water, a decent diet, and an unpolluted atmosphere? Stepping back even further, do the relatively low incomes of Third World countries indicate a negligible budget available for public spending on clean water, sanitation, education and health facilities? Is not poverty the cause of susceptibility to diseases and is not economic underdevelopment the cause of poverty?

2.4 Strategies for controlling diseases

The debate between those arguing for a largely biomedical and technological approach to controlling diseases, and those arguing for more broadly based, public health programmes is not new, and not exclusive to the Third World. Indeed, a useful comparison can be made with diseases in nineteenth-century Britain.

Although the diseases vary individually (e.g. scarlet fever), the major diseases in nineteenth-century Britain fell precisely into the same categories prevalent in many parts of the Third World today: diarrhoea, respiratory infections, and other infectious diseases.

Apart from measles and diphtheria, these diseases all declined massively over the latter part of the century. In his book *The Role of Medicine*, Professor Thomas McKeown asks:

> "Are the improvements in health with which medicine is commonly credited determined essentially by medical science, or are they due largely to fortuitous changes in which biomedical research has played little part?"
>
> (McKeown, 1979, p.156)

Answering his own question, Professor McKeown argues that the decline in infectious mortality in Britain in the latter part of the nineteenth century was due, not so much to the development of medical science, but to:

- improvements in nutrition;
- improvements in hygiene, which were 'the predominant reasons for the decline of water- and food-borne diseases';

- the change in reproductive behaviour, which led to the decline in the birth rate. If this had not happened, the beneficial effects of improved diet would have been wiped out by runaway population growth.

Smallpox, for which a simple vaccine was developed, is the obvious exception, but Professor McKeown's overall thesis is now widely accepted. The implication is that lasting improvement in health in Third World countries will not come about as a result of medicines so much as through improved nutrition, clean water and sanitation.

Does this mean, therefore, that Third World countries should forget about drugs and vaccines and place their efforts into public health measures? What are the possible objections to such a strategy?

I have noted two:

1 Biomedical science has improved since the last century and now has the potential to make rapid inroads against infectious diseases.

2 Any public health programme delivering improved nutrition, clean water and sanitation would require enormous resources, and even then the resulting health improvements would be long term rather than immediate.

The first point was assumed and the second taken up by Professors Anthony Robbins and Phyllis Freeman in a paper in *Scientific American,* November 1988, entitled 'Obstacles to developing vaccines in the Third World', in which they wrote: 'many countries cannot yet afford the capital investment needed to improve housing, sanitation and drinking-water purity' (Robbins & Freeman, 1988, p.90).

Their thesis is that the Third World cannot wait for improved public health measures to come about. Immediate vaccination is seen as a kind of *crisis management*, designed to achieve immediate results.

The problems associated with the implementation of a vaccine programme, however, are themselves formidable, both economically and politically. Robbins & Freeman claimed in the same paper:

"The obstacles to the development and distribution of the needed vaccines are many, but among the most important is the fact that the decision to develop new vaccines is left almost entirely in the hands of a few institutes or commercial manufacturers in the developed world. The engineering knowledge and skills are concentrated in these few institutes and firms..."

(ibid., p.93)

High development costs result in vaccines only affordable by the industrialized countries, at least until these costs have been recovered. This in turn means that research and development tends to be into the ailments of the industrialized countries rather than those of the Third World. As *The Financial Times* reported in an article on drugs companies operating in Indonesia:

"While research efforts in the West concentrate on chronic conditions such as heart disease and cancer, drug companies in Indonesia are grappling with simple solutions, with simple antibiotics accounting for 30% of sales, which compares with 10% in the US."

(*The Financial Times*, 16 February 1989)

The situation is exacerbated by an emphasis in the West on improving existing drugs and vaccines (so they have fewer side effects or are slightly more effective) which involves new and expensive production technology.

To overcome these problems, Robbins & Freeman outline various possible approaches:

1 The United Nations raises money to buy existing vaccines at their full market price or to pay manufacturers the development costs of new vaccines essential for the Third World.

2 The United Nations creates a public institute to develop and manufacture its own vaccines, by-passing commercial makers.

3 Development and production units are established in countries or regions of the Third World with large populations (i.e. where there is a large demand for vaccines), transferring the necessary technology and expertise from the industrialized countries.

However, a basic problem with all these approaches is lack of finance. As Robbins & Freeman themselves note:

"A major roadblock...is the fact that the World Bank and other agencies that might provide loans to establish such units are increasingly concerned with meeting bankers' standards (seeing a good return on investment) when they finance new industrial enterprises. The development and production of vaccines to meet the health needs of the Third World is unlikely to turn a profit."

(ibid., p.95)

Furthermore, the 'public health lobby' might well respond by pointing out that the breakthroughs of biomedical science are not a solution. There is little evidence that even the 'wonder drugs' of modern medicine really have the capacity to turn around mortality trends. A child from a poverty-stricken family may be cured of measles, but is likely to remain chronically malnourished, and may well die of something else a few months later (Figure 2.7).

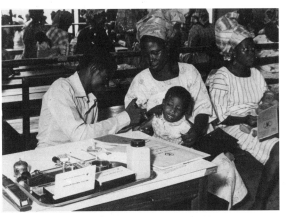

Figure 2.7 A health clinic in Lagos, Nigeria. Vaccinations can be effective against particular diseases, especially tuberculosis, measles, polio, tetanus, diphtheria, whooping cough, and smallpox, but are no guarantee that the child will survive other poverty-related diseases.

2.5 Integrating prevention and cure in primary health care

It is often said that public health measures are essentially aimed at *prevention*, medicines at *curing*. Leaving aside the fact that vaccines are also a preventive measure, the debate between biomedical intervention and public health is invariably posited in these terms.

But it does not have to be either/or. The two do have things in common. Take aid, for example: I have so far talked about it exclusively in terms of biomedical intervention, but many individual aid projects have been aimed at providing clean water in rural areas. I have also referred to biomedical *technology* but clean water and sanitation involve technology too. In fact, whatever approach is adopted, technology is required.

Nevertheless, any strategy aimed at providing permanent improvements in health for all has to go beyond the distribution of biomedical technology. It has been argued that such a strategy must be a social process that encourages the community to involve itself in preventive public health measures such as personal hygiene, improved diet and health education. If this is combined with an infrastructure of decentralized health facilities providing an essential vaccination programme and basic curative measures based on early diagnosis, the strategy is known as primary health care, or sometimes as **comprehensive primary health care** (Figure 2.8).

Cuba is an example of a country that has placed a strong emphasis on preventive measures together with decentralization of facilities throughout the island. Improvements since Castro came to power in 1959 in housing, sanitation, nutrition and health education (made possible by the eradication of illiteracy) have all contributed to health statistics comparable to those of Western Europe, Japan and the USA (see Table 2.1). It was envisaged that the 4500 family practitioners operating in 1989 (out of a total of 26 000 doctors) would rise to 20 000 by 1995, roughly one for every 500 members of the population.

Cuba now sees itself as a 'medical power' (see Chapter 16).

> **Comprehensive primary health care:** A broadly based strategy which explicitly links the prevalence of infectious diseases with poverty. It maintains the approach to primary health care, established at the Alma Ata conference in 1978, which aims at providing promotive, preventive, curative and rehabilitative services. It is often contrasted with *selective primary health care,* an approach promoted by UNICEF since 1983, which emphasizes growth monitoring of children, the use of oral rehydration salts for diarrhoea, the encouragement of breastfeeding for infants, and immunization. In practice, selective primary health care schemes concentrate particularly on the distribution of oral rehydration salts and on vaccinating as many children as possible. They are therefore more like biomedical interventions than public health programmes.

China is another country which has made enormous strides in controlling diseases via what are essentially primary health care policies. In its 1990 *World Development Report* the World Bank stated that in the previous 25 to 30 years life expectancy in the country increased from 52.7 to 69.5 years and the infant mortality rate dropped from 90 to 32 per thousand live births. The report adds:

"China's remarkable performance owes as much to safe drinking water, improved sewage disposal, and other sanitation measures as to broad immunization coverage and mass campaigns against parasitic diseases. It has much to do with the provision of basic health care and affordable drugs to even the most remote parts of the country. It reflects the successful drive to reduce

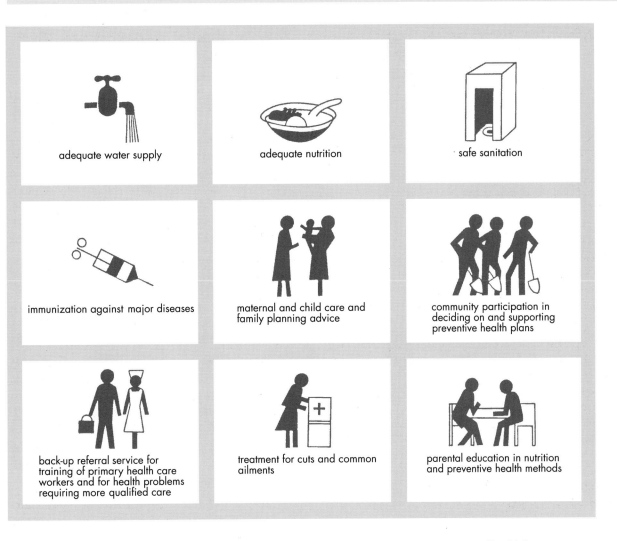

Figure 2.8   The comprehensive primary health care ideal, but can the poorest countries afford it?

fertility and to increase, through legislation, the age of first delivery, as well as great efforts to provide education on health and nutrition. And it would probably have been impossible without a safety net that, among other things, guaranteed food rations to even the poorest rural people.

China's performance...teaches an important general lesson: large improvements in the health of the population can be achieved if there is a broad and lasting political commitment, with a consistent emphasis on preventive measures and basic curative care. In other words, social progress is not merely a by-product of economic development. Policies matter."

(World Bank, 1990, p.74)

But ask yourself, do all countries have the capacity to implement this type of programme? What kinds of resources are needed? (Box 2.1)

The AIDS epidemic has become a threat throughout the world, but is particularly serious in countries which combine a high incidence of infection with inadequate preventive health care services. Unlike most other killer diseases, AIDS victims tend to be from the most productive age group. One study has estimated that in 10 high-incidence African countries, more than 10% of children will have lost at least one parent by the end of the 1990s. Limiting the spread of the epidemic necessitates changes in social behaviour through public health programmes. Many countries find these hard to implement because their present health care services concentrate the limited resources available on urban centred, curative interventions. Unless more resources become available and there is a shift into public health oriented programmes, or unless a new, cheap 'wonder drug' is produced within the next few years, it is likely that the demographic impact of AIDS in some countries will be enormous.

2.6 Infectious diseases, wealth and public policy: contrasting circumstances in Sri Lanka and Brazil

I want to end this chapter by briefly examining two other countries: Brazil and Sri Lanka. Look at Figure 2.9, which is a scatter diagram of under-five mortality rate and gross national product (GNP) per capita for selected countries. Why do you think Sri Lanka and Brazil might be particularly worthy of comment?

The scatter diagram shows that there is a strong correlation between a nation's average income, as measured by its GNP per capita, and its under-five mortality rate. This is a *negative* correlation because, generally, the higher the GNP per capita, the lower the under-five mortality rate.

Brazil and Sri Lanka are both *outliers* on this scatter diagram, albeit in opposite ways. In 1987 Brazil had a GNP per capita of US$2020, exceptionally high compared with other Third

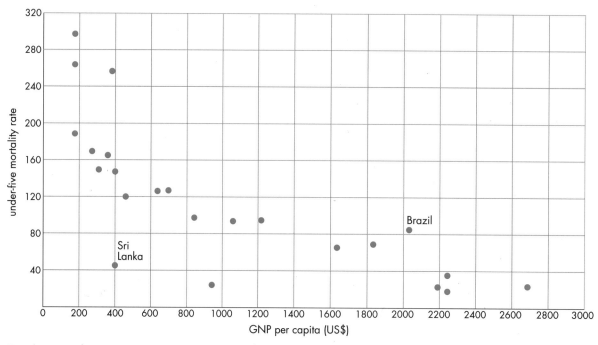

Figure 2.9   Under-five mortality and GNP per capita for selected countries.

World countries, and over five times that of Sri Lanka (US$400). Yet its under-five mortality rate of 85 is almost double that of Sri Lanka (43).

This is a good example of how broad-brush statistics can conceal as much as they reveal. The high average per capita GNP of Brazil tells us nothing about how that relatively high income is distributed. Now examine Table 2.8, which provides further economic, health and education data for the two countries.

Table 2.8 shows that, despite being a poor country compared with Brazil, Sri Lanka scores much more highly on its education indicators. Only 22%

of those who enrol successfully complete first grade primary school in Brazil compared with 88% in Sri Lanka. The percentage enrolling in secondary school is also much higher in Sri Lanka. Three per cent of government expenditure goes on education in Brazil compared with 8% in Sri Lanka.

You will remember that earlier in the chapter I noted the strong influence education has on susceptibility to diseases and that it is claimed by the World Bank probably to be a more important factor than access to safe water. This is borne out here as a much greater proportion of the population has access to safe water in Brazil

Table 2.8   Comparative indicators for Brazil and Sri Lanka (1988)

Indicator	Brazil	Sri Lanka
GNP per capita (US$)	2020	400
U5MR	85	43
Adult literacy rate (%)	78	87
Gross enrolment at primary school[a] (%)	103	104
Children successfully completing first grade of primary school (%)	22	88
Secondary school enrolment ratio (male/female)	32/41	63/69
Daily per capita calorie intake as % of requirements	111	110
Mothers breastfeeding (%; 3/6/12 mths)	66/58/34	94/92/81
Access to safe water (%; total/urban/rural)	78/85/56	40/82/29
Access to health services (%)	no data	93
Share of household income (%)		
top 20%	67	50
lowest 40%	7	16
Government expenditure (%)		
on education	3	8
on health	6	5

[a] The gross enrolment ratio is the total number of children enrolled in a schooling level (whether or not they belong in the relevant age group for that level) expressed as a percentage of the total number of children in the relevant age group for that level.

Source: UNICEF (1990) *State of the World's Children,* Oxford University Press, Oxford.

*Figure 2.10   Slums in Brazil.   The poorest live in
grinding poverty.*

than in Sri Lanka. One of the World Bank
reports I referred to estimated that maternal
education accounted for 34%, and increased
access to piped water 20%, of the mortality
decline in urban Brazil between 1970 and 1976.
It then added:

> "Obviously, piped water is a costly inter-
> vention; an analysis of the relative cost-
> effectiveness of various interventions (e.g.
> piped water vs female education) requires
> that their costs be estimated and matched
> to these estimates of relative effectiveness."

> (World Bank, 1983, Abstract)

Another striking contrast is the much higher
incidence of breastfeeding in Sri Lanka. Again,
this has long been associated with reduced infant mortality.

The final point to note about Table 2.8 is that,
whereas both Sri Lanka and Brazil show income
inequalities, the gap between rich and poor is
much larger in the case of Brazil, where the
poorest 40% of households receive only 7% of the
total private income and the richest 20% receive
67%. This is one of the largest gaps between rich
and poor in the world (Figure 2.10). The figures
for Sri Lanka, however, are roughly equivalent
to those in the western industrialized countries.

Brazil's average indicators, therefore, conceal a wide range between a wealthy élite and the great mass of rural poor and shanty town dwellers. Sri Lanka, on the other hand, has attempted a more equitable distribution policy, and this is one reason for its relatively good health indicators, despite its being a poor country. The government has long operated a food distribution programme, although its extent was much reduced during the 1980s by the ruling, market-oriented United National Party. The programme included issuing ration books or coupons for essential commodities, running a state system of food subsidies for low-income sections of the population, and providing special protein-enriched supplements to schoolchildren. The World Bank has noted that, despite major differences in priority between the two main political parties (the Sri Lanka Freedom Party and the United National Party), all governments in Sri Lanka since independence in 1947 have been committed to 'welfare, equality and the alleviation of poverty'. These extensive welfare expenditures plus a democratic tradition in the country have been linked to 'a level of literacy and advanced education unusual for low income Asian countries' (World Bank, 1987, pp.3–4).

This brief comparison shows that good health in a nation is not just a matter of creating economic wealth. It is also about the distribution of that wealth in terms of food, of water and sanitation infrastructures, and of education, as well as providing conventional health services. In other words, it is about alleviation of poverty.

This makes decisions concerning health policy fundamentally a political matter, where choices have to be made concerning limited resources. Look back at the quotation from the World Bank (1983) above. Here the choice is presented starkly between allocating resources to women's education or to piped water. Yet both will almost certainly bring about lasting improvements in a nation's health, as will sanitation and equitable food policies. They are essentially long term, however. For quick results, it is claimed that preventive vaccines and curative drugs are needed, but delivery of these depends primarily on multinational companies and the market place or on international aid, which do little to deal with underlying causes.

Finally, the dilemma of how to control infectious diseases raises general questions about prescriptions for development. Should development rely primarily on private enterprise and market-oriented incentives which reduce welfare subsidies but create economic wealth that may eventually filter through to the poorest sections of society? Or is it about state interventionist strategies aimed at distributing what wealth there is?

Tragically, there are no easy answers. Everything is much more complicated than is implied in *The Observer* article with which I began this chapter. We can make headway in some areas, and there have been success stories, but the diseases of poverty will be with us for some time to come.

Summary

1 The under-five mortality rate has been steadily decreasing in almost all Third World countries, but is still considerably higher than in the West. Nor does the gap between Third World countries and the West seem to be closing at present. In many cases the gap is widening.

2 There are considerable variations in patterns of disease prevalence within a country which the national statistics can mask. The poor and the very young in Third World countries seem particularly vulnerable to disease and there is evidence to suggest that in many countries the position of women in society relative to that of men makes them more prone to diseases.

3 Infectious diseases are the major killers in Third World countries. These result from dirty water and food, and polluted air, with malnutrition an important contributory factor. Poverty lies behind all these factors, but the alleviation of poverty is the subject of debate.

4 The experience of nineteenth-century Britain suggests that improved nutrition, water and sanitation are most likely to bring about lasting improvements in health in the Third World. While not denying this, others argue that biomedical intervention is an essential short-term measure. The problems associated with such intervention, however, are themselves formidable and involve large-scale aid or development of indigenous manufacturing facilities.

5 Biomedical responses to health in the Third World concentrate on drugs and vaccines, provided either via the market place or as aid. Public health measures emphasize proper nutrition, clean water, effective sanitation and education.

6 A policy that combines public health policies with basic bio-medical provision and which involves the active participation of the population is known as primary health care. This has been shown to be effective in Cuba and China.

7 Health policies require allocation of resources; such allocation is fundamentally a political question of priorities in development.

3

UNEMPLOYMENT AND MAKING A LIVING

DAVID WIELD

The issue of unemployment, like famine and diseases, seems simple but conceals great confusion and debate. There is complexity, for example, in the different understandings of the meaning of unemployment, as well as in the different views about what should be done about it. Here are simple statements of four views about the causes of unemployment. Which of them corresponds most closely to your own opinion?

- Overpopulation – too many people in the world (see the next chapter). The world's low-income countries cannot provide enough jobs to absorb the increased numbers seeking work.

- Overurbanization – too many people in the cities. If only people would move back to the land, problems of unemployment would decrease.

- Inappropriate technology – overinvestment in large scale capital-intensive industry that requires few workers.

- Inappropriate education – too many secondary school leavers who shun lower level and manual work and thus 'make themselves unemployed'.

You may not agree with any of the above. For example, you may think that unemployment is not such a serious problem – employment over time has risen and there must be a natural balance between supply and demand for employment. You may suggest that problems to do with lack of paid employment are a 'red herring' and that the official statistics overemphasize paid employment at the expense of the myriad means by which Third World peoples manage to make their livelihoods. Alternatively, you may argue that the problem is much worse than the statistics suggest since much of the poorer quality and worse-paid work is not registered (principally the work of women). Whichever view you take now, I hope that by the end of this chapter you will understand better the assumptions behind such different views and have the basis for a much more informed opinion.

Q What are the special characteristics of unemployment in the Third World? To what extent do the ways people make a living in the Third World differ from the rest of the world?

Q In the Third World, is there a sense in which people can be unemployed? Is there a sense in which they can be fully employed, or fully engaged in work?

Q How can we use an understanding of work and unemployment to assess policies designed to improve the work situation of Third World people?

The rest of this chapter is structured in three sections aimed at engaging with these questions. In the course of the chapter the focus moves from unemployment to the issue of human capacities, since attempts to increase human capacities to make a living lie at the heart of issues of economic development. And vice versa, since the kinds of economic development which emerge shape the potentials for human development, whether individual or collective.

3.1 What is unemployment? And what are work and employment?

"First Person: Ah looking work, mam…

Second Person: I have no work for you…

First: Ah can wash your car, mam…

Second: My husband has that done downtown…

First: Ah can tek care of the garden.

Second: We have a service for…

First: Ah can do anyt'ing, anyt'ing, mam.

Second: There is nothing you can do for me, nothing.

First: Well den, beg you a ten cent, mam…

Second: I don't believe in young healthy boys begging – that's what's ruining this country. Beg. Beg. Beg. You should be ashame. Go try to make something of yourself."

(Thelwell, 1982, p.109)

There are strong perceptions that the Third World is a world of high **unemployment** and low **employment**. In practice, the situation is more complex. In First World countries, unemployment means not having paid work. The vast majority of those with paid employment work for others for wages. Others, the 'self-employed', are also likely to take a salary out of their businesses.

The notion of employment being for money is fairly basic. Being unemployed is less simple; it is normally understood as lacking the means of *earning* a living – but this discounts the fact that unemployed people have to keep themselves (and their dependants) alive somehow.

> **Unemployment:** A concept generally restricted to the wage economy. It means being without work, i.e. not in paid employment, nor in self-employment (performing 'some work for profit or family gain') but currently available for employment and seeking it. This is the official meaning used in statistics.
>
> **Employment:** Either (1) paid employment for others, or (2) self-employment performing 'some work for profit or family gain'. There are three major attributes of employment (the income aspect; the production aspect; and the recognition aspect).
>
> The *unemployment ratio* is the number of unemployed people expressed as a proportion (usually as a percentage) of the total employed and unemployed population.

But unemployment has a more general meaning than that. We can get a sense of this from the quotation above, which is clearly about someone looking for work from someone else. We do not actually know whether the person is unemployed – but it seems likely. Any type of work will do for him, it seems, providing it brings money. Money, we can safely assume, is needed to survive in this environment (actually Kingston, Jamaica). The second person neither holds out any hope of employment, nor seems sympathetic to the fact that the first person needs to get paid work to earn his living. There seems to be an assumption that work can quite easily be found – blame is attached to the unemployed person for his predicament.

Two dimensions of unemployment are apparent here. The first is the economic, the second social and cultural. To be unemployed is generally to lose self-esteem, and often to be judged as not wanting to work, to be lazy (Figure 3.1).

Figure 3.1 Images of employment and unemployment: (left) from a Zimbabwean comic, 1991; (below) gold-miners coming off shift, Ghana.

The official usage of the concepts of 'employment' and 'unemployment', important as it is, does not cover all key ideas associated with making a living. In particular, these concepts focus on work for wages. The gap is an issue even in the First World – some work is not paid for. Examples are domestic work and voluntary work. But in countries of the Third World, where non-market (particularly agricultural) production is significant, the polarization between paid work and unemployment is nothing like as straightforward. The idea of self-employment put forward in Chapter 1, for example, is much broader than the official usage of the term.

Data on unemployment

One way of unpacking the conceptual problems is by looking at available statistical information. Statistics on unemployment show remarkable variations from country to country. Column 6 of Table 3.1 shows the overall unemployment rate for a selection of 10 countries in Asia. You can see that it varies from 1.8% in Bangladesh to 14.1% in Sri Lanka. Even the latter figure does not seem as high as expected, to me at any rate; 14.1% is a little under one in seven people.

Table 3.1 has two other types of data. Rural and urban unemployment are separated, as are male and female unemployment. Can you pick out any significant trends?

The only country where rural unemployment is higher than urban is Malaysia. Rural unemployment rates reported are relatively low, except in Sri Lanka. The national differences between male and female unemployment are less clear cut. In six out of ten, women's unemployment rates are higher than those for males. In Sri Lanka and Bangladesh the differences seem dramatic!

Table 3.1   Incidence of unemployment in selected Asian countries by rural–urban residence and sex, recent years

Country	Year (reference period)	Age of population covered	Unemployment rate All areas			Rural areas			Urban areas		
			Male	Female	All	Male	Female	All	Male	Female	All
Bangladesh	1984–85	10+	1.4	5.6	1.8	1.2	6.0	1.6	3.1	4.0	3.2
India	1987–88	5+	2.7	3.0	2.8	1.8	2.4	2.0	5.2	6.2	5.4
Indonesia	1985	10+	2.2	2.0	2.1	1.2	1.1	1.2	5.3	5.6	5.4
Malaysia	1980	15–64	4.7	7.2	5.6	4.7	7.7	5.8	4.5	6.2	5.0
Pakistan	1986–87	10+	3.3	1.1	3.1	2.8	1.0	2.5	4.7	2.0	4.5
Philippines	Oct. 1988	15+	7.7	9.9	8.5	5.0	8.7	6.3	12.7	11.5	–
Singapore	June 1987	15+	5.1	4.0	4.7	–	–	–	–	–	–
South Korea	1988	15+	3.0	1.7	2.5	0.5[a]	0.4[a]	0.5[a]	3.6[b]	2.2[b]	3.0[b]
Sri Lanka	1985–86	10+	10.8	20.8	14.1	9.5	21.3	13.2	15.8	27.9	19.5
Thailand	May 1986	11+	4.0	5.3	4.6	3.6	5.2	4.3	6.2	5.7	8.0

[a]Farm households; [b]non-farm households; –, no data.

Source: Visaria, P. (1990) *Concepts and measurement of unemployment and underemployment in ESCAP countries: a comparative study*, The Gujarat Institute of Area Planning, Ahmedabad, Working Paper No.31.

It is quite difficult, even with these relatively sophisticated data, to get a clear sense of the meaning of unemployment. Can it really be, for example, that in a country as poor as Bangladesh, there is such a low rate of unemployment as one in 50? Is there any significance that in India unemployment is registered for people aged over five years whereas in the Philippines and three other countries it is for those over 15?

There are a number of possible explanations. It could be that the statistics 'lie'. To be registered unemployed, for example, one first needs to be counted. If there are no labour offices in the rural areas or in the shanty urban areas there may well be a zero or low count in those locations. Also, to be counted as unemployed one usually has to be without employment, currently available for it, and 'seeking' it. There are good reasons why many people may not bother to 'seek' employment. If there are few jobs to be had

then it is well known that the number of people seeking them goes down. If employment rises then more people 'come out' to find it. If at the same time there are few jobs and no social security or unemployment benefit, then there is little reason why anyone should register. It is specifically the case that most Third World countries have little or nothing in the way of social security systems. There are no benefits for most who lose their waged jobs, quite apart from those without waged work in the first place. The unemployment rate may well be very low in poor, predominantly rural countries precisely because the poorest cannot afford to be without work of some sort.

The age figures in Table 3.1 are also suspect. Statistics over long historical periods show increased national income associated with lower use of child labour (Figure 3.2). So India, as one of the poorest countries measured in terms of

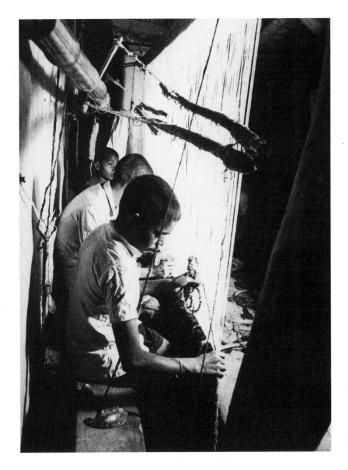

Figure 3.2
Child labour: carpet weaving for export is promoted in India as a 'village industry'.

national income, may well have large numbers of child workers – but does it have more than other countries in South Asia such as Bangladesh?

All in all, unemployment statistics by themselves do not allow us to get a good picture of how people do make a living nor (crucially if we want to avoid hunger, ill-health and famine) how people stop being able to sustain themselves.

Data on employment

Can we get any further by looking at data on employment, instead of unemployment? Table 3.2 compares total employment and total population for a selection of countries. Crude though such data are, they again give a striking sense of diversity. Employment as a source of livelihood seems much more important in some countries than in others. Do any patterns emerge?

Paid employment is clearly important in many countries of the Third World. For those who think of Third World countries as predominantly peasant societies made up of those producing mainly for themselves (subsistence producers – see below and Chapter 1), Table 3.2 is quite a surprise. In some Third World countries paid employment is dominant for men. Such countries, in the main, are those with a high level of urbanization (proportion of the population living in towns). Thus Latin America has a higher proportion of paid work than Asia and Africa. (In fact, some countries in Asia have a relatively high level of paid labour: for example, the newly industrializing countries of Taiwan, South Korea, Hong Kong and Singapore, which are not shown on the table.)

But Table 3.2 also shows that some countries more closely fit the stereotype of having predominantly non-waged producers. Burundi and Niger are particularly clear examples but most African countries in the table fit the pattern of places where most work is not waged work, even for males.

Data on male and female economic activity

To get at the idea of **work** and 'making a living' more broadly, we can try looking at another set of statistics, on **economic activity** (Figure 3.3). Table 3.3 gives data on the economically active population for a range of countries. What can be seen from the table?

Figure 3.3   Men's work tends to be counted in economic activity statistics whether it is formal waged work or not: (top) craft production of cars in an Indian factory; (above) informal craft work.

Table 3.2 Employment as a proportion of population, selected countries (1986–89)

Country	Employment (in millions) (year)	Population (in millions) (mid-1988 unless noted)	Employment as proportion of population (%)
Africa			
Algeria	4.418 (1989)	23.8	18.6
Angola	0.368 (1988)	8.8[a]	4.2
Botswana	0.170 (1988)	1.2	14.1
Burundi	0.054 (1988)	5.1	1.1
Kenya	1.311 (1988)	7.3	5.9
Niger	0.026 (1988)	22.4	0.4
Zambia	0.361 (1988)	7.6	4.7
America			
Bolivia	1.662 (1989)	6.9	24.9
Costa Rica	0.987 (1989)	2.7	36.5
Cuba	3.445 (1988)	10.1[a]	34.1
Chile	4.425 (1989)	12.8	34.6
Guatamala	0.801 (1989)	8.7	9.2
Jamaica	0.845 (1987)	2.4[b]	35.2
Mexico	8.291 (1989)	83.7	9.9
USA	117.342 (1989)	246.3	47.6
Asia			
India	25.745 (1988)	815.6	3.2
Malaysia	5.984 (1987)	16.5[b]	36.2
Sri Lanka	0.799 (1986)	15.8[a]	5.1
Philippines	21.849 (1989)	–	–
Europe			
Netherlands	6.155 (1989)	14.8	41.6
United Kingdom	26.212 (1988)	57.1	46.0

[a]Data for mid-1987; [b]data for mid-1986.

Sources: International Labour Organization (1989–90), *Labour Statistics*, ILO, Geneva; World Bank (1988, 1989, 1990) *World Development Report*, Oxford University Press, New York.

Table 3.3 Total economic activity rates, recent years (%)

Country	Men	Women	Year
Algeria	42.4	4.4	1989
Botswana	38.1	36.0	1984–85
Nigeria	40.7	19.7	1986
Tunisia	46.5	12.7	1989
Zambia	46.0	17.4	1984
Bolivia	45.3	33.0	1989
Chile	51.5	21.8	1988
Cuba	55.4	31.7	1988
Jamaica	44.1	23.1	1988
Mexico	53.7	23.1	1988
United States	57.4	44.7	1989
Bangladesh	53.6	6.5	1983–84
China	57.3	47.0	1982
India	52.7	19.8	1981
Indonesia	50.6	34.3	1988
South Korea	46.8	30.6	1986
Syria	40.3	6.8	1984
Greece	50.6	27.5	1987
Hungary	51.3	40.5	1987
United Kingdom	57.5	39.5	1986

Source: International Labour Organization (1986, 1987, 1989–90) *Yearbook of Labour Statistics*, ILO, Geneva.

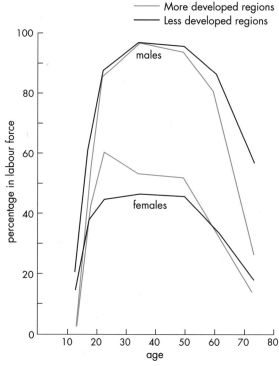

Figure 3.4 Labour force participation rates by sex and age, 1970.

First, one sees immediately that the economic activity rates are higher for men than for women, everywhere. Overall, the reported activity rates for women are a little over half those of men. Countries where female activity rates approach those of men include China, Eastern European countries and the USA.

Second, the activity rates for men are rather similar for men across all countries even between developed and less developed countries (Figure 3.4).

Work: Expenditure of energy for a purpose. Thus work includes both paid work (employment) and unpaid work, so-called 'formal' as well as 'informal' work, and domestic work, work done in kind, even voluntary work.

Economic activity: The *economically active* population includes 'All persons of either sex who furnish the supply of labour for the production of economic goods and services... [The] production of economic goods and services should include all production and processing of primary products whether for market, for barter or for own consumption,...the production of all other goods for the market and, in the case of households which produce such goods and services for the market, the corresponding production for own consumption.' (ILO, 1989–90)

Third, although the unequal division of domestic work could perhaps help explain differences between male and female activity rates, it cannot easily explain the widely differing activity rates of women. Compare Algeria with Bolivia, Bangladesh with Indonesia. Why should women's activity rates be so different and apparently so low? Beneria (1982) queries the low female activity rates reported in several parts of the world, including Arab countries as well as Africa and Latin America:

"A census taken in Sudan in 1966…resulted in a female labour force participation rate of almost 40%… In contrast with other official statistics that report rates barely above the 10% level. In a survey…in the Andean region, it was found that the proportion of women participating in agricultural work was 21% instead of the 3% officially reported. This type of underestimation is very common across countries, and especially in agricultural areas… To the extent that the amount of agricultural work performed by women is greater in the poorer strata of the peasantry…it implies that this underestimation differs according to class background and affects women from the poorer strata to a higher degree."

(Beneria, 1982, p.125)

Beneria gives three main reasons for the underestimation of women's work. First, the problem of defining who is an unpaid family worker.

"Given that involvement in home production is not considered as being part of the labour force, and to the extent that women's unpaid family work is highly integrated with domestic activities, the line between the conventional classifications of unpaid family worker and domestic worker becomes very thin and difficult to draw."

(ibid., p.123)

She is referring here to the difficulty of differentiating between 'domestic work' (such as preparing food for the household) and unpaid work done 'helping the head of the family in his (sic) occupation' (ILO, 1976, pp.28–9) as in agriculture.

Second, most censuses classify workers according to their main occupation. This will exclude women who are classified as 'housewife' but who are also working outside the household. In India it was estimated that this single exclusion lowered the female activity rate from 23% to 13% in the 1971 census. There are various reasons for underreporting of women's 'secondary occupations', according to Beneria.

"They range from the relative irregularity of women's work outside the household – that is, the greater incidence among women of seasonal and marginal work – to the deeply ingrained view that woman's place is in the household… In many countries it is considered prestigious to keep women from participating in non-household production; when asked whether women engage in such production, both men and women tend to reply negatively even if that is not the case."

(ibid., p.124)

Third, some activities are performed by women at home even though they are clearly tied to the market, as when they produce food and drink for sale, make handicrafts and clothes inside the household for sale outside, and so on. The integration of these activities with domestic labour makes them highly invisible (Figure 3.5).

The limited picture from statistics

So, in summary what can we deduce from the statistics?

1 Unemployment statistics do not give a reliable estimate of much other than those who expect paid work and are without it. The statistics are thus particularly problematic in those Third World countries with low numbers of wage workers.

2 Employment rates do give an idea of the level and extent of payment for work (commoditization of work, as Chapter 1 describes it).

3 Economic activity rate statistics do give a better idea of the extent of unpaid work. These data, even though they exhibit significant

Figure 3.5   Much women's work is statistically invisible: (left) formal waged work for women does get counted; (right) but work such as collecting firewood, whether for domestic use or for sale along with other items produced at home, is probably not counted.

underreporting, illustrate how important such work is to the ability of most households to make a living.

What are these statistics *not* able to help with?

Much economic activity in developed countries cannot easily be counted – that in the 'hidden economy' of second jobs, cash payment for work to avoid tax and so on. In the Third World, the very nature of much work as being outside 'formal' wage work makes it difficult to count. More generally, the issue of work and economic activity goes beyond the problem of measuring it, to questions of perceptions and social conditioning.

Examples of the kinds of work not adequately described by the statistics include:

1   Work in the *informal sector*.

2   Women's work, some of which can, of course, be categorized as informal. Time spent in unpaid labour in the industrialized countries of Europe has been estimated approximately to equal the time spent in paid work. In Third World countries the ratio of unpaid to paid work is likely to be considerably higher, given the time spent on agricultural activities (Ginwala *et al.*, 1991, p.7).

3   Agricultural work, which is considerably underestimated in most calculations of its contribution to national wealth.

4   Perhaps most important, the relationship between different types of work. For example, the statistics cannot give an idea of the varied ways in which households (and societies) combine activities to make their livings.

Informal sector: A common description, encompassing petty trading, self-employment, casual and irregular wage work, employment in personal services or in small-scale enterprises in manufacturing and services. Those unable to find (or retain) regular wage employment (the 'marginals') swell the ranks of the informal sector, characterized by its relative ease of entry with low capital investment requirements, and by being relatively labour intensive and unregulated. By contrast, work in the 'formal' or 'modern' sector refers to larger scale enterprises and employers with relatively stable employment, higher wages and more regulation of work conditions, and where workers can organize themselves more easily.

Examples of making a living

If statistics cannot deal with all the work in the Third World, are there other ways we can understand what is going on? Detailed research at a micro level can throw considerable light on the myriad ways in which people make their livings and relate to each other while doing so. Below is a description (from 1977) of a day in the life of one person, Domitila Barrios de Chungara, a women's leader and a miner's wife in a tin-mining community in Bolivia.

How many different types of work does Domitila describe? And in how many ways does she relate to others?

Domitila's day

"My day begins at four in the morning, especially when my *compañero* is on the first shift. I prepare his breakfast. Then I have to prepare the *saltenas* [a Bolivian small pie, filled with meat, potatoes, hot pepper and other spices] because I make about one hundred *saltenas* every day and I sell them in the street. I do this in order to make up for what my husband's wage

doesn't cover in terms of our necessities. The night before, we prepare the dough and at four in the morning I make the *saltenas* while I feed the kids. The kids help me: they peel potatoes and carrots and make the dough.

Then the ones that go to school in the morning have to get ready, while I wash the clothes I left soaking overnight.

At eight I go out to sell. The kids that go to school in the afternoon help me. We have to go to the company store and bring home the staples. And in the store there are immensely long lines and you have to wait there until eleven in order to stock up... So all the time I'm selling *saltenas*, I line up to buy my supplies at the store... From what we earn between my husband and me, we can eat and dress. Food is expensive: 28 pesos for a kilo of meat, 4 pesos for carrots, 6 pesos for onions... Considering that my *compañero* earns 28 pesos a day, that's hardly enough, is it?

Clothing, why that's even more expensive! So I try to make whatever I can. We don't ever buy ready-made clothes. We buy wool and knit... Well, then, from eight to eleven in the morning I sell the *saltenas*, I do the shopping in the grocery store, and I also work at the Housewives' Committee, talking with the sisters who go there for advice.

At noon, lunch has to be ready because the rest of the kids have to go to school. In the afternoon I have to wash clothes. There are no laundries. We use troughs and have to go get the water from the pump.

I've also got to correct the kids' homework and prepare everything I'll need to make the next day's *saltenas*.

Sometimes there are urgent matters to be resolved by the committee in the afternoon. So then I have to stop washing in order to see about them. The work in the committee is daily. I have to be there at least two hours. It's totally voluntary work.

The rest of the things have to get done at night. The kids bring home quite a lot of homework from school. And they do it at night, on a little table, a chair, or a little box. And sometimes all of them have homework and so one of them has to work on a tray that I put on a bed... So that's how we live. That's what our day is like. I generally go to bed at midnight, I sleep four or five hours. We're used to that.

I think that all of this proves how the miner is doubly exploited, no? Because, with such a small wage, the woman has to do much more in the home. And really that's unpaid work that we're doing for the boss, isn't it?"

(Johnson & Bernstein, 1982, pp.234–5)

It is possible to count the number of different types of work that Domitila did, and to see that only one of Domitila's activities involved payment for her work: selling food. It is clear, of course, that there was another source of household income – that of her *compañero*, a miner. Her last point is very simple, very important, and often forgotten: that her miner *compañero's* paid work depends on her unpaid work, such as food preparation for her family and washing clothes. The example illustrates also that low levels of wage employment do not equate with a simple life. Making a living here involves a complex and exhausting series of interactions, inside and outside Domitila's family. The implications are enormous. To give just one example, earlier we found that employment statistics give a better idea of those 'earning a living' (through paid work) than those 'making' a living without regular wages. In fact, it seems impossible to divide 'paid' workers from 'unpaid' in this way, given the ways in which they interact.

Domitila's workload is not uncommon for women. A study in Java showed women working an average of 5.5 hours of unpaid labour and 6.7 hours on income-generating activity. One study in Burkina Faso showed women working a 15-hour day in an area with poor access to fuel and water. A north Indian case study showed an average for women of 4.9 hours on 'domestic' work, not including childcare, and 4.5 hours on agriculture for household use and sale (Rogers, 1980 in Ginwala *et al.*, 1991).

At the household level, such mixed and complex work arrangements are common. The archetypal peasant household, with enough land and tools to produce for its own consumption and to market crops for those goods it cannot produce itself, is even less of a generalizable family than the proverbial two-parent, two-child household in Europe. Below is a description of the work of two households in Sri Lanka. As you read it you could make two lists: one of paid work; the other of unpaid work.

"One household with 13 members had seven sources of income: (1) operation of 0.4 acres of paddy land by the adults, (2) casual labor and road construction by the head and eldest son, (3) labor in a rubber sheet factory by the second son, (4) toddy tapping and jaggery by the wife, eldest son and daughter, (6) mat weaving by the wife and daughter, and (7) carpentry and masonry work by the head and eldest son. Another household with 11 members had six sources of income, mostly agricultural: (1) home garden by the family, (2) a one acre highland plot operated by the wife, (3) labor on road construction on weekdays and on the plot on weekends by the head, (4) seasonal migration to the dry zone as agricultural labor by the daughter and son, (5) casual labor in a rice mill in the dry zone by the eldest son, and (6) casual agricultural labor in the village by the head and his wife."

(Tinker, 1979, p.20)

These households were engaged in unpaid work in the household to produce food crops for their own consumption (subsistence production), in the production of a variety of goods in the household for sale outside it, and in paid work outside the household. The paid work includes full time work in a factory, seasonal migrant employment in another area, and casual labour of various kinds.

Overall, such examples show that we need to be concerned not just with 'earning a living' but also with 'making a living'. Conceptual distinctions are required over and above employment and unemployment. Two fundamental distinctions (see also Figure 3.7 below) are:

1   between paid (i.e. waged) and unpaid (i.e. unwaged) work;

2   between work which is remunerated (either indirectly through wages or directly through marketing the product) and work which is unremunerated (i.e. non-market-orientated subsistence or domestic work).

Although these are important distinctions, it is a mistake to regard different types of work as completely separate. For this reason, the idea of *duality* or *dual economy*, in which the typical Third World economy is seen as comprising two separate sectors, the modern/formal and the traditional/informal, is not very useful. It is certainly the case that large-scale modern industries and government bureaucracies offer formal paid employment alongside traditional and informal economic activities of all kinds, but in practice the sectors are not unconnected. The modern sector tends to depend on the informal and traditional for various inputs, and the majority of individuals and households combine work in the two sectors. Thus there are also conceptual issues surrounding the relationship between different types of work and different types of production. Much work is not done alone. Production necessarily involves relations with others.

3.2 Unemployment and making a living: processes and concepts

In this section, I shall delve into some of the conceptual difficulties surrounding unemployment, employment and work. I begin with processes that may lead to unemployment, and then introduce concepts of production through which emerge the incredible diversity of ways of making a living.

What causes unemployment?

Unemployment is not a single process. It can be caused by a range of different phenomena that affect the nature of unemployment.

Closing down socio-economic activity

Perhaps the clearest mechanism is the closure of an organization, whether a factory or a whole company, office, school, or hospital. Closure may not necessarily cause unemployment because, in principle, employees can be moved from one site to another, but it usually does. Unemployment can also be caused by the closure of part of a unit – a hospital ward, part of a factory.

In the period 1950s to 1970s, rapid industrial growth in the Third World (see Chapter 11) meant few closures. Even so, closures have dramatically affected some industrial regions. Closures and rationalizations increased in the 1980s.

For example, in Bolivia, following the collapse in the price of tin in the world market and the election of a right-wing government devoted to neo-liberal market economics, all state mines were closed in 1986 with 23 000 miners laid off including the *compañero* of Domitila Barrios de Chungara (Box 3.1).

Relocating socio-economic activity

The relocation of activity can cause unemployment. Although there may not be an overall global decrease in employment, this is hardly any comfort to those made redundant. Companies can threaten to move from one part of the country to another in search of cheaper costs. They can also threaten to move to another country. Although we have heard more about relocation from First World to Third World countries, threats to relocate from one Third World country to another have become more common as companies look for cheaper labour, better terms from governments, and so on. As with the closing down of socio-economic activity, such relocations from and within Third World countries are not that widespread. However, there are examples of multinationally owned production units relocating from 'higher wage' newly industrializing countries (NICs) to lower wage NICs.

Intensification of labour

So far, the processes causing unemployment have been straightforward. But 'intensification' is more complicated. Unemployment can be

Box 3.1   Extract from 'Domitila – the forgotten activist'

In three years, the district, which includes the Catavi mine, has lost over 2500 of the original 3000 workers. Those who remain are employed in maintenance and repair work. The miners' houses…were abandoned and rapidly becoming derelict. The hospital and the school, hard-won gains of the miners, lacked patients and pupils, and a few people stood idly around the union building, former hub of the district. The only activity of any note was on the strip of land leading from the mouth of the mine. Like convicts, co-operatives of ex-miners and local campesinos spend day and night there breaking rock in the hope of finding a little mineral that they can sell.

I asked Domitila about her own experience, and what had happened…

'In 1986 they shut down the mine and offered all the workers a pay-off. Well, of course we protested… The new government's economic policy was fatal for us. They wanted to break the strength of our union. They stopped providing the subsidized food which had made it possible to survive. At the beginning we all resisted and put up with the hardship, but it was impossible. In the end it's the hunger that wears you down. It's the children crying to be fed, and having to get up in the morning and pawn whatever you've got left to buy four rolls for breakfast or a kilo of sugar. In the end, despite our efforts and those of friends, we just had to go.

It was terrible. Most of us had been born in the mining community. It was the only life we'd ever known. When you want to move somewhere you plan it in advance, you prepare yourself for the change. But it wasn't like that for us. We were just thrown out from one day to the next. It was so sad. We were bewildered. We felt as if our mother was dying. When your mother dies you don't know where to go or what's going to happen to you. We didn't know when, or even if, we were going to meet again. We couldn't understand why we had to go. We stayed and stayed until we were forced to leave.

So we came to Cochabamba. My husband stayed at the mine. I suppose we're typical of so many families that split up over the changes. It's been difficult. Finding the younger children schools took up all our time to begin with, and then we had to find a way of surviving. It's been tough on the kids. We've all tried to get work, but none of us has anything permanent. You try selling pasties in the market – something we always did to supplement our income at the mine – but it's impossible. Oh, there are plenty of people who want to eat, but no-one has the money to pay. My son did get a job in a button factory and an optician's but he was laid off from both because of lack of demand. It's more expensive to manufacture things here in Bolivia than it is to buy contraband. So we make out as best we can. Sometimes we have enough to eat, sometimes we don't.'

(Sophia Tickell, *New Internationalist*, October 1989)

caused by changing the organization of work without increasing investment – for example, by getting employees to work harder. By this means, output can be maintained with fewer workers, or increased with the same workforce. Thus **labour productivity** is increased, perhaps causing unemployment. Labour intensification has been historically important in the Third World (see Chapter 9 on colonial labour regimes).

Labour productivity: The quantity of goods and services that someone can produce with a given expenditure of effort, usually measured or averaged out in terms of time spent working or labour time. It is the ratio of the amount produced to the amount of labour put in, measured as product per person-hour or person-year.

Technical change

The unemployment of employees (and their unemployment problem) does not necessarily correlate with problems for their employer. The productivity of labour can depend to a great extent on the tools or technology that the producer uses. Unemployment can result from increased capital investment – investment in new ways of making things or doing things that increases labour productivity and thus requires fewer workers per unit of output (Figure 3.6).

Technical change can occur at the level of one factory, and also more generally. Sometimes a new investment by one company can result in new jobs in that company, while wiping out a whole other way of producing the same goods or services. One major historical example was the introduction of the weaving machine that was estimated to put half a million handloom weavers out of work at the beginning of the industrial revolution. There are many such contemporary processes. For example, the increase in production and sale of plastic footwear has destroyed local leather industries in some countries.

Similarly, plastic household wear has lowered the demand for metal pots and pans. Negative reactions against 'technology' have been important in some political movements – Gandhi-ism in India is a well-known example since it served as a model for the appropriate technology movement (see Chapter 16).

Demographic forces

The mechanisms mentioned so far are examples of social processes that can be important causes of job loss. But there are other processes that have been very important in explanations of Third World unemployment, though they may not lead to job loss *per se*.

For example, a major cause of unemployment in many Third World countries is a growth in the number of young people needing employment at a faster rate than employment possibilities for them. There are at least two policy implications. One is that policies are required to stop populations growing so fast (see Chapter 4). The other is that other types of work creation could be prioritized over permanent paid employment,

Figure 3.6   Work in capital-intensive industry: (left) compare this South Korean factory with top of Figure 3.3; (right) compare these women's work with Figure 3.5 (left).

for example the hand-over of land for people to make their own living in rural areas rather than migrating to the city.

Land alienation

But policy and practice do not always converge. In practice, the opposite trend is more common – namely the privatization and enclosure of common lands, concentration of land ownership in fewer hands, and loss of land by many peasant producers.

The alienation of land was an early act of most colonizers of territory in what became the Third World. This act of land alienation is not exclusive to the Third World – on the contrary. However, what is different is the extent to which the contemporary history of the Third World has been shaped by colonial land alienation, and by land alienation after the colonial period.

In Africa, for example, the early years of colonialism were years of labour shortage. The President of the Chamber of Mines of South Africa complained in the 1890s:

> "The tendency of the native is to be an agriculturalist, who reluctantly offers himself or one of his family for just so long as the hut tax be earned, and expects the industrial demand to expand to give him work when his crops are bad. He cares nothing if industries pine for want of labour when his crops and home-brewed drink are plentiful."
>
> (Bundy, 1977, p.3)

Note the perennial comment about workers and alcohol!

Gradually, however, the level of employment rose. As an example, in one country, Tanzania, the first few decades of colonialism structured the colony into three main zones: (1) areas where agriculture changed towards the production of cash crops for export together with food crops for subsistence; (2) areas requiring wage labour (towns, plantations); and (3) labour reserve areas from where most males travelled as migrants to the areas requiring wage labour, leaving the rest of the household remaining on the land they owned or worked. The processes in

each area were different, but together they added up to a new labour regime linked to the new colonial economy: some people without land needed to earn wages and became 'proletarianized' (see Chapter 9); others without sufficient land to produce all their needs became partly proletarianized, working for a part of the year, or on contract for a part of their lives; some were able to make a living mainly from their agricultural production for the market as well as for themselves.

Even in a country like Tanzania, with relatively low colonial investment and industry, the process of proletarianization, slow and incomplete though it was, was under way and an 'unemployment problem' with it.

Some concepts

Analysis of the causes of unemployment leads us to the nature of work itself and thus to basic notions of production and reproduction.

Production

Production at its simplest involves interaction between people and nature. At one extreme, raw materials can appear 'ready made', as in the gathering of natural products. More typically, the raw materials used in a particular production process are the product of another production process. For example, the cotton used in the manufacture of textiles first has to be grown and harvested, then the harvested cotton has to be processed to separate the fibre from the other parts of the plant, and finally the cotton fibre has to be spun into thread before it can be used in making textiles.

A general definition of production is a process in which human energy is expended in changing nature (through various means) to produce goods for consumption (see Chapter 1).

I emphasized earlier the idea of productivity of labour – and how it depends on the capital invested.

> "For example, a farmer in the United States using a tractor and a combine harvester can produce, say, a ton of wheat with much less expenditure of time and human effort

than a farmer in India using an ox-plough. In turn the latter can produce a ton of wheat using less time and physical effort than a farmer in Africa who lacks a plough and has to cultivate with a hoe and other hand tools."

(Bernstein, 1988, p.55)

This quotation contrasts different tools and machines, but one crucial 'input' to production is human labour in the shape of those engaged in production. Human labour consists of a variety of physical and mental capacities and skills. We can think of the 'quality' of labour in terms of its possession of the capacities demanded by certain kinds of tasks. If those capacities are not fully available, this affects the productivity of labour adversely; for example, a producer might lack the training or experience to use a complex piece of machinery efficiently, or people's ability to perform arduous agricultural jobs in Africa or India might be undermined by their low levels of nutrition and health more generally.

The division of labour

It is an obvious assumption that American farmers like the one above do not make their own tractors, and also a realistic assumption for the majority of farmers of India and Africa today who use factory-produced ploughs and hoes. They have to obtain them from somewhere – ultimately from others whose work is to produce these different kinds of tools. This provides a simple example of the **social division of labour**, meaning that there are producers of different kinds of goods and services whose activities are complementary, and who are related to each other through the exchange of their products (even if the producers do not meet each other directly). As the social division of labour increases in complexity, it makes available a more diverse range of goods and services.

We can still assume for the moment that the three farmers are working alone on their own farms (even though they depend on tools produced by others). Such an assumption would be nonsensical, however, in the case of a car factory, which is a very different kind of *unit of production*. A car factory gives a good example of

a **technical division of labour**: the combination of different operations and tasks performed by a number of waged workers in the manufacture of a single product.

> **Social division of labour:** The degree of specialization *between* different units of production, and how they are related through the exchange of their products.
>
> **Technical division of labour:** The degree of specialization and combination of activities *within* any single unit of production or production process.

Work then, is more than what one individual does. The work of one individual affects that of others. This simple fact has important implications. For example, policies to improve the quality and quantity of jobs will affect a wide range of related types of labour.

Reproduction

It has been implied that the various elements or 'inputs' of the production process first have to be produced. Even the land used in agriculture, while originally a 'gift' of nature, is changed through people's interaction with it – its fertility can deteriorate, be maintained or even be enriched. Tools and machines used in production become worn out after a time; raw materials tend to be used up more quickly – for example, stocks of seeds and fertilizers used in each cycle of agricultural production. Therefore, these elements of production have to be replaced: that is, they require **reproduction** for production to continue in future. In the case of land, tools, seeds, etc., this is called *social reproduction*. It is also the case that the most vital element of the whole process – namely the producer – needs reproducing.

So far we have discussed production without using the 'gendered' terms 'he' or 'she', 'his' or 'hers'. Did you notice? I have signalled gender because we are entering an area of what many people think of as self-evidently 'women's work'.

Reproduction: All the processes by which the inputs of production are themselves produced. *Social reproduction* replaces the inert elements of the process. The 'production of the producer' involves *biological reproduction* (childbearing), *generational reproduction* (childrearing) and *daily reproduction* or *maintenance* (provision of human needs like food, shelter, etc.).

"The first and indispensable step, and also the most obvious, in the 'production of the producer' is that of *biological reproduction*. People have to be born for society to continue, and biology determines that only women can bear children. But this is only the first step. A child has to be cared for and raised until it reaches that stage of maturity when it can become independent, economically and in other ways. We will call this *generational reproduction*. A third aspect of reproduction is *daily reproduction* or *maintenance*. Adults as well as children need to replace or restore the physical and mental energies that they use up in the course of their daily lives. They have to eat and to rest, which requires the provision and preparation of food and drink, the maintenance of somewhere to live, clothes to wear which have to be washed, and so on."

(Bernstein, 1988, p.60)

Even childbearing is a social practice, conditioned by social relations and ideologies. While it is 'ordained by nature' that *only* women can bear children, there is nothing 'natural' about *whether* all women bear children, *when* they bear them, *how many* children they bear, nor that in many cultures there is a particular pressure on women to bear sons. There is nothing 'natural' about the fact that responsibility for bringing up children devolves on their mothers (or grandmothers, or aunts, or older sisters, or female servants, in different societies and social groups). Nor is there any 'natural' necessity that women should carry out the task of maintaining the current generation of producers as well as their children, who will provide the next generation. Earlier, I have called these tasks *domestic work*. In so far as domestic work is practised as 'women's work', it provides a special case of the division of labour – a *sexual division of labour* (see Chapter 15).

The specific ways in which the labour associated with generational reproduction is women's work clearly have significant implications for employment policies. In particular, it involves 'double' or even 'triple' days for many women. We have seen that women often have longer working days on average than men, they do different types of work, and often (like Domitila) their remuneration is low or zero.

The sexual division of labour extends beyond the domestic sphere. In many areas of Africa and elsewhere the process of agricultural production is divided into tasks by gender, with men tilling the soil and women performing the arduous work of weeding and harvesting. Generally, sexual divisions of labour are accompanied and justified by ideologies of men's 'place' and women's 'place' in society, which go beyond the different kinds of work regarded as fitting for them. The private or domestic domain tends to be seen as the province of women, and the public domain that of men. The latter may include wage employment outside the home (though in practice women also do wage work) and participation in public affairs more generally, whether particular religious and ceremonial roles and activities, deliberations of village councils, holding office in a co-operative, or active membership of a trade union.

What do you think are the implications of sexual divisions of labour for a society's or a group's productive capacity? Chapter 15 gives a more detailed answer. For the moment, note how, at the very least, if a particular sexual division of labour restricts a whole category of people to certain kinds of work and disqualifies them from acquiring other kinds of knowledge and skills which enhance their productive capacities, then this is likely to inhibit economic development.

Subsistence and commodity production

You have already encountered the next (and my final) key concepts. **Subsistence production** is production for direct consumption; **commodity production** is for sale and consumption by other than the producer. The distinction here is not in the physical properties of what is produced (tea, sugar, maize) but in the social relations of its production, distribution and consumption.

> **Subsistence production:** Production for the producer's own (or household) use.
>
> **Commodity production:** Production for sale through the market and consumption by other than the producer. *Full (capitalist) commodity production* is completely integrated into the market. Producers buy all inputs, including waged labour, and sell all the output. *Small (petty) commodity production* is only partly integrated into the market and is based on family not waged labour (see Chapter 9).

Subsistence production makes up a more significant proportion of goods and services in the Third World than elsewhere. In agriculture for example, a higher proportion of crops produced are consumed in the household. There are important implications for work.

1 Subsistence production is both unwaged and unremunerated.

2 Although it should count as 'economic activity', such work is likely to be underestimated in statistics, and thus its importance 'hidden' from policy makers.

3 In general, more such work is likely to be undertaken by women and children.

However, the use of the term 'subsistence production' can be confusing because it is often assumed that all peasant production is subsistence production. This is far from the case (see Chapter 1): some household members must either sell some product or work at least temporarily for wages in order to buy any basic essentials not produced in the household itself – tools, fuel, school fees, certain foods. Thus, there may be no completely 'subsistence household' in practice, even though a significant proportion of Third World producers engage in some subsistence production, together with production for markets or paid work.

Production for the market takes two major forms: *small (petty)* and *full (capitalist) commodity production*. The latter is defined simply as production which is completely integrated into the market – producers must buy all their inputs (including labour) and sell all their output. It is called capitalist because it involves as the key social relationship that between owners and wage workers. Since full (capitalist) commodity production is most important in First World economies it is perhaps unsurprising that our initial ideas of employment as 'wage work' correspond most strongly to it (Figure 3.7).

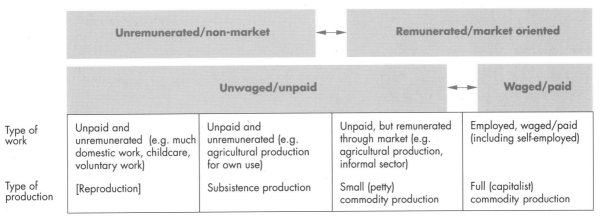

	Unremunerated/non-market		Remunerated/market oriented	
	Unwaged/unpaid			**Waged/paid**
Type of work	Unpaid and unremunerated (e.g. much domestic work, childcare, voluntary work)	Unpaid and unremunerated (e.g. agricultural production for own use)	Unpaid, but remunerated through market (e.g. agricultural production, informal sector)	Employed, waged/paid (including self-employed)
Type of production	[Reproduction]	Subsistence production	Small (petty) commodity production	Full (capitalist) commodity production

Figure 3.7 Work and production: how they fit together.

It is the in-between category of small (petty) commodity production that corresponds most closely to those most difficult categories of work: those that are unwaged but remunerated through marketing (Figure 3.8).

Petty commodity producers (see Chapter 9):

1   have access to means of production (land, tools, technologies);

2   operate on a small scale;

3   use household labour;

4   produce to satisfy basic needs;

5   have links to markets (they buy and sell) and thus operate in a wider social division of labour.

Some petty commodity producers have, overall, rather more control and ability to manoeuvre than others, or than wage workers. For example, peasant households that produce food crops can make decisions on the proportion to sell. On the other hand, a peasant household producing predominantly a permanent 'cash' crop of tea or coffee will be much more reliant on the market, since they cannot just tear up their bushes. Some petty commodity producers are able to 'get by' on this mixture of production forms. Others cannot, and may need to sell land or increasingly undertake wage work.

Not only do we find, then, a diversity of ways of making a living, but the linkage and mix between different types of work can, and does, change. It can be a very dynamic system: some households 'getting by'; others 'more than getting by'; and others gradually losing their ability to 'get by'. Small commodity production plays an important role, being both in-between subsistence production and full commodity production, and a key means by which households retain some control over their livelihoods.

In terms of the question 'Is there a sense in which people can be unemployed?', it may be difficult to find huge unemployment in the formal sense. But it is possible to conceive of a lack of 'quality' employment. People with expectations of wage work may lose it, fall back on petty commodity work or subsistence work, but still expect to go back eventually to 'better work'. Sen puts the situation well:

> "The ILO mission to Sri Lanka discussed cases where people said they were 'unemployed' when they meant that they did not have a regular job offering security and some sort of steady income... the question remains whether joblessness is the best way of viewing unemployment, and because of its emphasis on 'matching employment opportunities and expectations', the mission was, in fact, forced to view employment in a wider perspective."

> (Sen, 1975, p.6)

And the linked question: Is there a sense in which people can be fully employed, or fully engaged in work? The answer to that seems to follow quite naturally. People may not be employed in the formal sense, but may be relatively fully engaged in work. For example, because of

Figure 3.8 Petty commodity production is not only agricultural: Indian blacksmith.

low productivity and poor remuneration they may be obliged to work tremendously hard just to survive, engaged in petty commodity and subsistence production. Unsurprisingly, there is strong evidence that such people want better quality work and employment.

3.3 Improving ways of making a living

What use are concepts of work and production? The purpose of this brief final section is to show that understanding of policy issues can be improved by use of these concepts.

Using the 'disguised unemployed' for development

One relatively early development policy idea was that there is a high level of 'disguised' unemployment in the rural areas of Third World countries that could be drawn off and used for new development projects. The argument goes like this:

Some (maybe even a significant proportion) of the population engaged in peasant agriculture is not strictly necessary for production to continue and reproduce itself. In other words, taking away a part of the labour force would not lower production output.

Let us look more closely at this proposition. First, it is unlikely that any member of the family will not be working. I described earlier the high workloads of women. The existing data show women working much harder than men but they show men working hard also. Second, households are usually involved in petty commodity production and sometimes wage work so that income is shared in some way within the household. Intra-household distribution of income, uneven though it is, usually gives some rights to all household members while they are living in the household. But rights usually change when members leave the household. Third, it is likely that those in the peasant households, at least in many parts of the Third World, have rights of ownership or control over land. Again,

such rights may change for those who leave the household. In many of the poorest countries, people hand over control or access to land at the peril of famine and death. For these three reasons, and others, the idea of relatively free individuals able to take up employment elsewhere does not quite seem the way to describe the situation, though underemployment is certainly common.

> **Underemployment:** Work that does not permit full use of someone's highest existing skills or capacities. This could mean, for example: working for shorter periods, less intensively than able or willing to work; working at a lower level of productivity than capable of doing; earning less than able or willing; or working in a production unit with abnormally low productivity (Standing, 1981, pp.38, 43).

The argument goes on to suggest that the 'disguised unemployed' or 'underemployed' can be transferred to more productive work. The implication is that such a transfer means rural to urban migration to provide the labour for economic development through industrialization.

Can you think of any problems with this argument? I introduced some problems above. Access to land can be the key to long-term survival, i.e. daily and biological reproduction of households. Thus we have seen above how households develop ways of linking employment in higher productivity units of production with continuing production on the land: for example, by some household members migrating for regular wage income at the same time as keeping a base on the land.

A second problem is that, in practice, opportunities for higher income wage labour have been easier to obtain in some Third World countries than in others. Overall, it is important to emphasize that there has been significant growth in wage employment over time. But it has been unevenly spread.

The third issue is the actual level of spare time. You will remember the two earlier descriptions: of Domitila, and the household with diverse activities. Many development projects have been criticized because they have targeted those who already work very hard. Put simply, they do not have much extra spare time; rather they need assistance in increasing their productivity of labour and thus their output, without further intensification of their work!

Overall then, our first look at policies to use the 'disguised unemployed' for development does not look too promising.

But our concepts can point to some useful openings for those interested in improving economic conditions. For example, underemployed people may not have spare time but work with very low productivity. Small interventions may improve their conditions. An example might be the installation of a water pump close by that would lessen the time spent collecting water and thus free time for more productive activities. Another example might be that people work hard but do not earn enough money. A development policy targeted at this problem might need to increase income-earning possibilities while avoiding loss of subsistence production and lowering time spent on domestic work.

Women's work and development

The well-meaning policy mentioned in the last example of targeting those who work hardest is especially important for women. Consider this quotation from a policy document of a political party in the Third World.

> "[We] will strive to ensure that any new economic policy will have as an important policy objective the rapid successful integration of women into economic activity."

I chose this quotation because it is from a group which has made considerable efforts to integrate women and to increase their participation in politics. The quotation illustrates how persistent and easy it is to assume that women are not already integrated into economic activity.

From your reading of earlier parts of the chapter, what kinds of women's work may be hidden from policy makers?

Childcare, production and preparation of food, possibly subsistence cultivation, may be 'invisible'. But there are other hidden elements that have been the key to failed development projects: the ignorance of the overall extent of women's work contribution – both paid and unpaid; the assumption that the household is male headed; not taking into account women's independent cultivation or other income-generating activities.

The fact that such policy statements can appear, even from groups cognisant of the dominant role of women in much economic activity, has serious implications for development projects. Women can be targeted for 'integration...into economic activity', on the basis of massive misunderstanding of the extent to which their unpaid labour allows paid labour. There may be little 'free' labour time available (see Chapter 15).

Development interventions can affect existing social relations of production in various ways. One example is the provision of training courses to increase the skills of women. But such training courses can be set up without planning or provision to substitute for the women's unpaid labour. Courses can fail because women just cannot get free time. Other interventions have been designed to increase labour-intensive low income-generating opportunities for women without thought about their already 'breaking point' working days.

Targeting women in development policy more sensitively could lead to increasing the priority given to unpaid over paid labour and would involve social as well as economic policy. Measures like those mentioned earlier (improving water provision under women's control or improved access to cheap fuel supplies for cooking) could dramatically free women's labour time in rural areas in many countries.

Finally, there is strong evidence that the structural adjustment policies pursued in the economic crisis from the 1980s have tended unequally to push women out of the development process and to increase the proportion of domestic, subsistence and low paid commodity production. Chapter 15 will deal with this in more depth.

Summary

1   Preconceptions of the Third World include those that there is much unemployment and little employment. We found rather that work in the Third World is characterized by its complexity and diversity and, importantly, by the relationship between paid and unpaid work. There are few, if any, situations where households are either all waged or where none receive income.

2   Unemployment statistics, by themselves, do not give an accurate picture of those in work and those who need increased income to make a living.

3   Statistics do not adequately portray the larger proportion of unpaid labour done by people in the Third World, nor that much of this is done by women.

4   In Third World economies, there is a higher proportion of subsistence production, and also a higher proportion of petty commodity production.

5   Rather than people being unemployed, a major issue is that many work extremely hard but at levels of low productivity, receiving low financial recompense, and thus remaining in relative poverty. They require opportunities for better quality and better remunerated work.

6   Understanding work involves understanding the relationship between different forms of production:

- What is produced?

- How is it produced?

- For whom (division of labour)?

- What is the relationship between production and reproduction?

Answering such questions helps us assess development projects and development ideas to improve peoples' work capacity.

4

IS THE WORLD OVERPOPULATED?

TOM HEWITT AND INES SMYTH

"The population bomb about to explode"

(Headline, The Guardian, 23 February 1990)

"There is now solid scientific proof that population growth does not hinder economic development, and does not cause resource depletion."

(J. Simon, *The Guardian*, 21 July 1990)

"Let us act on the fact that less than five dollars invested in population control is worth a hundred dollars invested in economic growth."

(President Lyndon Johnson, speech to the United Nations, 1965)

"Development is the best contraceptive."

(Bucharest slogan, 1974)

There is clearly some disagreement over whether there is a population problem and what this problem is. Are there too many people in the world or is this an exaggeration? Should family planning be a priority in developing countries or should development be top of the list, leaving population growth to take care of itself?

Your immediate reactions may well be that there are too many people in the world; the greater the population, the less there is to go round; more people means less income and less food. In short, high population growth is a problem in itself and the cause of many other problems in the world. This is the traditional Malthusian view that population growth will sooner or later run up against the limits of the Earth's finite stock of resources. This is the message from the press and television. It has also long been the message from the major Western aid donors.

Alternatively, you may believe that population growth is not a problem at all. You may agree with Julian Simon in the above quotation and think that population growth is a crucial impetus behind human progress and that any control of individuals' reproductive choices is an infringement of fundamental human rights.

These are the two most extreme views of the population issue. But you may realize there must be other, more complex, factors at play, and we need a more comprehensive set of questions.

Q Is the world overpopulated and, if so, in relation to what?

Q Whatever the consequences of population growth – and these are by no means clear – what are the causes?

Q Is the majority of the world's population poor because there are too many people and not enough resources to go round or does the world's population continue to grow as a result of the poverty of the majority of people?

Crude birth rate: The number of births per 1000 population in a given year. Not to be confused with *growth* rate.

Demographic transition (theory of) and **demographic trap**: see Box 4.3.

Fertility: The actual reproduction performance of an individual, a couple, a group, or a population.

Fertility rate: The number of live births per 1000 women aged 15–44 years in a given year.

Infant mortality rate: The number of deaths to infants under one year of age in a given year per 1000 live births in that year (see Chapter 2).

Population distribution: The patterns of settlement and dispersion of a population.

Population growth rate: The rate at which a population is increasing (or decreasing) in a given year due to natural increase and net migration, expressed as a percentage of the base population.

Replacement level: The level of fertility at which a group of women on average are having only enough daughters to replace themselves in the population.

Total fertility rate (TFR): The average number of children that would be born alive to a woman (or group of women) during her lifetime if she were to pass through her childbearing years conforming to the age-specific fertility rates of a given year. In the US today a TFR of 2.12 is considered to be replacement level. (If there were no deaths before childbearing age it would be 2.0.)

Q Is family planning the ultimate solution to population growth, and, if so, why does the world's population continue to grow after 30 years and US$3 billion per year invested in population control? Are there other solutions to population growth?

By the end of this chapter, you should be able to give reasoned answers to these questions. First, let's see what the problem is and examine its dimensions. You may find it helpful to have a look above at the definitions of some demographic terms that will occur in the chapter.

4.1 The size of the problem

What does Table 4.1 tell us about the size and distribution of the world's population?

First of all, note that the total world population in 1990 amounted to over 5 billion people, about twice as many as 40 years earlier in 1950. You will also notice that in 1990 nearly 80% of the world's total population was to be found in 'low'

Table 4.1 Population size and growth rates by region (1950–90)

(a) Population size by region (in millions of people)

Date	World	High-income countries	Low-income countries
1950	2515	832	1683
1970	3698	1049	2649
1990	5321	1214	4107

(b) Average annual population growth (%)

Date	World	High-income countries	Low-income countries
1950–55	1.8	1.3	2.1
1970–75	2.0	0.9	2.4
1985–90	1.8	0.5	2.1

Source: UN (1989) *World Population Prospects 1988*, United Nations, New York.

income countries. This is hardly surprising since these countries comprise a large part of the world's square mileage. Yet the proportion of the population which is in low-income countries has been increasing over the years. Why do you think this is?

Look at Table 4.1(b). The population growth rate is much higher in low income countries. But you should also note that growth has *declined* in high-income countries and *fluctuated* just above 2.0% in low-income countries. So the growing gap between the two groups is due as much to the declining rates of growth as it is to the high rates of growth.

From the current level of 5.3 billion people, the world's population is projected to increase by 90 million per year to the year 2000. Most of this increase will be in developing countries. The United Nations Population Fund (UNFPA, 1989a) projects that population will be 6.25 billion at the end of the century, 8.5 billion by 2025 and may stop growing at about 10 billion perhaps a century from now. We should stress, however, that projected population data which go further than the following decade should be treated with great caution (Figure 4.1).

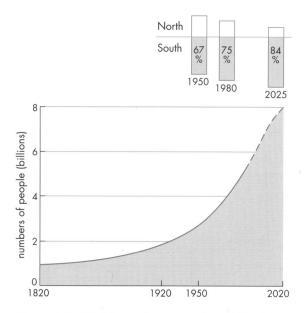

Figure 4.1 World population trends and North–South distribution.

Whatever the precise figures, the foreseeable future will produce many more people. But why does population grow and why do LDCs have higher growth rates than developed countries?

From a demographic point of view, two elements account for this: *mortality* and *fertility*. In most developing countries, while mortality rates have declined substantially, fertility rates have not. In developed countries, there is a much closer match between these two rates. When mortality rates are low, high fertility generates a 'population momentum', such that an increasing proportion of women in the total population is of childbearing age who will reproduce the next generation. This is why, even if growth rates in LDCs continue to decline, they will have ever greater shares of the world's population.

Improved health, as we have seen in Chapter 2, is a major factor in mortality decline in developing countries. Nevertheless, health and sanitary conditions are still far from satisfactory, and fatal disease is still prevalent in children. This has a direct bearing on fertility. Why? Because the more the under-fives die from curable diseases or causes related to poverty and malnutrition, the more likely are parents to have more children to compensate for possible losses. Note, though, that this is not an unproblematic explanation nor is it the only one to explain persistent high fertility, as we will see later in this chapter.

The third and often forgotten element of the population equation is the *distribution* of population. Density of population in the world is very uneven. Urban agglomerates are in stark contrast to rural regions in many countries. By the end of the century it is calculated that 48% of the world's population will be urban (see Chapter 11). Whole continents show a remarkable variation in population density. Africa, for example, is a continent where we find underpopulation and overpopulation side by side (Box 4.1).

The causes of unevenness in the distribution of population range from ecological endowments, colonial history and forms of production to national and international migration and resettlements (or any combination of these).

Box 4.1 Africa: overpopulated, underpopulated or both?

Historically, many areas of Africa actually suffered from *depopulation*, as the result of the slave trade, exploitative colonial labour policies and the introduction of new diseases from Europe. In the eighteenth century, 20% of the world's population lived in Africa; by the year 2000 the figure is expected to be less than 13%, despite recent high rates of growth. Only since the beginning of this century did Africa begin to rebuild its population.

Today, the problem is not so much absolute numbers of people, but their distribution. The average density of sub-Saharan Africa is only 16 people per square kilometre. In interior central Africa there is a belt of relatively low population growth, where women suffer from high rates of infertility associated with sexually transmitted diseases. The continent has millions of hectares of potential rain-fed cropland in the humid tropics, most of which is uncultivated and underpopulated.

Yet there are also areas of very high population density in the cities, along the coast, and in the highlands, where population pressure has contributed to environmental degradation. In Ethiopia, for example, the highest densities are in the drought-prone, environmentally vulnerable highlands, while there are thousands of acres of uncultivated arable land in the south and the east.

In West Africa the demand for labour is the crucial determinant of population densities and high fertility. There is a very low level of economic development and an undersupply of labour in many parts of the region. This can be traced back to the colonial era when the forced recruitment of labour, forced growing of cash crops, taxation, and military reprisals by colonial troops compelled African peoples to 'produce as many children as possible to increase labour supply and reconstitute as much of their local economy as possible under colonial conditions.' Added to this was the large-scale male migration to plantation zones and coastal cities for employment. With husbands absent, women depended even more on children as a source of agricultural labour and security.

Today this same migration pattern continues. In Burkina Faso, for example, during certain times of the year when young men migrate to raise cash for taxes and consumer needs, there is not enough labour to clear bush for new farms or maintain wells, and food production suffers accordingly. A number of observers have in fact noted how low population densities serve as a brake on agricultural production in Africa.

(Adapted from Hartmann, 1987, pp.17–18)

From the perspective of development, migration is as important as size and growth of population. First, it is an integral component of the demographic equation: population change is a combination of natural increase (births minus deaths) and net migration (immigration minus emigration).

Second, it has a complex, two-way relationship to development. The level and the nature of development in a given country or region determine migration patterns, while the latter in their turn impact on development. Finally, migration decisions have emotional and practical repercussions of great significance, affecting the quality of people's life in a way that is comparable to fertility- and mortality-related events.

People migrate in large numbers within and across international boundaries. The substantial migrations after the Second World War to Western Europe from many developing countries (frequently ex-colonies) are well known, as is the movement of Spanish-speaking labour from Central America to the southern states of the US, but there has been little consideration of the development impacts of other migrations. For example, it is estimated that in the early 1980s there were nearly seven million migrant workers in the oil-producing Middle Eastern countries from the Indian subcontinent and South-east Asia (Khan, 1990, p.88).

Other examples of substantial local and international population movements would include: the seventeenth-century slave trade, European migration to North and South America at the turn of this century, migration induced by war and famine such as from Indo-China and sub-Saharan Africa, the urban evacuations imposed by Pol Pot in Cambodia, the Indonesian transmigration programme (Figure 4.2). There is a long list of forced and voluntary migration around the world. Can you think of others?

Figure 4.2 'Transmigration – for a brighter future'. Under this Indonesian government programme, thousands of families have been moved from densely populated Java to other Indonesian islands.

Already you can see the complex nature of population issues. Table 4.1 provided only limited and aggregated information. Table 4.2 gives you a (numerical) taste of the variety within this, as a point of reference for the rest of the chapter. For now, you can check what the column headers mean by looking at the key definitions at the beginning of this chapter; and then look at the individual country data to see whether they are above or below the average for their region.

4.2 Views on population

Given declining mortality rates and generally high fertility rates in many countries of the world, we now return to the original question: is the world overpopulated? Already we must take a more considered approach because of the patterns of distribution of populations around the world. So we must ask: in relation to what is the world overpopulated? Is it, as some believe, in relation to available resources such as food, or is the problem the distribution and, therefore, access to such resources? We know, for example, that food production has outstripped population growth in the last decades, as Figure 4.3 indicates.

So why do so many children live below minimum nutritional levels? Think back to Chapter 1. It is a question of whether the mouths to be fed have access to the food they need. This can be summed up as follows:

> "...overpopulation is not a matter of too many people, but of unequal distribution of resources. The fundamental issue is not population control, but control over resources and the very circumstances of life itself."

> (Michaelson, 1981, p.13)

So is there a problem of overpopulation in the world? The question has to be asked not in absolute terms but in relation to distribution of, access to and use of resources. Nevertheless, this has not stopped people from speculating on 'overpopulation' as a problem in itself. In what

Table 4.2 Population indicators for selected countries and regions (1988, unless specified)

	GNP per capita (US$)	Infant mortality rate	Total fertility rate	Urban population (%)	Population projection mid-1990 (millions)	Population projection 2020 (millions)	Population growth rate (%)	Population under 15 years old (%)
Bangladesh	170	120.0	4.9	13	114.8	201.4	2.5	43
Brazil	2280	63.0	3.3	74	150.4	233.8	1.9	36
China	330	37.0	2.3	21	1120.0	1496.0	1.4	27
Cuba	n/a	11.9	1.9	72	10.6	12.8	1.2	25
India	330	95.0	4.2	26	853.4	1375.0	2.1	39
Indonesia	430	89.0	3.3	26	189.4	287.3	1.8	38
Kenya	360	62.0	6.7	20	24.6	60.5	3.8	50
Sri Lanka	420	22.5	2.3	22	17.2	24.0	1.5	35
Thailand	1000	39.0	2.6	18	55.7	78.1	1.5	35
Uganda	280	107.0	7.4	49	18.0	42.2	3.6	49
Africa	600	109.0	6.2	31	661.0	1481.0	2.5	45
Asia	1430	74.0	3.5	29	3116.0	4805.0	1.9	34
Latin America	1930	54.0	3.5	69	447.0	705.0	2.1	38
Europe	12 170	12.0	1.7	75	501.0	516.0	0.3	20
North America	19 490	9.0	2.0	74	278.0	328.0	0.7	22

Source: PRB (1990) 1990 World Population Data Sheet of the Population Reference Bureau, Population Reference Bureau, Washington DC.

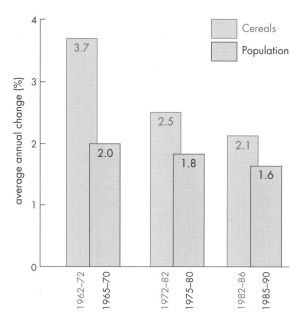

Figure 4.3   Growth in world food production compared with population growth. If the food supply can be increased at the current pace (or even a slower one), there will be enough food for a stable world population of 10 billion in 100 years.

follows, we present some of the most influential ideas concerning the world's population.

New Malthusian view

One of the most pervasive views in contemporary thinking on population is New (or Neo) Malthusianism, derived from the arguments of Malthus (Box 4.2). It shares Malthus' original view that population growth is a major cause of poverty but differs from Malthus in the belief that human intervention can put a check on population growth (through birth control). There are several variants of New Malthusian arguments, but the common feature of the more modern version is demographic determinism, i.e. that the weight of population itself is the cause of problems. This is the 'people versus resources' perspective (Lappé & Schurman, 1988) – epitomized by Paul Erlich's *The Population Bomb* published in the 1960s – which states in no uncertain terms that there are (or will be) too many people for the available resources (Figure 4.4).

Figure 4.4   Is population growth the cause of these conditions? Evicted squatters in Dhaka, Bangladesh.

Box 4.2 Malthus on population

Malthus (1766–1834) was an economist, most famous for his pessimistic *Essay on Population* (1798). In this he maintained that the human race tends to reproduce in geometrical progression (2, 4, 8...) while food supplies can only grow arithmetically (1, 2, 3...). Because of this law of Nature, humanity would breed up to the limits set by the supply of food, checked only by the inevitable consequences of its own growth: extreme poverty, famine, wars, pestilence etc. Thus, as population grew, there would be a fall in average output of food per head, which would create growing misery and eventually could only be 'resolved' by famine or war. Malthus was renowned for his support of the landed gentry, which blinded him to the view that extreme misery was caused not so much by diminishing returns to labour as population grew but by the lack of political bargaining power of the peasantry *vis-à-vis* the landlords.

In this view it is argued that rapid population growth is the primary cause of the problems of the Third World since it results in widespread poverty, economic stagnation, environmental destruction, rapid urbanization, unemployment, and political instability. During the years of the Cold War and the Vietnam War, in particular, the view prevailed that overpopulation, resulting in mass hunger, was a breeding ground for revolutionary activity. Central to such views is that it is the poor who produce more children, because of their ignorance and lack of foresight.

The most recent variant of the New Malthusian argument comes from some of those concerned with the pace of environmental change in the world. Look, for example, at this statement from an editorial in the prestigious medical journal, *The Lancet*:

"Human population has exploded five-fold in the past century and a half. For the first hundred years this explosion did not impose irreversible pressures on the biosphere. Over the past fifty, however, the inexorable growth of human population and increasingly numerous industrial offshoots have come to threaten the health of the planet."

(*The Lancet*, 1990, p.659)

The debate over environmental degradation is taken up in detail in Chapter 5. Suffice to say two things here. Arguably, it is those countries which are near or below population replacement level (as in Western Europe; see Table 4.2) which are the greatest consumers of non-renewable resources. Here, too, the worst environmental dangers are created, in the guise of acid rain, nuclear 'accidents' etc., and it is here that economic interests motivate the practices of deforestation and overexploitation of the soil and the sea, from which threats to the world's environment are generated.

Second, in certain parts of the world, the large numbers of people hungry for woodfuel, pasture or food crops result in the overuse of land. But again, this appears to be due to population distribution and an unequal access to resources rather than the sheer weight of humanity. In other words, there may be a link between population size and lack of resources in a given region but there is not always a direct relationship.

In the New Malthusian view, to set matters to rights, it is imperative that people, and especially the poor, should be persuaded (or forced, if this is necessary) to have fewer children. Improving the impoverished conditions in which they live is a secondary concern, reducing birth rates being the first. This can be achieved through family planning programmes, which have been promoted for over 40 years by international agencies as an efficient and cost-effective way to tackle the problems of development.

The New Malthusian orthodoxy sees the tendency to population growth as a constant factor in human history which needs no explanation in itself. When we ask *why* populations grow, the difficulties with New Malthusian argument begin to appear. The causes of population growth are precisely those aspects of socio-economic life which they put forward as its consequences. Equally, we could ask why fertility decline has occurred in industrialized countries, and find answers in a complex set of socio-economic changes, not in population control policies.

Being able to understand the causes of high population growth (and of fertility decline, where it has occurred) would, logically, give us strong clues on how to reduce fertility rates more effectively in those countries or regions for which population pressure mounts by the decade.

Specific explanations of why fertility rates are high or low

There is no agreement on what causes fertility rates to be high or low. Farooq & De Graff (1990) group many of the explanations into three types.

1 *Proximate variables*. This type of explanation recognizes that there are a series of social, cultural and economic factors which have an impact on fertility. However, between these and fertility outcomes there are a number of 'proximate variables' which affect fertility rates directly in a given social, cultural and economic context. These variables include:

- the proportion of people married;
- contraceptive use;
- the prevalence of abortion;
- post-birth infertility.

The explanatory power of such variables is not great. They provide a link between broader socio-economic 'factors' and fertility, but questions remain about what are the underlying causes.

2 *Mechanisms of demographic transition*. In demographic transition theory (Box 4.3), no mechanisms are given precisely, but the following factors are suggested (individually or in combination) which all relate to a general process of industrialization and development, and which could all cause fertility to fall following an earlier reduction in mortality:

- urbanization, which increases the cost of raising children;
- a fall in the production value of children due, for example, to education, the introduction of labour laws, increased demand for skilled labour;
- increased education and age of marriage for women;
- the changing role of women and increased earning opportunities outside the household;
- intervening religious and cultural values;
- a shift to greater individual control of fertility decisions;
- lower infant and child mortality;
- higher returns for education of children;
- more welfare and insurance schemes for the elderly;
- greater availability of contraception.

Box 4.3   Demographic transition theory

This theory (much cited in books on population) is based on what was observed to occur in the industrialized countries. In the wake of the Industrial Revolution, death rates dropped dramatically, mainly as a result of improved sanitation and public health. Then, after a generation or so of higher population growth, the birth rate also began to fall.

In other words, there was a historical shift from a situation of high mortality and high fertility to one of low mortality and low fertility.

Today, in large parts of the Third World, as previously in Europe, death rates have been reduced. Demographic transition theory expects that after a period of transition (which according to World Bank estimates may be shorter than in industrialized countries), fertility levels and thus birth rates will also fall.

In LDCs, the experience is much more varied than that of the industrialized countries in the past. The theory would hold that with development and lowering mortality, fertility should follow suit and also reduce. But this has by no means always occurred. This has led some to a discussion of a 'demographic trap' where LDCs become stuck in the transition phase of low mortality/high fertility (King, 1990).

We should note that demographic transition is a model and like any other is open to question. The fact that the model proposes that mortality and fertility rates will take a certain path, does not mean that this will happen in practice. In fact the model is one based on a particular view of development – modernization theory (see also Chapters 6 and 11), which assumes an inevitable progression towards a 'modern' society, though at very different rates.

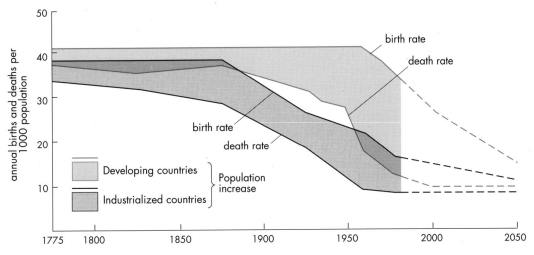

Trends in birth and death rates, actual and projected, 1775–2050 (World Bank, 1980).

3 *Micro-level explanations.* Here there is strictly no single theory, but a range of analyses of specific situations. Their common denominator is that, one way or another, they are criticisms of grand theories, such as demographic transition. Among this category of explanation are the following:

- explanations based on the economic value of children which is high in societies with low levels of social security for the elderly or where the value of children's labour is relatively high;

- the notion of intergenerational wealth flows: fertility is high where wealth flows from children to adults (as in different forms of child labour) and is low when it flows from adults to children (as in paying for school fees).

The social view

None of the above constitutes a broad coherent alternative view to Malthusian orthodoxy. They are too specific and partial as explanations. However, in 1974, at the Bucharest World Conference on Population set up by UNFPA, the New Malthusian argument was challenged by representatives from the Third World, on the basis that it diverted attention and money from the real causes of underdevelopment. Representatives from developing countries also reclaimed the right to define for themselves what they perceived their population problems to be and to resolve them, free of pressures from more powerful international agencies. Since 1974 a new wisdom has become accepted, incorporating some of the partial explanations of fertility rates mentioned above, that can, perhaps, be called a *social view*.

Central to this is the notion that rapid population growth is not the cause of the development problems of the Third World, but rather a symptom. This also means that, rather than being poor because they have many children, people may have many children because they are poor. In fact, the economic value of children to the poor is one reason that fertility rates are high in some developing countries. We saw above that, when infant mortality rates are high, women tend to have more births to ensure that a sufficient number of children survive to adulthood. However, a 'sufficient number' is a variable quantity dependent on economic and security considerations such as the provision or not of pensions, and the level of savings. Poor women have children because children's labour is valuable in itself and also may release their mothers for work (Figure 4.5). In addition, in economies with little or no welfare provision, children provide parents with security in old age. Proponents of this view would point to remarkable declines in fertility in a few countries with low per capita

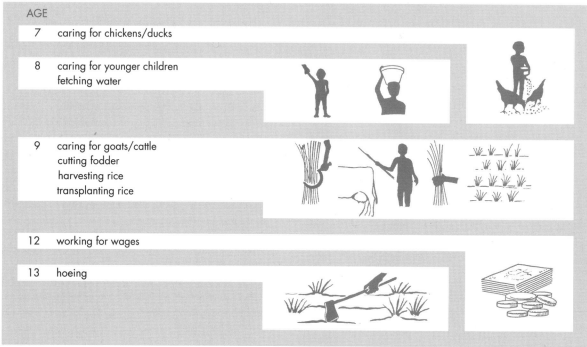

AGE			
7	caring for chickens/ducks		
8	caring for younger children fetching water		
9	caring for goats/cattle cutting fodder harvesting rice transplanting rice		
12	working for wages		
13	hoeing		

Figure 4.5   Children's labour is valuable: (above) survey of child activities in Java, 1977, showing age from which different activities start; (top right) boy tending water-buffalo, Indonesia.

incomes (Sri Lanka, Thailand, Cuba, the Indian state of Kerala) and point out that in those cases there has been greater access to social resources such as health care and education, particularly for women. It is lack of access to such resources rather than material poverty *per se* that tends to mean high fertility.

Can you imagine parents sitting at the kitchen table with a calculator, working out the future economic benefits of their children? Of course, individuals do not take decisions only on the basis of economic calculations. Other, non-economic considerations are equally important (Figure 4.6). These include relations between women and men, social pressure, dominant norms concerning the family and the role of women, individual (men's and women's) desires and circumstances. Calculations over numbers of children are further influenced by the widespread phenomenon of son preference.

Thus, it is argued that increasing living standards, lowering child mortality and improving the position of women via more equitable social and economic development will motivate people to have fewer children.

Family planning services, in this approach, are simply part of the necessary provision of health and other social services to the entire population, and a response to a genuine need for voluntary fertility reduction.

Figure 4.6 Why have children? Non-economic considerations are important.

The female autonomy view

Another angle on population is centred on the role of women. It is not entirely a separate approach, since the 'social view' has included a recognition of the importance of women in both development and demographic change. However, this 'female autonomy view' goes further in underlining the importance of the women's dimension of the population issue.

The close relationship between women's status and fertility is now accepted by demographers and policy makers. More egalitarian gender relations in and outside the household are known to have a positive impact on fertility reduction. Women who have access to better education and employment opportunities tend to have less need to rely on their children for economic security and social recognition. Potentially, better educated women are more able to safeguard the health of their children, thus contributing to the reduction of infant mortality. They also have better access to contraceptive information. Can you think of any other ways in which women's status and fertility are related?

The international agencies have also put women centre stage. For example, the UNFPA chose women as their focus of the 1990s.

"In many societies, a young woman is still trapped within a web of traditional values which assign a very high value to childbearing and almost none to anything else she can do. Her status depends on her success as a mother and on little else. Increasing a woman's capacity to decide her own future – her access to education, to land, to agricultural extension services, to credit, to employment – as an individual in her own right has a powerful effect not least on her fertility."

(UNFPA, 1990a, p.4)

Though this recognition of the centrality of women was inevitable because of the obvious role of women in bearing and rearing children, for some it has unsatisfactory connotations. There is a too narrow emphasis on statistical correlations between different aspects of women's life (e.g. education, employment) and fertility. Although these correlations are significant, they tend to hide the social mechanisms which are at the root of demographic behaviour. For example, why would you think there have been observed correlations between electrification of rural villages and declining fertility? What has also not been sufficiently realized is that:

1   Women are not always able to control their fertility because of their lack of decisional and formal powers in their households and in the community.

2   High fertility may indeed be advantageous to some women (despite the possible health risks) as a route to higher status, economic security or access to financial and other resources.

Criticism has gone further. For example, in the work of G. Sen (1989), Hartmann (1987) and Lappé & Schurman (1988) there are two powerful messages. The first concerns the difference between rhetoric and practice and the second is about women's empowerment.

The means through which women would be able to regulate their fertility according to their own needs have been available in most countries of the world for several decades. Their availability, however, is determined by the national and international agencies which are concerned with population control rather than birth control (Box 4.4).

While at least some of the implementers of population policy now recognize women's role in population dynamics, there is still a reluctance to attack the problem practically from this angle. Interest in women may still be confined to reducing fertility, rather than their welfare as such. Other UNFPA documents acknowledge the positive value of women's self-determination and even their need for choice in childbearing (UNFPA, 1989b, pp.12, 36). However, there is no indication that this element of choice should apply to the services of the family planning programmes themselves, to which methods of contraception should be developed, to which type of information should be available to allow

> **Box 4.4   Population control versus birth control**
>
> The distinction between population control and birth control goes beyond semantics. Birth control refers to the right of individual couples or women to control childbearing (timing, spacing and total number of children) on the basis of individual choice. Population control refers to policies aimed at controlling the same for the population as a whole on the basis of demographic 'imperatives' such as the perceived need to cut the population growth rate (Rogow, 1986, p.74). The fundamental difference between these two approaches is in who is empowered to make decisions over fertility.

women to make a free and informed choice, to how and when to use specific contraception methods, and to where and from whom contraceptives should be available (Figure 4.7).

The female autonomy view argues that, yes, family planning should be available but it is doomed to failure unless individual wants and needs are first taken into consideration. This, it is argued, is achieved through the **empowerment** of women and is the second message. Women's lack of social power and autonomy is an explanation of high fertility. However, women's empowerment should be an aim in its own right. Fertility reduction is viewed as a secondary aim to be adopted by women themselves according to their needs and desires. In other words, fertility decline comes as part of a bigger package (G. Sen in *Wuyts et al., 1992*).

"The question of reproductive choice ultimately goes far beyond the bounds of family planning programmes, involving women's role in the family and in society at large. Control over reproduction is predicated on women having greater control over their economic and social lives and sharing power equally with men."

(Hartmann, 1987, p.34)

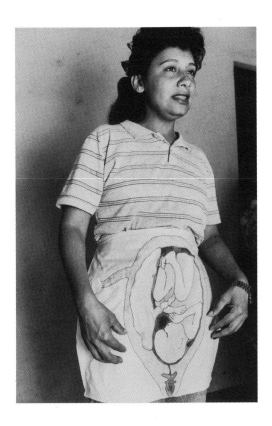

Figure 4.7 (Left) Instructor at ante-natal class, Nicaragua. (Above) family planning campaign image from Cuba, January 1991: 'For your health and your children's, plan your family'. Compare these posters with those in Figure 4.8.

Empowerment: Having or being given power or control. It is generally used to describe the desirable state of affairs in which individuals have choice and control in everyday aspects of their lives: their labour, reproduction, access to resources, etc. However, there is an immediate contradiction within the idea of individual empowerment, since people tend to be restricted in their lives or to have power over others as a result of social relations and structures outside their own control. Paulo Freire, the Brazilian radical educator who promoted the term, argues that empowerment should be thought of in social class terms. 'The question of social class empowerment…makes 'empowerment' much more than an individual or psychological event. It points to a political process by the dominated classes who seek their own freedom from domination, a long historical process' (Freire & Shor, 1987).

In sum, the female autonomy view stresses women's control over reproduction as opposed to population control. This is not only a reaction to the excesses of population control programmes (which have been notoriously negligent of human rights and of the severely disadvantaged conditions of many women's live) but also a prescription for the way forward. Women's autonomy is seen not only as a right but as a precondition for future fertility decline. Exponents of the female autonomy view see the social and economic welfare of women as the prime target. This includes birth control but does not give it priority. It is a long-term view requiring substantial social change. The New Malthusians would charge that the long term is too far away and action has to be taken now. The response to this could be that 40 years of population control to date is also long term and the results have not been altogether favourable.

4.3 Seeking solutions

It may by now seem obvious that what is seen as a solution depends on what is perceived to be the problem. A government believing its territory to be relatively underpopulated may put forward policies aimed at increasing the birth rate. Conversely, most international agencies, and many Third World governments, believe the problem is one of overpopulation, or at least of too high a rate of population growth, and their policies attempt to reduce the birth rate by reducing fertility rates. You should also realize from the previous discussion that different views on causes and consequences will have different implications for deciding on how it is thought solutions can best be achieved. It is the views adopted by those agencies and governments which dictate what policy measures are actually implemented.

Over the last four decades, increasing amounts of financial and human resources have been pumped into 'population control' in developing countries. In 1960 international population 'assistance' was US$6 million. By 1970 this figure was US$172 million (a 20-fold increase in real terms) and in 1988 it was US$420 million. This comes mostly from US-AID, The Population Council and UNFPA (UNFPA, 1990b).

In the 1950s and even more in the 1960s the activities of such international institutions and the countries funding them (donor countries) were clearly informed by the New Malthusian view. Massive support was given to direct population

control initiatives as well as to contraceptive technology. However, the perceptions which regulate the approach of such agencies may have changed over time.

Does this view still inform and guide the present-day ideas and activities of donors in the field of population?

The World Bank's *World Development Report 1984* remains probably the most comprehensive official document on population in recent years and remains an influential statement of the official position on population issues. So it is worth examining it to see what official view still prevails. Does this reflect the new 'social view' or is it a restatement of the New Malthusian position? As you read the following extract from the concluding section of the report, you will see elements of both positions. Which do you think is the dominant one?

"This Report has shown that economic and social progress helps slow population growth; but it has also emphasized that rapid population growth hampers economic development. It is therefore imperative that governments act simultaneously on both fronts. For the poorest countries, development may not be possible at all, unless slower population growth can be achieved soon, even before higher real incomes would bring down fertility spontaneously. In middle-income countries, a continuation of high fertility among poorer people could prolong indefinitely the period before development significantly affects their lives. No one would argue that slower population growth alone will ensure progress: poor economic growth, poverty and inequality can persist independently of population change. But evidence described in this report seems conclusive: because poverty and rapid population growth reinforce each other, donors and developing countries must cooperate in an effort to slow population growth as a major part of the effort to achieve development."

(World Bank, 1984, p.185)

Although it is not always made explicit, the position is evident from the themes stated at the beginning of the report. First, 'Rapid population growth is a development problem.' Then, 'There are appropriate public policies to reduce fertility.' While there is welcome recognition that poverty can be a direct cause of high fertility, it is argued that poverty alleviation (improving income opportunities and social insurance, expanding education, employment and health opportunities for women, and so on) 'takes time to have an effect' on population. The World Bank argues that population growth can be stemmed more quickly and directly with 'some appropriate combination of development policies geared to the poor, family planning and incentives' (p.9). It then goes on to say that: 'Experience shows that policy makes a difference.' At this point, family planning programmes take centre stage.

In other words, despite some acknowledgement of the importance of social factors, the official view is still largely the New Malthusian one which argues for direct population control measures (alongside development policies). You should have been able to understand this also from the quotation above from the report's conclusion.

Indeed, the international population agencies and donors continue to back top-down programmes aimed at reducing fertility directly, rather than simply extending the availability of contraception and family planning services as one of a range of improvements to living standards (Figure 4.8). In practice, however, despite the huge investments mentioned above, results have been mixed. For example, in Indonesia fertility rates have dropped while in Bangladesh they have not, despite long-term, expensive population control programmes in both countries.

Figure 4.8 All over the Third World the message is the same.

This is not to argue that top-down policies cannot be effective, but that their outcomes are difficult to assess. In any case they are only partial solutions. The different successes of such policies in different countries is likely to be due to the very kinds of social factors that the 'social view' proponents emphasize.

Top-down population policies vary considerably in terms of how much external control is exerted over fertility and how much room is left for individual autonomy in family planning decisions.

In practice, most attempts at fertility reduction do not involve the imposition of direct means of population control. Instead, programmes use systems of incentives and disincentives aimed at increasing or decreasing fertility levels. For example, high fertility may be encouraged by increased child benefits, maternity grants, child care allowances and special family loans (as in many Eastern European countries). Disincentives may range from loss of tax exemption, low priority in access to education and housing for families who have more than a set number of children (as in China), or even penalties, such as delay in payments of salaries for civil servants (as in Indonesia).

These are examples of indirect incentives and disincentives. Many family planning programmes also include direct incentives in the form of sums of money or clothes given to health and family planning workers for reaching set targets, or to the 'acceptors' themselves. Such payments have been internationally condemned and so they are often presented as 'compensation' for lost work time or for transport expenses. The problem with direct incentives is that, particularly in situations of poverty, they lend themselves to abuse by both health workers and acceptors.

Despite the promotion of the 'social' and 'female autonomy' views, recent trends in fertility control do not augur well for individuals' autonomous control over fertility decisions. One such is the increased use of long-lasting, provider-dependent contraceptives. These include IUDs, injectables (e.g. Depo-Provera) and implants over which the user has no individual control. These are selected on the grounds of efficacy (for target reaching), costs, ease of delivery and duration, rather than considerations of people's ability to determine the size of their families.

A related trend involves vested interests in such contraceptive methods. Most research and testing on these methods used to be carried out in the US. Since the Thalidomide case and, more recently, the Talcon Shield case, and with the rise of a strongly conservative political lobby in the US, pharmaceutical companies have become reluctant to invest in reproductive technologies which bear the risk of liability to heavy payments for damages or costly indemnity to the firms. Testing in developed countries has also become restricted. Research and testing has been taken on by international agencies such as UNFPA and WHO which contract out production (Population Reports, 1987; Nair, 1989). These are the very agencies which administer many of the population programmes of developing countries. As a result the agencies have a vested interest in one particular kind of reproductive technology, i.e. long-lasting and provider-dependent contraceptives.

These trends are particularly worrying for women: first of all because most contraceptive methods are directed at women who have no part in determining research priorities and standards (IWHC, 1986, p.8). Second, some of the trends have serious implications for women's health. For example, women in developing countries have been used frequently for testing contraceptives (Atkinson, 1989). The women of Puerto Rico, Haiti, Guatemala and Chile were among the first to take part in the tests of contraceptive pills and sterilization injections in the 1960s and 70s (Mass, 1976). More recently, women in India and Bangladesh have been used in the trials of hormonal contraceptives (Nair, 1989; Hartmann & Standing, 1989). In all cases, such experiments were carried out with little or no information given to the women themselves or, at times, to the local personnel involved in the trials.

It may seem incredible that despite all these initiatives to control population, mostly in the direction of reducing it, not all people have access to means of contraception. In many countries there is still a large 'unmet demand' for contraception (*The Economist*, 1990, p.21) from people who wish to control their fertility but do not have access to reliable means of doing so. For example, unmarried women may be barred access to contraception because of their marital status.

Thus, on one side we have enormous efforts, in money and organization, to reduce the number of children born, especially in developing countries; in many cases these efforts have overstepped the boundaries of what is considered acceptable in terms of human rights, personal safety and self-determination. On the other side we have regions and groups of people (e.g. unmarried women) for whom the means of controlling fertility are still outside their reach. What do you think are the reasons for these apparent contradictions and what would be the most appropriate means to solve them?

Summary

To summarize, let us reconsider briefly the questions posed at beginning of the chapter.

1   Is the world overpopulated? If so, in relation to what? Overpopulation is relative: it depends on where you are talking about and depends on what you measure population against: food, the environment, wealth, employment, health and so on. One needs to look at the population of specific regions relative to specific factors.

2   What are the causes of population growth? Here the New Malthusians, who blame the world's ills on population growth itself, cannot help much. In fact, we have argued, they confused consequences with causes. Continued high levels of fertility in many countries are real, but they will not be reduced without first identifying their causes. The 'social view' and 'female autonomy view' point to a variety of social factors largely related to poverty, and to lack of female autonomy, as at least partial explanations for high fertility.

3   What is the relationship between poverty and population growth? Certainly it is too simple to say the latter causes the former. On the contrary, poverty can be identified as a key variable in continued high fertility. Poverty induces poor health which in turns endangers the lives of infants. Where the under-five mortality rate is very high, individuals reckon more births are needed to ensure the survival of one or two children who can care for the parents in old age. But material poverty in itself is not always a cause of high fertility. Some countries with low per capita incomes but good access to social resources such as education and health care, especially for women, have recorded big declines in fertility. A fuller definition of poverty as the cause of overpopulation should include questions not only of distribution but also of women's autonomy and power to choose family size and timing.

4   Finally, are population control policies the solution to population growth? The international donors have followed the New Malthusian line, and put their money where their mouth is for the last 40 years. But women and men, even when coerced, do not limit the number of children they have without good cause. In the industrialized countries, fertility decline occurred as a result of a complex set of socio-economic changes. In developing countries, population policies have proved most unsuccessful (for example, in Bangladesh and parts of India) where the causes of high fertility have not been taken into account. In other words, population control by itself does not work, even under coercive conditions.

Recently, however, the 'social view' and 'female autonomy view' are beginning to be noted even among the large donors. Education and health provisions have entered the official vocabulary of international population agencies, as have women's employment and status. But still the bulk of finance goes towards population control. One may conclude that such programmes will continue to have a marginal impact on fertility until women and men have greater power of decision-making over the circumstances of their lives.

> "Economic and social development is a precondition for slowing down and eventually halting population growth. Therefore...it is in the common interest of humankind to ensure that the developing countries are enabled to make sustainable economic and social progress as rapidly as possible... jobs and better wages, health care and better nutrition, better education, emancipation of women, access to family planning, these are the conditions in which family size will fall, as has happened in the developed countries."
>
> (Thomas, cited in Ekins, 1986, p.12)

5

ENVIRONMENTAL DEGRADATION AND SUSTAINABILITY

PHILIP WOODHOUSE

The second half of the twentieth century has witnessed a growing concern with the negative impact of human activity on the physical environment. By the 1990s concern with 'the global environment' has become one of the most potent factors shaping politics in industrialized countries, engendering a debate that concerns not only the environmental consequences of economic development through industrialization, but also environmental degradation in non-industrialized economies. As a result, the question of environmental change has become a key element in discussion and analysis of Third World development.

> **Q** What factors determine the environmental outcome of development?

> **Q** Are there certain forms of environmental degradation specific to the Third World? If so, what are their causes and main features?

In addressing this question, I shall first trace briefly the growth of an awareness of the 'global environment' and of a perception of a specifically 'Third World' aspect to the degradation of that environment. Secondly, I shall examine the ways that Third World environmental problems are related to Third World livelihoods. Finally I shall examine ideas of 'sustainable development' to see what implications they have for Third World development.

5.1 The Third World in the 'global environment'

A changing awareness: from 'conquest of Nature' to 'managing the commons'

Transformation of the physical environment has been an integral part of the development of human society since its inception. The 'domestication' of plants and animals for agricultural production was achieved by modification through human intervention of the genetic composition, numbers and distribution of biological species. The construction of irrigation systems and other systems of water control, such as terracing of hillsides and dykes for flood control, constituted drastic transformations of natural ecosystems in order to improve conditions for human social existence. Even environments that seem natural have usually been modified by a degree of human intervention not apparent to the outsider. The pattern, common to semi-arid parts of West Africa, of scattered leguminous trees in land used for shifting cultivation and pasture, is beneficial to livestock and crops but is no more 'natural' than the traditional English hedgerow and field pattern.

Until comparatively recently human transformation of the environment was considered a necessary and creative activity: 'the conquest of Nature'. Perception of a negative impact only began with the advent of industrial production and concerned the visible threats to health posed by 'the two great groups of nineteenth-century urban killers – air pollution and water pollution, or respiratory and intestinal disease' (Hobsbawm, 1968, p.86).

The study of ecology in the post-war period, however, brought a perception of less visible dangers. Rachel Carson's book *Silent Spring*, published in the early 1960s, documented the way in which the insecticide DDT accumulated in successive stages of the food chain and reached toxic levels in the higher stages, thus killing large numbers of mammals and birds. The case of Minimata in the 1970s demonstrated a similar process in which a population of Japanese fishermen were the ultimate victims of an accumulation through the food chain of mercury which entered rivers from local paper-processing mills.

This evidence of invisible but systemic damage, demonstrating interconnectedness of apparently separate processes and the cumulative damage caused by local activity, laid the basis for a questioning of the whole pattern of development through industrialization. A typical conclusion is found in *The Limits to Growth*, published in the early 1970s: 'If the present growth trends in world population, industrialization, pollution, food production and resource depletion continue unchanged, the limits to growth on this planet will be reached sometime within the next one hundred years. The most probable result will be a rather sudden and uncontrollable decline in both population and industrial capacity' (Meadows *et al.*, 1972).

In the 1980s, international concern was expressed through the setting up of the UN World Commission on Environment and Development (the Brundtland Commission). Its report stated:

> "Over the course of this century, the relationship between the human world and the planet that sustains it has undergone a profound change. When the century began, neither human numbers nor technology had the power radically to alter planetary systems. As the century closes, not only do vastly increased human numbers and their activities have that power, but major, unintended changes are occurring in the atmosphere, in soils, in waters, among plants and animals, and in the relationships among all of these. The rate of change is outstripping the ability of scientific disciplines and our current capabilities to assess and advise. It is frustrating the attempts of political and economic institutions, which evolved in a different, more fragmented world, to adapt and cope. It deeply worries many people who are seeking ways to place those concerns on the political agendas."
>
> (WCED, 1987)

Shortly after publication of the Brundtland Report, the 'deep worries' had indeed made their way onto the political agenda in the form of the Montreal Protocol to protect the ozone layer (Box 5.1).

Box 5.1   Managing the global commons: the atmosphere and stratosphere ozone depletion

Atmospheric pollution from coal burning has always been one of the more visible consequences of industrial development. As clean air legislation enforced measures to remove soot particles from flue emissions, the pollution effects of the invisible gases produced by burning coal and other fossil fuels became more apparent. In the late twentieth century vehicle exhausts overtook industrial plants as the main source of such gases in some urban areas. The principal gases produced are: carbon dioxide, carbon monoxide, sulphur dioxide, nitrogen oxides, and (in less efficient burning) hydrocarbons such as methane. During the 1970s it was recognized that these pollutants, although comprising less than 0.1% of the atmosphere, were involved in chemical transformations which created hazards. In particular, at a local level, in areas receiving large amounts of sunshine, ultraviolet radiation in sunlight causes nitrogen dioxide and hydrocarbons to react chemically, producing 'photochemical smog' close to ground level, where one of the reaction products, ozone, causes irritation to eyes and respiratory system. At a regional level, 'acid rain' resulted from the reaction of nitrogen oxides and sulphur dioxide with water droplets in clouds to form nitric and sulphuric acids respectively. However, the first attempts to establish binding international agreements to limit levels of atmospheric pollution were triggered by concern in the 1980s over two more far reaching effects of atmospheric pollution: depletion of ozone in the upper atmosphere (stratosphere), and accumulation of infra-red absorbing gases in the lower atmosphere (troposphere) – the greenhouse effect.

Let us look at the first of these in more detail. Ozone in the upper atmosphere (the stratosphere) normally absorbs large amounts of ultraviolet radiation from the sun which would otherwise reach the Earth's surface and damage biological organisms through its effect on genetic material. Ozone depletion was first discovered over the Antarctic in 1985 by the British Antarctic Survey and has since been associated with compounds known as chlorofluorocarbons (CFCs) which have found their way to the stratosphere. These compounds do not occur naturally, but have been manufactured since the 1930s for use as refrigerant gas, aerosol propellants and cleaning fluids.

The discovery of the ozone 'hole' revealed a gap in scientific understanding of the atmosphere, since the mechanism involved did not form part of previous models of the atmosphere. Response was rapid. In 1987 the United Nations Environment Programme convened a meeting of 62 governments in Montreal to set out measures to halve the use of CFCs by the end of the century. By the time a second meeting took place, in 1990 in London, to ratify the 'Montreal Protocol', the aim had been changed to eliminate CFC manufacture completely by 2000. One factor behind this more ambitious target was the discovery by ICI in the UK and DuPont in the US that the compounds known as hydrofluoroalkanes (HFAs) and hydrochlorofluorocarbons (HCFCs) could be used instead of CFCs for the principal applications in refrigeration and aerosols. After 1987, ICI invested £100 million in research on HFAs, and by late 1990 had started manufacturing the chemicals at a new plant costing £30 million at Runcorn on Merseyside. However, not only were the new HFAs estimated to be about five times more expensive than the CFCs, but the production processes were protected by patent, unlike those for CFCs, on which patents had expired. The willingness of the UK to ratify the protocol was not shared by China or India, who pointed out that though they used few CFCs at present, they would ordinarily have expected to use CFCs in order to increase standards of living. By banning CFC use they were effectively having to accept lower living standards in order to clean up a mess created by richer nations. Indian industry had recently invested in CFC production technology in order to build up output to 200 000 tons per year – to supply 300 million fridges. The Indian representative at first refused to ratify the Protocol unless the new technologies were made available to Third World industries, and finally signed only on the understanding that if the technologies were not transferred, India would not be obliged to stop CFC manufacture.

Indian representative Maneka Gandhi at the London ozone conference, 1990. An unusual example of an environmentalist activist given a government position.

Environmental degradation in the Third World

In the discussions of global environment by the Brundtland Commission and others, two specific concerns with environmental degradation in the Third World stand out.

First, there is a fear that the industrialization of the Third World will greatly increase the amount of pollution, and particularly the atmospheric pollution which it has been claimed will cause catastrophic climatic change (Figure 5.1). This fear is derived from comparison of consumption of resources, and particularly energy, in industrialized and non-industrialized countries.

The Brundtland Commission has claimed that in order to supply people in the Third World with a level of consumption equal to that now enjoyed in industrialized countries, global energy consumption would need to increase fivefold,

entailing a similar increase in atmospheric pollution from burning coal, oil or gas. Others have argued further that, whereas technological change has increased the efficiency of energy use in industrialized countries, large-scale industrialization in the Third World using older (and more polluting) technology will cause a disproportionate increase in global pollution levels. Thus, of China's plans to industrialize by using its coal reserves to generate power (as Britain did in the nineteenth century), Greenpeace's Director of Science said 'It is an energy plan which spells death-by-climate for millions and disaster for all.'

The second type of environmental problem frequently associated with the Third World is typified by extracts from a 1988 newspaper article reproduced in Box 5.2. Note what the author sees as the specifically 'Third World' element of 'planetary problems'.

Figure 5.1 Atmospheric pollution then and now: (above) contemporary painting of Manchester, 1830; (right) Shanghai 1980.

Box 5.2 Extract from 'You can't see the future for dust'

From now until the year 2000 there are fewer than 4400 days, probably just enough for us to come up with responses to counter some planetary problems. But we'd better not wait too long.

In population terms, we have just passed the five billion mark. But in 1998 we shall number six billion people. Bangladesh will no longer have 107 million people, mostly farmers, packed into a space the size of England, but 145 million...

Yet this dismal scenario can be relieved somewhat, and the longer term's scenario can be relieved a lot by vigorous family planning campaigns implemented forthwith...

The projected total for Nigeria, for instance – not to be reached until early in the 22nd century, is 594 million but could be cut by 162 million...

By the year 2000 too, many countries will be further affected by the spread of deserts. Today some 900 million people suffer the effects of desertification, and by the century's end they may well have increased to 1250 million...

Again, there is a premium on countermeasures. Much more can be done now, and at much less cost, than if we put off until the year 2000. Yet since the UN Anti-Desertification Plan was announced in 1977, with its $4.5 billion annual budget, governments have chipped in a mere $26 million.

Also by the year 2000 we shall have lost at least another 45 000 square miles of tropical forest destroyed outright, and a similar amount of forest degraded, out of remaining forests that today total 3.3 million square miles... [The] prime agent of deforestation, the slash-and-burn cultivator, is building up his numbers by 3 to 6 per cent per year, due to natural increase of 2.5 per cent or so, the rest being accounted for by landless peasants who migrate into the forests. Again we know roughly what to do.

We have a Tropical Forest Action Plan prepared by the World Bank and the World Resources Institute in conjunction with UN agencies, with a budget of $1.3 billion per year, or only twice as much as has been spent annually on tropical forestry to date. Governments are pledging support in principle, and perhaps some funds in practice. Third World governments are more inclined to do so than they used to be, now that they have learned their revenues from hardwood exports are projected to slump from $7 billion a year today to $2 billion by the year 2000. Today 1.6 billion people must gather fuelwood. Within the next dozen years their numbers will swell to 2.3 billion. And if these throngs use crop residues and livestock manure as fuel instead of fertilizer, the decline in crop output could be worth $8 billion a year...

As tropical forests go, so do their species. At least half and possibly three-quarters (or even more) of all species occur in tropical forests. Of these, one third or more occur in just 7 per cent of the biome – 'hot-spot' areas that are both ultra-rich in species and ultra-threatened by the machete and matchbox. In these areas alone we may well lose half a million species by century's end. In other parts of tropical forests, together with wetlands, and other species-rich zones, we shall be fortunate if we do not lose another half million species. Among these 'in memoriam' species there could well be several plant species with promising anti-cancer properties, other plants that could serve as sources of new foods, and dozens of animal species that could help with medical research. Some scientists consider that one of the best bets for AIDS research lies with the creature from which the virus is thought to have originated, the green monkey of Zaïre's forests...

Yet in the main we know what to do to save species. We need many more parks and reserves. But setting up protected areas is rather like establishing fragile islands in face of the oncoming tide. So we need far greater efforts to deflect the tide, to stem it, eventually to stop it. That is, to eliminate the motivation of shifted cultivators and others who feel a compulsion to head for tropical forests or semi-arid zones or montane area. All these are marginal environments where 'marginal' people – the landless, the workless, and hopeless – wreak undue damage to wildlands and species' habitats.

We must devise ways of packing more people into existing farmlands through intensified agriculture. Yet this prospect fades as farmlands lose ever more topsoil through erosion...

According to some estimates preliminary and approximate as they are, we can expect to lose at least 275 billion tons of the stuff by the year 2000, or roughly eight per cent of remaining topsoil…

Declining food output per unit area will leave a lengthening list of countries unable to feed themselves from their own farmlands by the year 2000. For oil-rich Saudi Arabia, that is no deal at all… As for Ethiopia, the World Bank believes that with its fast-growing population, the country will need to increase its imports of food from 214 000 tons in 1980 to more than three million tons in the year 1990 – and still more mega-amounts by the year 2000.

All these problems will be complicated by the onset of the greenhouse effect. About 100 years ago the carbon dioxide content of the atmosphere was some 265 parts per million. Today it is 345 ppm and rising by 1.5 ppm per year. At least as much greenhouse effect is caused by other trace gases. By the year 2000 we shall surely be feeling the climatic repercussions, possibly an extra 0.3 degrees C warmer (perhaps more still) within a further two or three decades. A Mediterranean-style climate for us might not sound so bad. But food-growing capacities may be sharply reduced for those many territories that become not only warmer but drier…

The time to confront the challenge of climatic change is now. We have a window of opportunity, roughly till the end of the century… We could increase the energy efficiency of our fossil-fuel use by two per cent per year, at a very rough cost in the order of £20 billion a year. This would defer the full greenhouse effect by several decades (and even reduce the overall scale of its ultimate impacts), thus giving us all the more scope to adjust. Large a sum as £20 billion sounds, it would be peanuts as compared with some costs of adjustment…

(Norman Myers, *The Guardian*, 8 January 1988)

What did you find? You probably got the impression that the main cause for concern is increasing Third World rural population, which is regarded as causing widespread deforestation, which in turn results in declining land productivity and destruction of habitats of potentially useful (as yet unidentified) plants and animals which may then become extinct. The problem of this population is not that they have high levels of consumption, but that they are poor. Their poverty is held to drive them to seek survival through means which degrade the resources on which they depend.

I shall return later to consider a number of aspects of this article, but first I want you to consider whether you have encountered similar ideas in earlier chapters, particularly those on population and food. In doing so, you will probably identify the standpoint of the author of the article as a 'New Malthusian' one. Here let us note two particular problems with this view that you should be able to relate to criticisms in previous chapters.

First, aggregate statistics tell us very little about individuals' economic circumstances, which vary widely. In Chapter 1 we saw, for example, that famines are not caused by a decline in total food availability (and in Chapter 4 we noted that world food production has consistently increased faster than world population for the past 30 years), but by a failure of 'entitlement' of individuals to that supply.

Second, the original Malthusian analysis rested on the assumption that the only way of increasing the production of food on a fixed land area was through the investment of more human labour, a process that Malthus considered ultimately self-defeating as an ever greater proportion of the total production would be needed to feed the labour force. As the economist Edward Barbier (1989, p.8) has observed: 'this constraint on growth in the Malthusian system would be broken through introducing the substitution of capital for labour in agriculture', i.e. by increasing the productivity of labour. Similarly, the modern idea of 'carrying capacity', implying a fixed relationship between a human population and its immediate environment, is undermined wherever technical or social change alters that relationship.

In fact, a variety of technical and social factors have a central role in determining the environmental impact of human populations. Thus, where environmental degradation is concerned, numbers or even densities of people can be misleading. We should rather be concerned with how they gain their livelihood. It is livelihoods that establish the relationship of human populations with their environment. If we wish to identify a characteristically 'Third World' element of environmental degradation, we need therefore to consider the nature of economic activity in non-industrialized countries and how it might have a different relationship to the environment from that in industrialized countries.

5.2 The Third World environment and the global economy

Relating livelihoods to the environment

All human economic activity exploits the environment:

- through the use of natural resources as raw materials for production ('renewable' resources in the case of plants and animals, wind and water power; 'non-renewable' in the case of minerals for production and fossil fuels);

- through the use of the environment to accept waste products. The capacity for waste to be transformed into harmless forms in the environment can be regarded as a 'renewable' environmental resource.

The intensity of exploitation of these resources determines the environmental consequences, in terms of the depletion of non-renewable resources and degradation of renewable resources.

An important distinction we need to make in economic activity is between primary commodity production and manufacturing. The former

refers to agriculture, timber and other tree products, fisheries, mining, energy production; the latter includes food processing and relatively simple transformations to make paper, glass or textiles, as well as extremely complex processes to make pharmaceuticals or assemble machinery. The important point here is that complex manufacturing, which forms its products from a combination of different raw materials, offers more flexibility of resource use than primary commodity production. Whereas in primary commodity production the composition of the commodity (e.g. copper or groundnuts) is fixed, and therefore so are the environmental resources used to make it, there is considerable scope to alter the composition of manufactured products: copper piping can be substituted by plastic or even steel; cooking oil can be made from a variety of different vegetable oils besides groundnut oil.

We should note here that the capacity to change the composition of manufactured output is critically dependent upon the availability of capital to invest in technological change, and cheap energy to allow transportation of raw materials from distant sources.

The importance of the distinction between primary commodities and manufacturing from an environmental standpoint is that an increase in output of primary commodities is more likely to be achievable only through increasing the intensity of exploitation of local resources, with attendant risks of degradation of those resources (exhaustion of mineral reserves, depletion of fish stocks, reduction in soil fertility etc.). By contrast, manufacturing output may be increased, if the capital and cheap energy supplies are available, by drawing on a multiplicity of more or less distant sources of raw materials. The local environmental impact from manufacturing is the waste energy and materials that cause pollution.

Table 5.1 has been assembled from World Bank statistics on traded (exported) output, in which primary commodities are divided into two categories: 'fuels, minerals and metals' and 'other primary commodities' (mainly agricultural products).

Table 5.1 Percentage share of merchandise exports in selected areas

	Fuels, minerals, metals	Other primary commodities	Machinery and transport equipment	Other manufactures
OECD countries	7	12	41	40
East Asia	10	16	23	51
(South Korea)	2	5	39	54
(Malaysia)	18	37	26	19
Latin America	35	29	14	23
Sub-Saharan Africa	45	38	2	14

South Korea and Malaysia are included in East Asia.

Source: World Bank (1990) *World Development Report 1990*, Oxford University Press, Oxford.

Use the data in the table to calculate a figure for the percentage of primary commodities and manufactured products exported from each area. What are the main differences?

The table shows clearly the way that primary production dominates the economic activity of sub-Saharan Africa, and to a lesser extent Latin America. The category East Asia is less informative, as it aggregates countries like Korea and Taiwan, in which manufactures form more than 90% of exports, and others like Malaysia and Indonesia, in which primary commodities make up 55% and 71% of exports respectively.

We can see, therefore, that while some parts of the Third World, notably in East Asia, have important manufacturing sectors, much of the Third World's export earnings are derived from primary commodities. This division of the world into countries producing manufactures and others producing primary commodities dates from European expansion and colonialism (see Chapters 8 and 9).

Implications of dependence on primary commodity production

During the latter half of the twentieth century prices of primary commodities have undergone a long-term decline relative to those of manufactures, with the exception of the period of economic boom following the Second World War (Figure 5.2). During the 1980s the fall in commodity prices became precipitate, reaching the lowest levels recorded in the twentieth century. This fall in commodity prices can be traced to four main changes in the pattern of demand, production, and trade in the industrialized countries.

1 Demand for primary commodity exports is closely linked to overall economic activity in industrialized countries, and this has slowed down in the 1970s and 1980s relative to the boom of the post-war period.

2 The structure of consumption in industrialized economies has altered, with consumption of manufactured goods growing less quickly than consumption of services, which require little or no input of primary commodities.

3 Technological changes in manufacturing processes involving greater *efficiency* in the use of materials, increased *recycling* of scrap materials and increased *substitution* by synthetic materials has reduced dependence on primary commodity inputs.

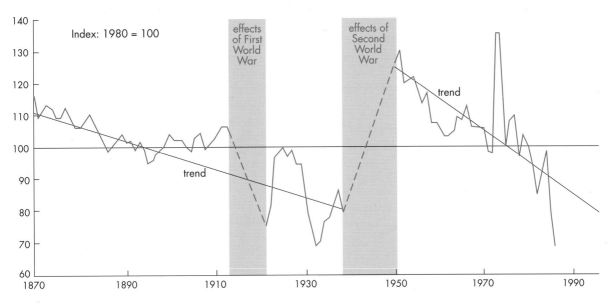

Figure 5.2 Real commodity prices deflated by price of manufactures, 1870–1986.

4 Agricultural protectionism has increased in industrialized countries. In Japan, the United States, and the EC countries, a willingness to pay high prices to local agricultural producers has stimulated overproduction of agricultural surpluses, which then need to be sold, often at much lower (i.e. subsidized) prices on the world market. This practice not only effectively closes markets for agricultural commodities in industrialized countries to producers in the Third World, but also generally lowers the world price for such commodities.

These trends have two main results. First, the consumption of primary commodities is either declining or rising less fast than economic growth in general. Whereas environmentalists in the 1970s saw depletion of mineral reserves as an important limit to world economic growth, few were so preoccupied in the 1990s. Secondly, for countries whose export income depends on primary commodities, falling prices has meant reduced income. Each unit of output of primary commodities can be exchanged for a decreasing number of units of manufactured goods.

Governments have responded to this situation by cutting imports, by borrowing, and by increasing the output of primary commodities and

reducing the cost of such output. There is evidence that these responses have not improved matters, however:

- Reduction of imports has reduced the capacity of Third World manufacturing industry to maintain or replace equipment, thus further diminishing local manufacturing output.

- Borrowing was favoured in the 1970s as a means of 'recycling' the large surplus cash deposits of oil-exporting countries. In the 1980s, however, the drop in commodity prices and the rise in interest rates increased the debt liability of many primary commodity-producing countries, while simultaneously reducing their capacity to repay. In Brazil and Indonesia in 1988 the interest due on outstanding debt was equivalent to 42% and 40% of export earnings respectively.

- For governments of countries dependent upon exports of primary commodities, the need to 'service' their debts may present little option but to maximize output of these commodities in order to try and recover lost revenues through greater volume of production. Indeed, this was the strategy advocated by the World Bank in its prescriptions of 'structural adjustment' for indebted nations. However, increased production of the same commodity by different countries tends to depress prices further, and in some cases has precipitated a collapse in the market.

For example, in 1989 the Indonesian government was planning to double cocoa production within five years and to take its share of the world market from 3% to 12% by the end of the century. At the same time though, falling prices were wrecking the livelihoods of established West African growers, and causing an exodus of people from the countryside in Côte d'Ivoire and elsewhere (Box 5.3). The main advantages of Indonesian producers were not only that their productivity was high, but also that the wages of rural labourers were extremely low. The decline in commodity prices therefore places a premium on low production costs, and the need to achieve these may require damage to the environment as well as to workers' standards of living.

A similar process occurred in tin mining in the 1980s. In 1985 the price of tin collapsed and in the following three-year period production remained static or fell in all the principal producing countries except China and Brazil. The IMF observed:

"World mine production of tin is estimated to have risen in 1988 by 11%. A large part of the increase occurred in Brazil where mine output was raised by over 50% to 44 thousand tons. As a consequence, Brazil accounted for over 20% of world mine production in 1988 compared with less than 4% in 1982 and is now the world's largest producer. The increase in 1988 came largely from the Amazon state of Rondônia where production from a new low-cost mine began in September 1987."

(IMF, *Primary Commodities. Market Developments and Outlook*, July 1989)

In 1989, British Petroleum, which owned 49.5% of the open-cast tin mine in Rondônia (Figure 5.3), was defending itself against newspaper reports of devastation of thousands of hectares of rainforest by its mining operation:

"The company admitted that reafforestation had not been attempted, because the land that had been used for extraction of prime-grade tin ore, and was now lying spoilt might be brought back into use for lesser grades if the price of tin rose. The mine management did have a plan for eventual reafforestation, the company said, but BP's stake was in any case about to be sold to another large British company, RTZ. There were no environmental restrictions on the sale, a BP spokesman said."

(*The Times*, 19 June 1989)

Figure 5.3 Maximizing output with low production costs: open-cast tin mining, Rondônia, Brazil.

Box 5.3 Extract from 'Ivorian cocoa, the cash crop nobody wants to buy'

Twenty sacks of cocoa are sitting in the store-room, but nobody can afford to buy them. Out in the fields the cocoa pods are bulging, the coffee beans gleaming red in the sun, but each day the farmers wonder whether they have either the manpower or the reason to pick them.

An Ivorian farmer examines pods bulging with cocoa beans on a plantation near Gagnoa

Massive arrears for payment of the 1988 crop hang over the heads of cocoa buyers who are now unable to buy the current crop, which is ready for shipping but is to be found sitting in store-houses throughout the country's cocoa belt…

The cash flow problems experienced by the buyers as a result of the dramatic fall in prices have left some unable to pay farmers for much of the 1988–89 harvest. This has created a climate of distrust.

'It is all over for us unless there is a price rise soon,' said one farmer. 'We are finished. Our village and the seven others which make up the co-operative we are part of are collectively owed 3m CFA francs (£6000) and most of us have not been paid for a year.

'Nobody can afford to buy meat to eat, nor can we afford tools for clearing ground around the crops.

Our children are going without school books and old people in the village cannot afford to buy medicines,' he said.

The cut in the producer price paid to cocoa growers in December reduced their income, at least on the official level, by half. The cut from last year's level of 400 CFA francs a kilogram to the current 200 CFA francs was the Government's response to the fall in world market prices which it had steadfastly refused to acknowledge in previous years.

But despite the official 200 CFA francs a kilogram price, farmers claim the market is so stagnant that crops are increasingly being sold at prices far below that figure: 'Secretly, in the night, buyers come and offer less than 200 francs. People who are desperate to buy medicines or send their children to school with books, are sometimes prepared to go as low as 150 francs,' one farmer said…

The co-operatives have long aimed to sell cocoa and coffee direct to exporters, cutting out the buyers…whose shortage of money has contributed to the current slow market. But this would require the farmers to provide transport, the expense of which makes up a significant proportion of the overall costs. With banks reluctant to lend, these funds are hard to come by…

The serious shortage of cash has forced mainly young people to leave villages, which makes current attempts toward diversification of crops more difficult. The shortage of labour, with 200 people having left one village in the past year, reducing the population to 900, adds to the difficulty of buying new seeds and tools caused by lack of income…

Everybody…is waiting for cocoa and coffee prices to recover from their current historic lows. But with increasingly powerful competition coming from neighbouring Ghana as well as Asian producers…they look like having a long wait.

(Mark Huband, *The Financial Times*, 22 November 1989)

The pursuit of increased output of primary commodities may intensify the exploitation of environmental resources to the point where the effect on livelihoods may be catastrophic. It has been argued that the Sahelian famine of the 1970s was an outcome of this kind, the result of reckless extension of groundnut cultivation into marginal rainfall areas (Box 5.4).

At the level of individual livelihoods, what is the response to low commodity prices? In Box 5.3 we saw two general products of a crash in agricul-tural commodity prices: a migrant labour force on the one hand, and, on the other, a residual, part-time, under-resourced agriculture. For those left on the land, lack of resources confronts the immediate need for subsistence and repro-duction. A more intense exploitation of local resources (land, trees, pasture) may present the only alternative, despite the environmental deg-radation which results. Here we have an agent of environmental destruction we can perhaps identify from the article in Box 5.2.

Box 5.4   Groundnut cultivation and environmental degradation in Niger

In the post-war period, European consumption of vegetable oils far outstripped local production and so US exports of soyabeans to Europe were vigorously promoted. The trade quickly incorpo-rated sales of seedcake (residue after extracting the oil) for incorporation into animal feed, and US soyabean processors established milling plants in Europe. As a result US exports of soya increased from 47 000 tons of soya seedcake in 1949 to 5 million tons of seedcake and 13 million tons of soya beans in the early 1970s. The near total dependency of the rapidly expanding in-tensive livestock (pigs and poultry) industry in Europe on American soyabean imports caused alarm to the French authorities who attempted to secure alternative sources of oilseedcake by stimulating the production of groundnut in its Sahelian colonial territories. In Niger, the area of groundnut cultivation increased rapidly, from 142 000 hectares in 1954 to over 300 000 hec-tares in 1957.

The years following formal independence of Niger from France in 1961 saw a further expan-sion of groundnut cultivation, which covered 432 000 ha in 1968. Research by Richard Franke and Barbara Chasin (1981) indicates resulting profoundly changed patterns of land use. Land was brought into cultivation that would have remained fallow, thus reducing recuperation of soil fertility and eliminating vegetation that would have served for grazing. Cultivation was also extended into marginal rainfall areas traditionally used by pastoralists for grazing livestock according to a 'transhumant' man-agement pattern (moving seasonally from one area to another to take advantage of seasonal vegetation growth). The overall effect was seri-ously to reduce the pasture available for pastoralists' herds and to generate conflict be-tween pastoralists and cultivators for control of land.

The onset of drought in 1968 found pastoralists vulnerable to shortage of pasture, therefore, and by the end of the drought in 1974, the Food and Agriculture Organization (FAO) estimates suggested some 39% of the total cattle in Niger had died through lack of water or lack of pas-ture. Price changes as well as the loss of ani-mals contributed to the destitution of many pastoralists, and the ensuing famine condi-tions established an imagery of African agri-culture that was to be reinforced throughout the following decade, providing many environ-mentalists with the evidence of the eco-catastrophe that they had been predicting.

Of the struggle by the French government to resist dependency on American soyabean im-ports, Franke and Chasin have observed: 'It was ultimately a losing battle: by 1971, France imported 1.3 million tons of soya, of which 773 544 tons came from the USA, as opposed to only 318 332 tons of groundnuts. Nonetheless, this losing battle was fought not only on the fields of France, but also in Niger.'

In this section I have argued that Third World environmental degradation has its roots in the structure of the world economy. It is not that environmental degradation in the Third World is worse than that in developed countries, but that it appears to many that it is less amenable to intervention because it is the result of economic desperation rather than profligate consumption.

5.3 Sustainable development

Whose development, whose environment?

If you return to the article in Box 5.2, you will probably find two agents said to be responsible for environmental destruction: the governments of Third World countries who have been slow to adopt 'vigorous family planning campaigns', and the 'slash and burn cultivator', who, together with 'the landless, the workless and the hopeless…wreak undue damage to wildlands and species' habitats'. By comparison, 'we' have the answers: 'a Tropical Forest Action Plan', 'more parks and reserves', 'ways of packing more people into existing farmlands', and since 'we could increase the energy efficiency of our fossil fuel use by 2% a year', 'we' are evidently resident in industrialized countries.

By looking at environmental problems such as deforestation in specific cases, we can quickly see the extent to which these views misrepresent the agencies at work in deforestation. Read Box 5.5, an account of deforestation in Malaysia constructed from press reports in 1989–90, and note all the different groups and interests involved.

Box 5.5   Developing the rainforest in Sarawak

Sarawak, in north-west Borneo, is one of the states of the Malaysian Federation. It is the largest source of unprocessed (log) tropical timber in the world. During the 1980s the annual output of logs doubled, reaching about 15 million cubic metres, worth about £2.5 billion, in 1990. The rapid increase in Sarawak's output accompanied reductions in supply from other important sources of timber in South-east Asia: log exports from Indonesia were banned in 1985; all logging in Thailand was banned in 1988.

Two thirds of Sarawak's log exports go to Japan, and Japanese companies Mitsubishi, Marubeni, and C.Itoh are active in logging operations, either directly or as owners or partners of local logging companies. Logging companies gain access to the Sarawak forests by obtaining 'concessions', of 10–20 years' duration, from Malaysian concession owners. In 1987, the State Chief Minister, Abdul Taib Mahmud, accused his predecessor of corruption in the distribution of concessions. It subsequently emerged that many state politicians had acquired concessions and other interests in logging. One of the most prominent was the Minister for the Environment, James Wong, who owned a logging company, Limbang Trading, with a logging concession of 300 000 ha.

Most of the concessions are for 'selective logging', in which a maximum of 10 mature trees per hectare are supposed to be cut, allowing the remaining trees to regenerate the forest cover sufficiently to allow a similar cutting after 30 years. Critics say that the selective logging process is poorly policed and that in practice the impact of logging roads and damage to other trees during felling and log removal results in a loss of 40% of the tree cover.

In 1990, a report commissioned by the timber traders' association, the International Tropical Timber Organization (ITTO) concluded that logging in Sarawak was not sustainable and that it was destroying the environment. It estimated the primary forests in the state would be 'logged out' within 11 years. The impact of the logging process was felt particularly by the 200 000 native inhabitants of the forest, whose livelihood is based on small-scale agriculture (2 ha cleared each year by each family, estimated to account for a total of 72 000 ha of cultivation per year), hunting, fishing, harvest of nuts, resins, and making rattan products.

Encroachment of logging activities on forest peoples' land frightened away wildlife, caused soil erosion which increased sediment loads in rivers, with negative consequences for fishing and drinking water supply, and damaged large numbers of nut- and resin-producing trees.

In 1988, one group of rainforest people, the Penan, started blockading the logging roads where they entered their part of the forest, and disrupted logging operations. Through Harrison Ngau, a member of Malaysian Friends of the Earth, and one of the protest organizers, the conflict received publicity in London and New York. In 1989 the World Rainforest Movement petitioned the UN General Assembly to defend the land rights of forest peoples against their governments.

The Malaysian government refused to acknowledge the Penan's claims, and by 1990 some 300 people had been arrested, detained and fined for disrupting logging in Sarawak. However, High Court judges have consistently upheld their refusal to pay fines, and released them. In 1989, the Malaysian Prime Minister was quoted as saying that the Penans were 'unfortunate people exploited by crusading environmentalists who wanted the tribe kept as ill-fed and disease-ridden museum pieces'. In 1990, Jewin Lehnen, chairman of the Sarawak Penan Association, said: 'It's simply wrong to say we don't want development. But by development we don't mean timber companies invading our land. We want the right to live here, and use our land, without disturbance. Then we want development in terms of schooling for our children and clinics to treat illnesses. But all these things we can only get when the logging stops and our rights are recognized.'

(Above) Logs awaiting shipment from Limbang in the north of Sarawak; (right) Penan spokesman Unga Paran and Mutang Urud of the Kelabit people at an international meeting in London.

As in most cases, a complexity of conflicting interests implies conflict over what kind of development is desirable. You should have noted the following: the environmentalists of the industrialized countries wishing to prevent destruction of the rainforest; the members of ruling groups in Malaysia wishing to generate wealth through the sale of timber concessions; multinational logging companies wishing to cut and sell tropical hardwood; the traditional inhabitants of the forest, such as the Penan, who see their livelihoods as hunters or shifting cultivators being destroyed by the incursions of the logging companies into the forest.

For another example, consider this description of the Brazilian government's goals in Amazonia:

"The roots of the environmental pillage in the Amazon can be traced back to the strategy for regional development elaborated by the Brazilian military whose influences in national and Amazonian policy expanded steadily from the first days of President Getulio Vargas's schemes of the late 1930s to develop the northern half of the country... The generals who drove out President Joao Goulart – with active US connivance – took power on April 1, 1964, with a long-nourished geostrategic vision. It was tersely expressed by their chief theorist, General Golbery do Couto e Silva... Brazil's manifest destiny, Golbery wrote, would be to occupy 'the vast hinterlands waiting and hoping to be roused to life'. Such a project would provide a disgruntled population with a sense of purpose and achieve the all-important settlement of empty lands and unguarded frontiers. The march to the west would integrate the Amazon into a new economic sphere. The central ambition, Golbery said, was 'to inundate the Amazon forest with civilization.' "

(S. Hecht &A. Cockburn, *The Guardian*, 25 November 1989)

On this interpretation, then, rainforest destruction has little to do with population growth, and much to do with a particular vision of national development and 'civilization' – arguably the same vision that operated among European immigrants to North America who also 'won the West' at considerable ecological expense. The agencies that benefit from this form of development are those with commercial interests in mining, timber and construction (particularly of transport infrastructure), and those favourably placed to take advantage of changing rights to land tenure. Those whose livelihoods are threatened by this form of development are existing users of the land and other natural resources, and those whose influence over land tenure is weak (Figure 5.4).

Views of sustainable development

The basic issue for the Brundtland Commission was whether or not the activities such as logging just described are sustainable in the medium to long term. The Commission's definition of **sustainable development** has since been widely quoted.

> **Sustainable development:** Development that meets the needs of the present without compromising the ability of future generations to meet their own needs.

Interpretations of 'sustainable development' diverge greatly, however, and in the following paragraphs I outline three: a neo-liberal view, a populist view, and an interventionist view emphasizing the need for international co-operation.

Neo-liberal view
The environment is natural 'capital'. The services derived from air, water, soil, biological diversity, and recreation (the countryside etc.). depend on maintaining those environmental 'assets' intact, or renewing them. If this is not done, those services will sooner or later decline. This evidently parallels the distinction between income and capital, whereby 'income' is the amount that can be consumed in a given period without being any worse off at the end of it. In these terms, if a forest can be used for various

Figure 5.4   Conflicting interests in the development of the Amazon: (above) 'Indian' women gathering forest products; (right) settlers beside the trans-Amazonian highway; (below right) the funeral of the rubber tapper union organizer 'Chico' Mendes, assassinated by gunmen in 1989.

purposes without reducing its long-term value in those uses, this can be regarded as 'sustainable development'. Similarly, discharges of waste are considered 'sustainable' if they are within the capacity of natural systems to transform them into harmless forms.

If values can be assigned to natural 'capital', then sustainable development can be secured by classical economics criteria, valuing the income from a particular course of development against any depletion of environmental capital. While considerable interest has been generated among economists over the possibility of incorporating environmental valuations in accounting procedures, there appear to be considerable unresolved difficulties in providing valuations of natural 'capital', particularly since these need to take account of future values of resources to future generations with unknowable lifestyles (livelihoods and consumption patterns).

The concept of pricing present-day consumption of environmental 'services', such as waste discharge, has found easier application in development policy, however, where it has formed the basis of proposals to regulate atmospheric pollution through the provision of tradable 'permits' to use (pollute) the atmosphere allocated to national governments according to adult population size. The principle of the market would allow non-industrialized countries the right to trade their 'right to pollute' to industrialized countries who would supply goods or services in return. Such proposals are unlikely to be welcomed by Third World governments, some of whom regard such proposals as a means of transforming the atmosphere into yet another primary commodity to be exported by non-industrialized countries.

Such trade in environmental commodities is the basis of moves to secure 'biosphere reserves' of the kind advocated by Norman Myers in the article in Box 5.2. The mechanism is known as 'debt-equity swaps' in which concerned agencies in the industrialized world 'buy' part of the debt owed by countries with 'threatened' ecological resources, in return for the right to determine the conservation of those resources, usually through the establishment of reserves. Advocated by international agencies like the World Wide Fund for Nature, such deals have also been denounced as (for example) 'neo-colonialism in its most stupid splendour' (Hayter, 1989).

Populist view

A *populist* definition of 'sustainable development' may be identified in the declarations of environmental movements in some Third World countries. The Inter-Regional Consultation on People's Participation in Environmentally Sustainable Development, held in Manila in 1989, stated:

> "The concept of sustainability is best understood in terms of the sustainability or non-sustainability of a community. Authentic development enhances the sustainability of the community. It must be understood as a process of economic, political and social change that need not necessarily involve growth. Sustainable human communities can be achieved only through a people-centred development."

A similar view of sustainable development can be found among Western writers who emphasize the need for priority in development to be given to securing 'sustainable livelihoods' for the poorest groups within communities (e.g. Chambers, 1988).

In this interpretation, sustainable development concerns local trade for local needs, and is often profoundly opposed to urban and industrial development, and the national government with which these are associated. Of the environmentalist peasant movements in India Ramchandra Guha (1989) writes: 'At one level they are defensive, seeking to escape the tentacles of the commercial economy and the centralizing state; at yet another level they are assertive, actively challenging the ruling class vision of a homogenizing, urban–industrial culture.' He traces the environmentalism of these movements as originating in resistance to the creation by the British colonial government of reserves for 'rationalized timber production'. The exclusion of the peasants from the reserves deprived them of their traditional use of the forest for pasture, fuel, herbs and medicines. Reassertion of such 'sustainable' use of the forest is more than an environmentalist statement, therefore: it is also a reclamation of economic control of forest resources. You can see how populist notions of this type relate strongly to ideas of empowerment, introduced in Chapter 4.

The major question mark over such an approach is the degree to which such 'sustainable' exploitation of forest products can survive integration into the world economy. Antonio Macedo, Coordinator of the National Council of Rubber Tappers in Brazil, following his murdered predecessor Chico Mendes (see Figure 5.4), emphasizes the use of local rubber extraction for local manufacturing to supply local markets, rather than international traders. However, entrepreneurs are active in exploring marketing opportunities for exotic rainforest products from cosmetics to ice cream. Observers are sceptical of the ability of forest dwellers to retain control of any trade with world markets:

> "Once botanists have taken their samples and repotted them or tucked away the seeds in gene banks, the forests are often no longer needed. Either, like rubber or cocoa, production is transferred to plantations, or the required chemical is synthesized in Western laboratories. Certainly neither Third World governments nor forest people receive any reward from the subsequent exploitation."
>
> (Pearce, *New Scientist*, 7 July 1990)

Interventionist view

This view emphasizes international co-operation. The Brundtland Commission, for example, envisaged the establishment of international environmental treaties, to be enforced, logically, by international agencies. The first such treaty was the Montreal Protocol to eliminate the use of CFC compounds (see Box 5.1). The discussions to implement the protocol were marked by intense conflict as governments sought to protect their domestic refrigerant industries. It is interesting to note how the advocacy of early CFC elimination on the part of US and UK governments contrasts strongly with their opposition to any specific targets for reduction in carbon dioxide emissions. The contrast clearly reflects these governments' perception of their own industrial capacity to exploit any proposed international regulations. This approach to environmental management has been termed 'technocratic' and doomed to founder on conflicting interests (Figure 5.5). For example, Redclift (1987) argued: 'it is impossible to arrive at the optimum mix of resource uses without preconceived, value-based criteria. Once you agree about the kind of society you want, then agreement about environmental goals may not be difficult, but a social consensus does not exist.'

Although for most of this chapter we have concentrated on specifically Third World environmental problems mainly relating to non-industrial production, all these views, but particularly this third one, purport to deal with the potential environmental problems of Third World industrialization as well.

Chris Patten, when UK Environment Secretary, stated at a London conference on Latin America: 'The environment is our new aid imperative,

Figure 5.5   Conflicting interests at the international level.

replacing perhaps some of the ideological and strategic arguments for aid which were often used to justify it in the Cold War years.' Remembering that those arguments were to preserve the world capitalist order from the threat of communism, one might ask whether the new environmental imperative is also to preserve the existing world economic order, or to change it.

Summary

Concern about environmental degradation and the Third World generally has two aspects: worries about the potential problems of pollution etc. likely to be caused by future industrialization, and present concerns related to specific Third World locations which are generally not industrialized. This chapter concentrated on the second. I have argued that to understand environmental degradation we need to consider people's livelihoods, for these establish the relationship between economic activity and environment. Characteristically, Third World forms of environmental degradation are generated by Third World livelihoods: participation in primary commodity production as wage labourers in mining, forestry, and agriculture, or as petty commodity producers (sometimes degraded to 'subsistence' producers). Finally, three alternative views of the environment and its role in development have been introduced: neo-liberal, populist and interventionist. All such views must be interpreted in relation to the interests of those concerned and the manner in which they participate in the development process.

6

DEVELOPMENT, CAPITALISM AND THE NATION STATE

ALAN THOMAS AND DAVID POTTER

So far, each of the chapters of this book has dealt separately with one of the enormous social, political and human issues which epitomize the so-called 'Third World'. However, these issues are clearly interrelated. Hunger, disease, unemployment and so on are aspects of poverty, and they all relate to questions of development. In fact, two phrases often used in place of 'Third World' are 'developing countries' and 'underdeveloped countries'. The use of such phrases indicates that what the countries in question have in common is their lack of development – though the first carries an optimistic implication that things are moving in the right direction, whereas the second seems more pessimistic.

But what exactly is this 'development' that is lacking in so many parts of the world? Certainly 'development' is a goal that virtually everybody believes in – the idea of the 'developed society' has effectively become a synonym for the 'good society'. All definitions of development contain the central notion of a process of change from a less desirable to a more desirable kind of society – in short, the notion of *progress* (Figure 6.1). Beyond this, however, lie some basic disagreements over questions such as: Development of what? How is what is desirable defined, and by whom? How is progress to be achieved? What if (as generally seems to be the case) progress for one group within a society is gained at the expense of other groups?

Of course we have progressed a great deal, first they were coming by bullock-cart, then by jeep – and now this!

Figure 6.1

Thus, just as there are many answers to the question of what constitutes progress, so there are many meanings given to 'development'. This chapter explains some of the most important of them. We will see that we cannot take particular definitions of development out of context but can only understand them in relation to various different views of how development takes place or can be achieved.

The context in which development occurs has many aspects, but we concentrate here on two. The dominant economic system of the contemporary world is a *capitalist* system, and from a political point of view the world is composed of *nation states*. Economics and politics are not the whole story, but they are certainly crucial. In fact, most influential frameworks for understanding development can be understood as sets of differing views on how development can be achieved, either within capitalism or in spite of capitalism, either through the agency of the state or against the state.

The rest of the chapter is arranged in three main sections. The first explores different definitions of development. The second sets out the context of development provided by global capitalism and national politics and the state. The final section uses the ideas of global capitalism and the state to introduce four contending views on development: neo-liberal, structuralist, interventionist and populist views.

> **Gross national product (GNP) and gross domestic product (GDP):** The World Bank defines GNP as 'the total domestic and foreign output claimed by residents of a country' in one year. What they 'claim' is also their *income*; thus GNP is a measure of national income and GNP per capita is a measure of the average income of each member of the population, including what they may earn or receive from abroad. GDP, on the other hand, is simply an *output* measure: the 'total final output of goods and services produced by an economy'. Thus GDP measures the size of the economy while GNP is the total income available for private and public spending. The two are of course closely related. The GNP of Nigeria, for example, is the output produced in Nigeria (its GDP), less whatever is 'claimed' by foreigners (repatriated profits, migrant workers' earnings, etc.), plus what Nigerians earn outside the country (remittances from abroad, returns on investments abroad).

6.1 Definitions of development

Economic well-being and GNP per capita

Perhaps the simplest way of thinking about development is that it means an increase in prosperity. At a national level, the most used measure of economic well-being is **gross national product (GNP)**. GNP uses market valuations, and is in practice a measure of national income; GNP per capita gives an indication of the average material living standard of a nation's people.

An increase in GNP per capita could mean development in that it implies better material living standards and less poverty. However, if you recall the approach to poverty via the idea of *entitlement* (as in Chapter 1), you should realize that a measure such as GNP per capita has limitations in this regard. GNP per capita is a measure of *average income* based on *market valuations*, and thus there are several ways in which the measure fails to give a full indication of the incidence of poverty.

First, as an average measure, GNP per capita tells us nothing about income distribution within a country. Income distribution is notoriously difficult to measure, both for political and for technical reasons. Where statistics can be found, however, they indicate that many Third World countries have greater inequalities in income than industrialized countries. A few rich individuals may so distort the picture that the GNP per capita figure corresponds neither to the low standards of the masses nor to the wealth of these few.

Also, even where income distribution figures exist, they will generally be based on surveys carried out using the household as a unit rather than individuals. This leads to two further difficulties. Since larger households may also tend to be poorer, income per head may be even more unevenly distributed than income per household. And inequalities within households will be completely invisible. For example, if men get the meat, if boys get better educated, or if girl babies tend to be weaned earlier, then statistics based on households will not give a correct impression of individuals in absolute poverty.

Second, using market valuations for measuring income per capita gives rise to at least two different problems with relating this measure to poverty levels. GNP is a measure of production and its value is given in currency which can be freely converted internationally, for example into US dollars. However, the wage represented by the average GNP per capita in a local currency does not have the same purchasing power for commodities at local prices. In fact, real purchasing power in poor countries may be relatively higher, so that, for example, a Sudanese on an income the equivalent of US$370 might subsist somewhat better than a US citizen with only US$370 per year in the USA, so long as local Sudanese prices for basic foodstuffs etc. were lower than international prices.

The other point is that well-being is not entirely a matter of purchasing power. The most obvious example of this may be production for use (i.e. direct entitlement) by peasants or petty commodity producers. One might be able to measure the potential market value of such items, but what about less concrete items? Education and health, for example, are much less tangible but no less basic needs. Education could be considered an inseparable part of life, with no price put on it; or it might be considered a commodity, with school fees charged and salaries paid to teachers, in which case it will form part of a country's GNP. Another, ironic, example, often quoted, is how a polluting industry with added pollution control could contribute more to GNP than a less polluting alternative.

Thus GNP as an indicator underestimates both subsistence and collective goods, whereas it overvalues the commercialized, the individualized and the organized.

Economic development

There is another sense in which an increase in GNP per capita could indicate development even though it does not relate precisely to individuals' improved well-being or reduced incidence of poverty. An increased GNP – or, more precisely, increased GDP – means *economic growth*. Economic growth is simply a continued increase in the size of an economy, i.e. a sustained increase in output over a period.

Since tools and other means of production are always required, economic growth cannot take place without investment of capital to obtain additional means of production. At a national level, this implies that a proportion of the total output must be set aside from consumption as *savings*, which can then be channelled into investment. The simplest model of economic growth shows that if a high rate of growth is to be achieved, either a high rate of savings (and thus of foregone consumption) is necessary, or else investment will have to be obtained from elsewhere, e.g. foreign investment or borrowing.

However, this does not tell us which social groups will be going without. The term 'savings' does not get at this so well as the notions of the **appropriation** of **surplus** and surplus **accumulation**. Also, although there can be no economic growth without investment, by itself investment is not sufficient to guarantee growth. As you will recall from Chapter 3, various aspects of the organization of production, both technical and social, will have a big effect on how productively any investment will be utilized.

Economic growth may do no more than keep pace with population growth, but in the case where the growth in output is greater than that of population – that is, labour productivity (see Chapter 3) has increased – then **economic development** in its simplest sense may be said to have occurred. However, for most economists development is more than this. Whereas growth means more of the same type of output, development implies more thoroughgoing changes, changes in the social and technical relations of production (see Chapter 3 again). Thus the productive capacity of a society as a whole has to increase, rather than just increasing productivity within its productive enterprises.

Surplus, appropriation, accumulation: That portion of what is produced at any given time that is not required for immediate consumption, including reproduction, is called the *surplus product* or simply the *surplus*. It may be preserved as a store of produce or implements, or converted through market relations into a sum of cash. There are three main possible uses for the surplus product. It can be kept as a reserve against future needs. It can be taken by dominant social groups or classes and used for 'luxury' or 'conspicuous' consumption and for maintaining their power (e.g. building palaces or temples, keeping armies). Or it can be used as a means of investment in expanding production and/or increasing its efficiency by improving productivity. In the second case the surplus is said to be *appropriated* by the dominant groups or classes concerned. The surplus may also be appropriated by dominant groups in the last case; the difference is that they then use it for what is termed productive accumulation, or simply *accumulation*.

Economic development: 'Raising the *productive capacities of societies*, in terms of their technologies (more efficient tools and machines), technical cultures (knowledge of nature, research and capacity to develop improved technologies), and the physical, technical and organizational capacities and skills of those engaged in production. This can also be expressed in terms of raising the productivity of labour: using the labour available to society in more productive and efficient ways to produce a greater quantity and a more diverse range of goods and services' (Bernstein, 1983, p.59).

Industrialization and modernization

Historically, the biggest change in the direction of economic development has been industrialization. The industrial revolution in Britain has been described as a process of 'total' change: 'a change of social structure, of ownership and economic power in society; as well as a change of scale' (Kitching, 1982, p.11).

There are two different ways of defining industry. The first divides all economic activity into sectors and defines industry as the production of all material goods not derived directly from the land. Industry thus comprises the mining, energy and manufacturing sectors, and does not include agriculture or services. The second definition emphasizes technical and social change: an industrial production process is one that uses advanced technology and a complex technical division of labour, and is linked to other forms of production through combining a wide range of raw materials, skills and sources of energy. Thus **industrialization** can mean simply an increased percentage of GDP from industrial sector outputs, or, more fundamentally, a general change of social structure, organization, scale, concentration and ways of thinking towards giving primacy to productivity, efficiency and instrumentality. Though harder to measure, the second definition is more useful for analytical purposes; it can cope with the idea of industrialized agriculture and differentiates fully industrial manufacturing from craft production.

Industrialization: The process by which production in the industrial sector becomes increasingly important compared with agricultural production; more fundamentally, a general change towards the use of advanced technology and a complex division of labour in production with associated changes in social structure and organization.

In a world dominated by advanced capitalist economies, all aspects of Western industrial society are elevated to represent the ideal of what development is trying to achieve. Our colleague Henry Bernstein encapsulated this in the idea of development as 'following in the footsteps of the West', which in effect, is to say 'If you want what we have (and have achieved), then you must become like us, and do as we did (and continue to do)' (Bernstein, 1983).

Figure 6.2

This view of development as *modernization* comes particularly from the 1950s and 1960s. For example:

"Historically, modernization is the process of change toward those types of social, economic and political systems that have developed in Western Europe and North America from the seventeenth century to the nineteenth and have then spread to other European countries and in the nineteenth and twentieth centuries to the South American, Asian and African continents."

(Eisenstadt, 1966, p.1)

Modernization (Figure 6.2) implies complete transformations in many aspects of life brought by economic development through industrialization. Another modernization theorist lists the following interrelated technical, economic and ecological processes':

"(1) in the realm of technology, the change *from* simple and traditionalized techniques *towards* the application of scientific knowledge;

(2) in agriculture, the evolution *from* subsistence farming *towards* commercial production of agricultural goods. This means specialization in cash crops, purchase of non-agricultural products in the market, and often agricultural wage labour;

(3) in industry, the transition *from* the use of human and animal power *towards* industrialization proper, or 'men aggregated at power-driven machines, working for monetary return with the products of the manufacturing process entering into a market based on a network of exchange relations';

(4) in ecological arrangements, the movement *from* farm and village *towards* urban centres."

(Smelser, 1968, p.126)

Smelser argues that all four processes give rise to the following changes in society:

"(1) structural differentiation, or the establishment of more specialized and more autonomous social units;

(2) integration, which changes its character as the old social order is made obsolete by the process of differentiation. The state, the law, political groupings and other associations are particularly salient in this integration;

(3) social disturbances – mass hysteria, outbursts of violence, religious and political movements, etc. – which reflect the uneven advances of differentiation and integration respectively."

(ibid., p.127)

For Smelser industrialization and capitalism were inseparable aspects of becoming modern and developed. In particular, *commoditization* – both the spread of production for exchange through the market and the spread of wage labour – was seen as one aspect of industrialization alongside technical innovation, urbanization and structural differentiation in society. However, in principle, industrialization, along with many aspects of modernization, does not necessarily imply a capitalist system. State socialist models of planned development, for example (see Chapter 12), generally include plans for industrialization. Indeed, when Kitching used the phrase 'the old orthodoxy' to describe the idea that 'If you want to develop you must industrialize' (Kitching, 1982, p.6), he was clearly pointing out how both capitalist and state socialist ideas of development take it to mean industrialization.

Human needs and conditions for development

Another approach to development is to start not from production but from human needs. Such an approach is exemplified in Dudley Seers' (1979) article 'The meaning of development' (first published 1969) which points out the importance of value judgements in deciding what is or is not 'development'. Seers suggests that 'the realization of the potential of human personality...is a universally acceptable aim', and development must therefore entail ensuring the conditions for achieving this aim.

The first three conditions are (Seers, 1979, pp.10–11):

- the capacity to obtain physical necessities (particularly food);

- a job (not necessarily paid employment, but including studying, working on a family farm or keeping house); and

- equality, which should be considered an objective in its own right.

Seers continues:

"The questions to ask about a country's development are therefore: What has been happening to poverty? What has been happening to unemployment? What has been happening to inequality? If all three of these have become less severe, then without doubt this has been a period of development for the country concerned. If one or two of these central problems have been growing worse, especially if all three have, it would be strange to call the result 'development'..."

(ibid., p.12)

Seers' formulation was designed to challenge the type of economic definition of development outlined above, with its emphasis on productivity, growth at any price, and increasing GNP per capita as the ultimate goal. As already noted, economic development of this type does not necessarily reduce the numbers in poverty, let alone meet other human needs such as those pointed to by Seers.

However, Seers underlines the importance of GNP per capita and economic development in a different way:

"Suppose that two countries start a decade with the same *per capita* income and one grows faster than the other over ten years, but that the increase in income in the former goes entirely to the rich, and that, because growth has been due to highly capital-intensive techniques, unemployment rates remain unchanged, while in the latter growth has been slower, but has meant lower unemployment and thus benefited the poorest class. Then, although the country with faster growth has, on my criteria, developed least – in fact, not developed at all – it has achieved greater potential for developing later.

From a long-term viewpoint, economic growth is for a poor country a necessary condition of reducing poverty. But it is not a sufficient condition. To realize the development potential of a high rate of economic growth depends on policy. A country where economic growth is slow or negligible may be busy reshaping its political institutions so that, when growth comes, it will mean development; such a country could develop faster in the long run than one at present enjoying fast growth but with political power remaining very firmly in the hands of a rich minority."

(ibid., pp.12–13)

Seers is here addressing an important debate in development, which is often referred to as the *growth versus equity* debate. We have already seen how GNP per capita, the indicator used to measure growth, cannot measure inequality. Seers' argument is that growth without equity might not lead to what he sees as development, though it might increase the potential for development later.

However, it is also important to understand the pressure for growth without equity, or *why* growth *with* equity might be difficult to achieve. The argument essentially is that the best way to achieve significant growth is by increasing the scale of production and concentrating capital investment in the larger and more productive enterprises. This also means industrializing. In the short run at least, the savings required to increase industrial productivity and output may have to come from the agricultural sector; in other words, part of the agricultural surplus will have to be appropriated for accumulation in the industrial sector. Trying to spread investment equitably in improving productivity everywhere at once may mean its impact is so diluted as to have no real effect anywhere. Hence the importance of political power and institutions; if some inequities are inevitable, it is very important that political controls are available to correct these inequities sooner rather than later or not at all.

Another strand of thinking on what is desirable for development includes non-economic factors directly in what is meant by development. One such approach (identified in Chapter 1 with Amartya Sen) starts with viewing poverty in terms not of poor material living standards but of lack of choice or of capability: poverty meaning the failure to be able to take a full part in human society. A particularly graphic description of what this means is given by Denis Goulet:

> "The prevalent emotion of underdevelopment is a sense of personal and societal impotence in the face of disease and death, of confusion and ignorance as one gropes to understand change, of servility towards men whose decisions govern the course of events, of hopelessness before hunger and natural catastrophe. Chronic poverty is a cruel kind of hell, and one cannot understand how cruel that hell is merely by gazing upon poverty as an object."
>
> (Goulet, 1971, p.23)

In these terms, if development means combating or ameliorating poverty, then development means restoring or enhancing basic human capabilities and freedoms. Or, as Edwards forcefully puts it:

> "development results from a long process of experiment and innovation through which people build up the skills, knowledge and self-confidence to shape their environment in ways which foster progress towards goals such as economic growth, equity in income distribution, and political freedom. At root then, development is about processes of enrichment, empowerment and participation, which the technocratic, project-oriented view of the world simply cannot accommodate."
>
> (Edwards, 1989, pp.119–20)

Empowerment, participation and freedom to choose involve politics. They are aspects of *democracy*, or democratization – a subject discussed further in Chapter 14.

Seers (1979, p.12) also recognized the political dimension and suggested further conditions for development in addition to those mentioned above:

- participation in government;
- belonging to a nation that is truly independent, both economically and politically (Figure 7.3);

'Here, Señor Carter, is the statue of Simón Bolivar, who liberated Latin America from foreign domination!'

Figure 6.3 'True' national independence – a condition for development?

Figure 6.4   Women's status and political participation: conditions for development? 1986 Muslim women's demonstration in India against divorce bill making concessions to fundamentalists.

• adequate educational levels (especially literacy).

Two further aspects of importance that have gained recognition since Seers wrote his article (in 1969) are *the position of women* and *safeguarding the environment*. Thus we might now add two further items to Seers' six:

• relatively equal status for and participation by women in society;
• 'meeting the needs of the present without compromising the ability of future generations to meet their own needs' (definition of sustainable development by the Brundtland Commission).

This makes a full list of eight conditions for what we might call **human-needs centred development** (Figure 6.4).

Any comprehensive definition of development is likely to combine a number of such dimensions and to be based on supposedly universal values. Many others besides Sen, Seers and Edwards could be quoted. We should note whether any such

Human-needs centred development: A term for development where the level of satisfaction of various dimensions of human needs is considered to have improved. Extending Seers' conditions for development to a list of eight gives:

1 low levels of material poverty;

2 low level of unemployment;

3 relative equality;

4 democratization of political life;

5 'true' national independence;

6 good literacy and educational levels;

7 relatively equal status for women and participation by women;

8 sustainable ability to meet future needs.

formulations go beyond the needs and aspirations of *individuals* and of *nations*. Groups, local communities, and social classes could also be mentioned, as one might well have development goals for these; equally there could well be an international dimension since international agencies are certainly involved in propagating values for development.

6.2 The context of development – global capitalism and the politics of the nation state

Whatever definition of development is taken, it has to be seen in its economic and political context, which means seeing development in relation to contemporary global capitalism and the politics of the nation state.

Capitalism is the dominant international economic system. Even before recent crises effectively discredited state socialism as a viable rival, it was capitalism which held sway over economic relations at a world level, even though the Soviet Union and its allies in Eastern Europe and elsewhere exemplified the possibility of an alternative system, at least within their national boundaries.

In an extreme or pure form of capitalism, there would be only a minimal role for states or any form of collective or public action. In practice, in contemporary or 'advanced' capitalism, states have an important role to play. Development in particular has to be seen not only within the economic context of global capitalism, but also in the political context of a world of 'nation states'.

The term 'nation' denotes a certain unity among a people which may include a common national language, a shared history, a common cultural heritage in terms of tradition, norms of behaviour, religious customs, and so on. By contrast, the term 'state' describes a political unit occupying a certain geographical territory and seeking to control it under a single jurisdiction. In practice, the test for what counts as a 'sovereign' state is that it should be recognized as such by other sovereign states – and by international agencies. The term 'nation state' thus depicts a supposed ideal where cultural and political boundaries coincide.

There is nothing 'natural' about the division of the world into nation states. However, there are several implications for development. The ideology of *nationalism* can provide an important impetus for economic development, as for example when the need for independent military capability justifies the sacrifices required to make resources available for investment in industry. Within and across state boundaries, differing ethnic identities may be blamed as the source of conflicts between interest groups, which at the very least complicate processes and programmes of development (see Chapter 18).

Although, as we shall see in Chapter 13, the nation state may be in relative decline and other political actors are of increasing importance, for the time being we will accept the world as one of nation states, and concentrate on the political context of development within such states. A discussion of the concept of 'the state' is included below in greater detail, but ideas of nation and nationalism are left aside.

The elements of global capitalism

Capitalism can be characterized as a system of production of goods and services for market exchange in order to make a profit. Capitalism has certain basic elements as an ideal system which, taken together, distinguish it from other systems: private ownership; regulation through the market via commoditization and competition; distribution of welfare through the market determination of wages; and enterprise management for profit and accumulation.

Ownership
Capitalism means *private ownership*: the ownership of the means of production is private and individual. The archetypal capitalist owns his or her land, buildings, tools, and equipment, and hires in labour and buys raw materials in order to use these means of production to produce goods and services for sale. In the case of large corporations, there are many owners, but they still hold shares as individuals. Of course, not all individuals own means of production, either directly or

Figure 6.5 Regulation through the market. The working of markets like this large grain market in north-west India sets prices for commodities. Such markets can exist within or outside capitalism; capitalism requires the combination of regulation through the market with other elements, notably production for profit and accumulation.

through holding shares in productive enterprises. The important point is that the main form of ownership is not a collective form, and in particular is not state ownership.

In practice, many prominent capitalist countries have large state sectors. This goes for Sweden, France, South Korea, and other Third World capitalist countries such as Mexico. Also, collective savings institutions such as insurance companies and pension funds own large amounts of shares in capitalist firms, but they tend to act as though they were individual owners and not to represent collective interests (see below).

In relating this aspect of capitalism to Third World development, the question of foreign investment is of prime importance. Capitalism would imply being completely open to overseas ownership through foreign investment – and also privatization of any previously state-owned industries. In the 1980s there has indeed been a big pressure for this, implying selling either to local entrepreneurs or to transnational corporations (TNCs).

Regulation
Capitalism means *regulation through the market*, not state planning or intervention.

In this context, regulation refers to how decisions are made about what is produced, how much of each product, at what price and what quality, and so on. It is clear how state planning within a country could imply regulation about these aspects of production. The state might impose price controls, set quotas for production from different factories, and employ inspectors to check quality. Under capitalism, such regulation is effectively imposed through the impersonal mechanism of the market, via *commoditization* and *competition*.

We saw in Chapter 3 the distinction between commodity and subsistence production. Under capitalism all goods and services are in principle turned into commodities for exchange rather than for the producer's own use. Individual producers and firms can then produce what they like, decide on quality and set what prices they like, but the assumption is that competition between firms, together with consumer choice between competitive versions of the same commodity, will force them to set the 'right' prices and produce what is actually required (Figure 6.5).

Once again, there is the proviso that in practice many capitalist states have a high degree of state intervention. Many public goods and especially public services are supplied directly by state agencies. States commonly impose direct price controls and quality standards; they also employ different types of incentives such as preferential tax treatment for reinvestment in certain areas of production.

In the Third World there tends to be a large number of state interventions regulating the relation between the national economy and the international. For example, Third World states typically maintain exchange rate and credit controls, tariffs, and import controls. International or Western agencies, such as the International Monetary Fund (IMF), or United States Agency for International Development (US-AID), that follow capitalist principles, often press for a reduction of state intervention in such areas.

Distribution

Under capitalism the allocation of resources and the distribution of welfare are done through the market determination of wages. In principle, there is no universal provision or rationing even of basic goods or welfare services.

This is the question of how it is decided who gets what. In capitalism all goods and services are commoditized and can be bought. Hence, how much a person gets (i.e. his or her entitlement, in the terms explained in Chapter 1) depends entirely on personal income – or on how a household's income is divided between income generators and dependants when it comes to consumption. For the majority of the population who work as wage labourers, this in turn depends on how much of the surplus derived from the sale of the products comes to them as wages, rather than going to the capitalist owners as profit. The level of wages is determined through the labour market, and competition for employment will tend to keep wage rates down. Thus workers' labour power is another commodity. Indeed one way of characterizing capitalism is to call it a system of *generalized commodity production*, implying that labour power as well as means of production and all goods and services are commoditized (see also Chapters 8 and 9).

Wages are not the only source of income under capitalism. Those who are capitalists, large or small, get income through dividends on shares or taking profit directly. Chapters 1 and 3 have discussed different forms of livelihood, and we can see that while the division into capitalists and workers may be basic to the notion of capitalism, there are many other groups only partially linked into the overall system but whose welfare is still determined mainly through market mechanisms. Peasants and craft producers, for example, may combine petty commodity production, deriving income from the sale of their produce, with a certain amount of direct production for own use (in terms of entitlements, combining trade with exchange entitlements).

In practice, almost all countries have at least some services provided universally, as with the National Health Service and basic educational provision in the UK. There is no doubt that Sweden, for example, where inequalities have been systematically reduced through state provision, is nevertheless a capitalist country.

There may be a greater variety of livelihoods – and entitlements – in many Third World countries than in the industrialized West. As discussed in Chapters 1 and 3, it is the inadequacy of many of these livelihoods that results in widespread poverty, and their failure that can lead to famine. Here again, state interventions such as the basic food provisioning policies of the Sri Lankan governments since independence, described in Chapter 2, would be opposed by agencies or governments with more capitalist orientations.

At an international level, there is the net flow of resources out of the Third World as a whole. This is not a question of market determination of individuals' incomes through bargaining for wages, but of general income levels which are affected by a country's international earnings and liabilities. Market principles apply at international level to set the interest rates and conditions on the servicing of international debts. Markets also regulate the changes in world commodity prices (and the generally adverse movement of the terms of trade from the point of view of the Third World, as noted in Chapter 5).

Enterprise management

Production under capitalism is run with the aim of making a profit in order to accumulate. Management is undertaken by or on behalf of the capitalist owners and in their interests, rather than directed primarily towards the interests of the workers in the enterprise, or of the local community or the state.

As noted above, the state itself, as well as collective institutions such as pension funds, owns shares in productive enterprises, but the management of such enterprises still treats these owners as individuals whose prime interest is to maximize the return on their investment. Indeed, in many capitalist countries it is enshrined as a legal principle that funds entrusted to others have to be managed in the owner's material interest – which is taken to mean maximizing profits first and foremost.

In the case of small-scale agriculture in the Third World, trying to achieve a profit leads to a different regime from that which attempts to safeguard all livelihoods against risk. It implies a polarization into a relatively small number of 'rich peasant' or capitalist farmer households and a larger number of landless wage labourers.

In larger agricultural or industrial enterprises, the idea of making profit for the owners leads to various questions. For example, will overseas ownership mean that managers who are local nationals will nevertheless ignore local interests in favour of the international focus of their parent corporation?

Legitimation

All these elements fit together to form a global system that also functions in an ideological fashion to legitimate actions taken in particular ways.

Examples have been noted where international and Western agencies have advocated new practices in line with capitalist principles, even where these involved dismantling previous welfare-oriented policies. The idea that the system works as a whole to promote efficiency and wealth creation is very powerful, and acts to legitimate actions which would otherwise appear simply to be favouring the interests of capitalists themselves.

Aspects of the politics of the nation state

In discussing the political context of development, we need first to be clear how we are using the term politics. Here we use it to refer to processes of ordering and steering society at different levels, including the means by which some people or groups attain positions from which they can effectively promote their own interests and values. Although the different levels range from the local to the international, we concentrate here on the *national* political context of development, broken down into a number of aspects grouped under four headings below. Thus development fits into wider national political processes which involve: choice and steering; material interests and values; a variety of political actors and institutions; and, perhaps most importantly, state power and the maintenance of order.

> **Politics:** Processes of steering and choosing through time at different levels and localities. The term expresses the idea of material interests and values; it fundamentally concerns power relations within and between political actors and institutions, including forms of popular struggle; and it involves state power and the maintenance of order in society.

Choice and steering

Politics centres on particular types of *processes* involving people engaged in interrelationships through time. Leaders come and go, the people acquiesce in, or enthusiastically support, or struggle against being ordered and steered in a particular way. The processes of politics are essentially embedded in and partly determine historical continuity and change.

Politics occurs at *different levels*, from the household, caste, hockey club or factory to the locality (local politics), the region (provincial politics), the nation state, and the international arena (global politics). The political process also has a spatial dimension; it can press down more or less firmly on different localities within any larger social and economic arena.

Politics is *steering*, or governing. It has been likened to steering a ship at sea ('the ship of state') – and has several similar requirements, like trying to stay in control, charting a course and trying to achieve an objective (reach a destination). All these are important aspects of political leadership, but the people being led are left out of the picture. Another analogy has been around in Asia for a long time: that of trying to ride a tiger in a particular direction, the tiger being the more or less obedient masses who must be coaxed along and who may, if goaded too hard, turn on the ruler and make big trouble.

Any deliberate action to bring about change in society in a desired direction always involves *choice*. To choose one development programme means not choosing another, and that choice is political because it benefits some people more than others. Thus development fits into politics in general in that development itself involves choices taking place through time, as part of broader processes of steering and at different levels – local, regional, national, global. Development choices at one level have consequences for choice at other levels or different effects in different places.

Material interests and values

Politics is not just about process; it always has content related to certain material interests and valued preferences. Those doing the ordering or steering usually have a political project or purpose of some kind, which may or may not be shared by those being ordered or steered. The project may be rather narrow – to enrich the leader and/or a dominant class. It may be more broad – to enrich and fulfil the lives of all the people through increased empowerment, the abolition of poverty, and the creation and optimal distribution of wealth. Political content produces conflict. It also inspires and mobilizes people in collective political effort.

Development plans and strategies also relate to interests and values. They express certain *values* in the sense that they rest on certain preferences and moral assumptions about what is good or desirable. For example, a preference for competition as broadly the right approach to development effort usually rests ultimately on some moral conception about the value of individual freedom.

A development strategy which gives strong emphasis to collaboration usually is grounded in a more egalitarian moral premise. *Interests*, on the other hand, are about material benefits. Any development choice or action is going to benefit certain interests more than others.

Usually, dominant interests and values in any development initiative reinforce each other. For example, development schemes which spring from those who value competition usually best serve the interests of those who are already dominant, with the result that there is little immediate change in the existing structure of rewards and benefits. This is frequently what happens with top-down initiatives by bureaucratic and political agents within the state. As Goldsworthy (1988, p.511) suggests, decisions and actions by such agents will commonly reflect either bureaucratic self-interest or the interests of friends, clients and class allies, or will at least 'objectively serve' those interests, resulting predictably in 'a pattern of development that both reflects and consolidates existing disparities'. But that may be too static a view. Over time, patterns of development can also change as the relative power of clients and classes changes.

Political agents and institutions

Politics is about people getting into sets of positions and their supporters pursuing their interests and purposes. People in such sets of positions may be called political agents. They make up political institutions that endure over time such as legislatures, courts, chief executives, councils, commissions, military junta, the police, the armed forces, civilian bureaucracies and political parties. Combinations of political institutions make up the state apparatus. Also part of politics are the processes by which some people get into these sets of positions; such processes include elections, appointment, inheritance, military coup, revolution. Here we consider how development relates to two important examples of political institutions: political parties and civilian bureaucracies.

There are different types of parties and party systems in the Third World, from (for example) those where only one party exists, to those where parties are ephemeral or non-existent, to multiparty regimes where there are decentralized patronage-based parties seeking votes. A conventional view of parties sees them as agencies representing certain views or interests in society and, when a party's leaders are in power, having a controlling or influencing say over government policy, including development policy. Political parties would thus provide a major *political input* into the development process (Figure 6.6).

This conventional conception of parties is not completely wrong, but in Third World contexts it can be misleading if too much emphasis is put on it. In much of the Third World, political parties are more important as *political outputs*, as instruments of government. They frequently are created and used by political leaders to help legitimize their regimes, as well as to provide a structure in society through which, generally via the distribution of patronage, a coalition of powerful political interests sufficient to sustain the government can be secured (Randall, 1988, p.184). Parties in the Third World can be used by governments to knit together alliances between certain groups in society at different levels, through which development efforts can be channelled.

One also has to be wary of conventional definitions when it comes to bureaucracies. They are usually thought of as organizations characterized by hierarchy, continuity, impersonality and expertise. They are also supposed to be distinct from the governing bodies which employ them.

This model in fact relates to an ideal type. Few if any bureaucracies in the Third World or elsewhere would fit it closely. Hierarchy? Formal organization charts frequently hide the reality of pyramids of informal, patron–client groupings. Continuity? Bureaucratic tenure for officials can be short or uncertain as determined by the rapidly shifting fortunes of ministers, generals or other political leaders. Impersonality? Bureaucratic decisions are frequently determined by particularistic loyalties rather than by impersonal application of general rules. Expertise? Training programmes are frequently inadequate or non-existent and selection can be influenced more by political loyalty than by the qualifications of the candidates. Indeed, some bureaucracies appear to depart so far from the model that it might seem misleading even to refer to them as bureaucracies.

But that goes too far. Most countries in the Third World and elsewhere do have organizations accountable to political leaderships, with noticeable elements of hierarchy, continuity, impersonality and expertise, which distinguish them from other types of political institutions. Development takes place in a political context in which such bureaucracies play a considerable role in several ways.

First, bureaucratic officials make an input into the making of development *choices*, by providing necessary information, suggesting likely political outcomes of different policy options, even framing the options. There are also a host of bureaucratic

Figure 6.6   Development involves political parties. Above are those present at the foundation of the All-India Muslim League at Dhaka, 1906. The League was eventually, as a political party, to lead the movement for an independent Pakistan.

Figure 6.7   Development involves local bureaucratic control. Officials of a co-operative credit agency in India getting the thumb impression of a peasant on documents before giving him a loan for agricultural operations.

decisions to do with finance, budgeting, scheduling, methods of recruiting and training necessary personnel, and so on, as well as the detailed allocations of resources at a local level.

Secondly, the *implementation* of any development effort involves power and control. The way bureaucracies are organized helps higher civil officials and political leaders to ensure that people working lower down in government departments are behaving appropriately. Also, people on the receiving end of development programmes normally must enter into a power relation with bureaucratic officials whenever government funds are involved. For example, conditions to be met before grants and loans can be released are enforced by local bureaucrats (Figure 6.7).

Thirdly, the *results* of development will be affected by informal political and social structures and alliances in society at different levels. Bureaucracies, like political parties, can help to knit together such alliances. Even when there is a comparatively rich assortment of local structures and alliances, bureaucratic and party institutions may together dispense development funds to local élites and others, thereby helping to sustain the alliances of those who benefit from development.

State power and the maintenance of order

Finally, politics concerns *power* relations in society, both authoritative power (perceived by those subject to it as legitimate) and coercion or 'naked power'. Within national boundaries, particularly, the maintenance of order in society depends on the power of the state, and development takes place within this context.

> "From one point of view, the state is society's way of controlling itself, and of making sure that others do not control it. From another, it is a sort of protection racket which claims a monopoly over the use of force."
>
> (Jordan, 1985, p.1)

Before looking at the different roles the state can play in development, we should be aware of some important general features of **states**.

1   Each state claims the right to regulate affairs within the boundaries of its own territory and aims to provide security from foreign intervention for people within its boundaries by conducting relations with other states. Such interstate relations can be peaceful (e.g. diplomacy) or warlike. It has been estimated that from 500 BC to the present day in the West, 'the average state has

engaged in at least one open, organized war with another state in about 50 per cent of years' (Mann, 1984, p.32). The military capability that states acquire can also help to make them exceptionally powerful and dangerous within their own territories.

2 The state's claim to 'a monopoly over the use of force' within its boundaries goes with its being 'distinguished from the myriad of other organizations in seeking predominance over them and in aiming to institute binding rules over the activities of other organizations within its boundaries' (Azarya, 1988, p.10). Notice the note of conditionality regarding the state's power – the state *seeks* predominance and *aims to institute* binding rules. The state's predominance and maintenance of binding rules might be strong and effective, but equally it might be weak and reflexive, so weak indeed that the state's power is really of no consequence beyond the walls of the political leader's compound.

3 The state also provides 'identity and cohesion' (Jordan, 1985, p.1). In order to enjoy the enormous advantage of having their rule accepted by the people (or at least some of the people) as legitimate, states continually promote a sense of national identity and common citizenship, and seek to define away, or suppress, competing ideologies subversive of their rule. This process of legitimation involves the reproduction of a common culture, a 'national interest', through the use of the mass media, the educational curriculum, religious organizations, and other interest groups and associations (Figure 6.8). The working of political institutions, the 'majesty of the law', and the pomp and circumstance that attend state ceremonies can also contribute to the process of legitimation so central to the reproduction through time of any state form.

4 The state is not coterminous with society; it is one of many organizations or spheres of activity *within* society. It is therefore both an agent and a structure. It acts as an agent within a broader social structure (and international arena) and is always influenced to some extent by this structure. At the same time, the state also provides a structure of binding rules (more or less) that influence or control to some extent the actions of other agents within society.

Figure 6.8 Identity and cohesion. The Chinese Communist Party and state leadership emphasized the importance of education and communicating its ideology.

5   The state sustains relationships with other agents within society; it 'coexists and interacts with' families, economic enterprises, religious organizations, and so on (Azarya, 1988, p.10) and presides over 'different spheres of the community' (Jordan, 1985, p.1). The most crucial of these relationships is the one between the state and the economy: broadly speaking, states participate directly in the processes of productive capital formation, they provide infrastructure, and they affect private sector resource allocation through monetary and fiscal policies.

6   Finally, although a state can be considered as a single cohesive entity or organization distinct from others in society, it is at the same time a set of organizations or an ensemble of political institutions – coercive, administrative, legal – which may not always act as one or in concert. The state in India in 1990, for example, had assemblies of political institutions at the centre and in each of 20 state governments, one of which in West Bengal was controlled by the Communist Party of India (Marxist) which was pursuing policies different from those in the other 19 states and at the centre. Even within West Bengal 'the state' behaved differently in different districts, development blocks and villages.

From these six features of states we can see that the state must always play a major role in development, though this can be in at least three different ways.

First, the state itself can be a *primary agent* for development initiatives in society. Political decisions can be taken regarding development as contained in national and local plans; development departments can be established employing trained personnel with development expertise; funds and other resources can be obtained and allocated; bureaucratic rules and regulations can be laid down regarding how the development programme is to be implemented; and so on.

Second, the state can provide an *enabling structure* for development by other agencies. That is, development can be taking place primarily through the activities of economic and other organizations outside the state. This is what occurs in the case of capitalist development, when the state provides essential infrastructural support, including the maintenance of order in society, establishing a set of economic policies favourable to capitalist accumulation, changing the domestic economy in response to the changing demands of global capital, controlling the labour force, providing necessary transport and communications, and so on. Actually, a strict separation between the state providing infrastructure and capitalist enterprises spearheading development is not easy to make in many countries where there is a long tradition of state involvement in the economy. For example, state guidance of capitalist development was central to the exceptional growth of the newly industrializing countries of South Korea, Taiwan and Singapore.

Third, the state can be a *structural obstacle* to development, resulting in development effort through collective struggle against the state. Workers and peasants may perceive the state and its allies as maintaining forms of exploitation and oppression that block any genuine development from which they might benefit. Development for them can only begin with collective struggles which 'can combine elements of *defence* against further exploitation and oppression; of *resistance* to the power of capitalists and landlords, and the legal, political and ideological forces that support them; and of *transformation* when such struggles

States: (1) Conduct peaceful and warlike relations with other states; (2) claim a monopoly over the use of force within their boundaries and generally seek to order society through power relations; (3) can provide identity and cohesion through processes of legitimation; (4) act as agents within society and structure the actions of other agents; (5) sustain myriad relationships with other spheres of activity and groups and classes – of which the relationship with the economy is most important; (6) are not unified organizations but rather ensembles of institutions and processes which are extremely various, conflictual and complex.

develop ideologies and forms of organization and solidarity that challenge existing structures of exploitation and oppression and point the way beyond them to alternative forms of society' (Johnson & Bernstein, 1982, pp.266–7). In such cases, the state is neither the main agent of development nor does it provide infrastructural support for development by others; it is central to the problem of development because it does not allow development to occur.

These three cases should not be seen as mutually exclusive. A state can act simultaneously as development agent, enabling structure, and obstacle.

Other important features of states and further consideration of the role of states in development are discussed in various contexts throughout the rest of this book. You should look especially at Chapter 10 on colonial states, Chapter 12 on socialist states and development, and Chapter 14 on development and the democratization of Third World states.

6.3 Competing views on development and social change

Let us now consider the question of *how* development occurs. As pointed out in the Introduction to this book, development can be seen both as an *historical process of social change* in which societies are transformed over long periods and as *what development agencies do*: deliberate efforts aimed at progress on the part of governments, organizations and social movements. Each of these two ways of looking at how development occurs gives an incomplete picture. Each gives different emphasis to **structure** and **agency** as factors in explaining development. These are terms which have cropped up above but which are worth considering in their own right. Both are necessary for a full appreciation of the process of development.

There are four major sets of competing views on how development occurs. These views differ in how they see development as social change in relation to the global capitalist system and the

Structure: A factor explaining development referring to relatively slow-changing sets of relationships between classes, economic activities and other general elements in society.

Agency: A factor explaining development referring to particular sources of action.

Explanations of social change based on structural factors emphasize shifts in relationships between classes, etc., whereas explanations based on agency factors look for specific individuals or groups as agents of change.

role of the state; they also differ in what are seen as the key agencies in development. Like all coherent views on any aspect of society, they each have both **analytical** and **normative** aspects: they attempt both to explain how development *does* occur and to suggest how it *should* occur. As you read the rest of the book you should consider how any of the views put forward at any point fit in with those outlined here.

Analytical: Such a view or theory attempts to explain or analyse some aspect of society, perhaps putting forward a conceptual framework for understanding.

Normative: Such a view or theory brings in value judgements and suggests how things should be rather than just explaining how they are and why.

In practice, a view or theory is bound to contain aspects of both the analytical and the normative. The use of certain concepts rather than others implies certain values, and conversely value judgements cannot be made without some view about how things work.

Neo-liberalism

In the 1980s neo-liberalism (or market liberalism) became the dominant view of development, at least in the industrialized West. Those who today promote this view are the direct descendants of the proponents of 'free enterprise' in the 1950s and earlier, and trace their theoretical ideas back to the classical economics of Adam Smith in the late eighteenth century – though modern neo-liberalism is perhaps more ideological rather than only economic. In this view the purest form of the above system of capitalism is the best. It is said to be both efficient and fair – and these two statements correspond to the analytical and normative aspects respectively of this theoretical view.

Let us look at the analytical side first. In neo-liberal thinking, the most important of the five elements of capitalism outlined in Section 6.2 is regulation through the market. It doesn't really matter who the owners are; they are always assumed to be acting rationally in accordance with their own material interests, which in turn is assumed to mean maximizing profit or return on investment in order to accumulate and reinvest. The important point about them is that they are viewed as individuals – and in fact individualism is a key aspect of this way of thinking.

However, market competition is even more important. It is seen as the main force towards economic progress – and hence development. Faced with market competition, the best ways to ensure continued profits are to grow and innovate. These both lead to increased labour productivity: growth does so through economies of scale, and innovation through capital investment in improved production processes. Thus successful capitalists are able to enter a positively reinforcing cycle: profit – accumulation – growth – innovation – increased productivity – increased profits; and then can use those increased profits to continue the cycle.

This system is seen as progressive because it allows enterprising individuals to thrive, and the benefits of their innovations and the increased productivity will eventually be benefits for all. This argument goes back to the famous phrases of Adam Smith: the 'hidden hand of the market' converts individual interests into 'the wealth of nations'.

The economic aspects of neo-liberal theory are generally underpinned by psychological arguments about the values, aspirations and motivations of individuals. The market presents a formal equality of opportunity to all who enter it, and distributes rewards objectively. What determines success is what individuals are able to bring; in other words, how well endowed they are before entering the competition. You will recall this idea of *endowment* from the discussion of poverty and livelihoods in Chapter 1, so you will probably take it to imply the whole bundle of a person's assets and liabilities (personal possessions, access to land and other means of production, education, family dependants, etc.). However, in this context it means simply that a person has a given psychological type. Some individuals have what it takes to succeed and others do not.

David McLelland, an American psychologist who claimed to have isolated the vital motivational factor necessary for economic development, suggested the following metaphor for market competition: 'The free enterprise system…may be compared to a garden in which all plants are allowed to grow until some crowd the others out' (McLelland, 1963, p.90).

Thus, in neo-liberal thinking, individual capitalist entrepreneurs linked through the market provide the dynamic for development. However, in the end these gradual changes initiated by many individuals are seen as leading to a total process of change in social structure, political systems and culture; in other words, to *modernization* as the term was used in the 1950s and 1960s (see Section 6.1 above and Chapter 11).

The most obvious distinguishing feature of those Western countries which are thoroughly modernized and developed is that they have undergone industrial revolutions and as a result enjoy high per capita income. Another feature claimed as equally important by many proponents of neo-liberalism is that these Western countries are liberal democracies: they combine the prosperity associated with industrialized economies with political systems based on parliamentary representative democracy.

Recall from the discussion in Section 6.1 above on modernization that the changes said to accompany development included 'social disturbances' (Smelser, 1968, p.127). Those countries which modernize successfully may then be seen as those where political institutions develop which allow people to give voice to their new aspirations, and which allow at least some of the wealth accumulated by the capitalist entrepreneurs to trickle down to other sections of society. Liberal democracy as a political form allows for 'freedom' in electing representatives to match 'freedom' in the market.

You might wonder why, if these processes of modernization and capitalist development are so inexorable and the result so desirable, not all parts of the world have developed to the same extent. Neo-liberals look for the answer to this question in the idea of obstacles – which in this context mean things that prevent the proper working of the market. Three main kinds of obstacle are put forward to explain different cases of lack of development:

1 *Tradition*. The continuation of non-market social relations and systems of obligation can be seen as preventing production for own use from being commoditized. Related, racist, notions such as that of the 'lazy native' were particularly prevalent under colonialism (see Chapters 3 and 9), but are by no means uncommon today. For example, Smelser's suggestion that modernization means increased social disturbance perhaps also implies that 'traditional societies' were somehow peopled by passive individuals whose lack of entrepreneurialism was mirrored by their fatalism. In fact, of course, there are plenty of examples both of resistance to colonialism and of innovative and creative methods of improving and safeguarding livelihoods, which continually transform so-called traditional societies without bringing about a transformation to a capitalist production system.

2 *Monopoly*. Under capitalism, the protagonists naturally try to minimize the regulatory effects of the market as a whole by finding a particular small market or market segment which they can either completely monopolize or at least partly dominate. There are two sorts of monopoly which can act as obstacles to market regulation: monopolies of capital, i.e. industrial monopolies; and monopolies of labour, i.e. trade unions.

3 *State regulation*. In general, any kind of collective or state action, except when the state is acting purely as a shareholder like any other owner, is seen as interfering with the proper working of the market. In the neo-liberal view, the role of the state should be a minimal one: guaranteeing political order, ensuring the conditions for capitalism (keeping a 'level playing field'), and 'policing' the casualties of the competitive system.

There is a real dilemma here for neo-liberal thinkers. While they favour 'rolling back the state' as far as possible, they also require its policing function, which in practice tends to be considerable. The dilemma is how to guarantee that this policing is done fairly, since it is necessarily done outside the market and hence outside the mechanism which this theory argues is the means of fair regulation. From what has been noted in Section 6.2 above about political institutions and values tending to match material interests, we might deduce that a capitalist state is more likely to act against a trade union monopoly than against an industrial one. Indeed, many Third World states compete to attract investment from transnational corporations – which are often industrial monopolies in themselves – by making a virtue of their tough anti-union legislation.

The main normative aspects of neo-liberalism are the positive values put on individual achievement and competition. In the Third World as elsewhere, the most important obstacle is usually seen as that of state intervention. Third World states commonly engage in various policies such as controlling exchange rates, food subsidies, imposing tariffs and quotas, which can all be regarded as 'distorting the price signals' that allow competition to work. Such policies are seen as counterproductive; for example, it is argued that without food subsidies there would be more incentive for farmers to invest in greater productivity and to produce more, and more production would in turn lead to cheaper prices through competition and hence obviate the need for the subsidies.

As already noted, neo-liberal thinking is often closely linked to ideas of modernization and the view that there is general progress towards not only greater prosperity but a single model for a modern prosperous state. The poorer peoples and countries of the world have no specially difficult structural position that makes their development problematic or impossible; indeed, they are governed by exactly the same principles of competitive relations between individuals as those who have succeeded in becoming rich. Thus, although in the short term countries whose comparative advantage is in agricultural commodities should concentrate on efficient production of such commodities and not try to force industrialization through state intervention, in time, according to this view, one can expect more and more Third World states to evolve into modern, industrialized, capitalist, liberal democracies.

Fukuyama gives a good example of how rigid this way of thinking can be:

> "There is a very tight relationship between liberal democracy and advanced industrialization, with the former following the latter inexorably. Industrial maturity is an interconnected whole, requiring higher levels of urbanization, education, labour mobility, and ultimately free communications, political participation, and democratic government."
>
> (Fukuyama, 1990)

Structuralism

The word 'structuralism' was first prominent in discussions on development in the work of Raul Prebisch and others in the UN Economic Commission for Latin America (ECLA) just after the Second World War. Here this heading is used to group several related but distinct strands of thought, of which two (Marxism and the dependency school) are set out further below.

In general, such views are concerned with underlying social and economic structures and see development as involving changes in these structures. They differ fundamentally from neo-

liberalism both in their view of history and in their approach to capitalism.

Thus, whereas to neo-liberals history is the sum of individuals' actions, including the actions of individual governments, firms and other organizations, structuralists see history in terms of political and economic struggles between large social groups, particularly classes, as new structures and systems replace old ones across the globe.

There is in fact a structural aspect to some of the views discussed above under 'neo-liberalism', particularly those of the modernization theorists, as well as to those discussed below under 'interventionism'. However, much structuralist thinking on development including the Marxist and dependency schools emphasized here, has in common a fundamentally critical view of capitalism. While global capitalism is seen as having a quality of dynamism that may be necessary for economic development, it is regarded mainly as a system of *exploitation* which should in the long run be radically altered.

Of the elements of capitalism set out in Section 6.2 above, structuralists regard ownership and management as most important, and do not regard regulation through the market as a defining feature. In practice, the most important markets, both internationally and within particular countries, tend to be either monopolies or controlled by the capitalist states in a way that safeguards the profits of the large capitalist enterprises and the interests of the capitalist owners.

Of the five elements, distribution is also highlighted by structuralists. As noted above, under capitalism the allocation of resources and the distribution of welfare are done through the market determination of wages. This is where capitalism's fundamentally exploitative nature comes in.

Marxism

Historically the most important of the two views grouped here is associated with the ideas of Karl Marx. Marx viewed capitalism as a particular type of *class society*, one constituted by antagonistic relations between different social classes, of which the most important are capitalists and workers.

Any class system is based on particular relations of production, and in the capitalist system those who own the means of production have the power to appropriate surplus, whereas those do not own means of production have to sell their labour power. However, one can also argue that the contradictions inherent in capitalism tend to bring about or accentuate other divisions as well, such as those based on gender, ethnicity, nationality, and so on.

Marx believed that industrial capitalism, in particular, represented a massive advance in the progress of society, particularly in the impetus it gave to the systematic application of science to methods of production. He also saw as very positive the way that capitalism brings people together in an ever-increasing scale of co-operation, with integrated production processes organized on the basis of socialized labour, as opposed to the small-scale 'privatized' labour of household production.

On the other hand, Marx saw class exploitation and oppression as essential features of capitalism. He vehemently condemned the conditions of life of the industrial working class in Victorian Britain and the Europe of the time, and the brutality perpetrated by British colonial rule in India.

In short, for Marx capitalism was profoundly contradictory, at two levels. First, the development of productive capacities under capitalism represents an enormous potential force for human emancipation and freedom from want, at the same time as the class relations through which the productive forces have developed deny their promise to the majority of people. Second, these class relations embody a contradiction between *private* ownership and control and increasingly *socialized* labour. Marx thought that in time private ownership would begin to obstruct the further development of productive capacities. Then conditions would be ripe for the overthrow of capitalism. This would not be automatic, but result from class struggle between capitalists and workers. The latter would organize in a political movement to dispossess the former and then utilize the productive capacities made available by capitalism to go on and form a different kind of society, i.e. communism.

As for the role of the state, during the class struggle it could be seen as both an enabling structure for capitalist development and a structural obstacle to development that would benefit the workers. Then, once in the hands of the workers, the state could be used to promote socialist development (see Chapter 12) in the transition to communism.

The dependency view

The other approach that we call structuralist is that of the dependency school. Here capitalism is still seen primarily as a system of exploitation (indeed, many dependency thinkers could be called neo-Marxists) but the important point is its international nature. In this view, the historical process which resulted in the development of the industrialized world was the same process in which the Third World did not become developed. Put simply, Western capitalist industrialization created structures in which Third World economies were dependent and which tended to lead to and maintain underdevelopment.

In its most crude version, dependency thinking simply substitutes countries for classes so that capitalism is not so much a system of class exploitation as one of exploitation of Third World countries by the First World. In less crude versions, the international capitalist class, together with allies from the ruling élites of Third World countries, is able to exploit workers and peasants in Third World countries, at the same time as 'buying off' its own working class with a mixture of material rewards and racist ideology.

As in classical Marxism, capitalism is viewed as having positive as well as negative aspects. For example, dependency thinkers tend to favour industrialization and to note the positive aspects of the dynamism of capitalism in raising productive capacities. However, the dependency view is not clear on how the contradictions of capitalism as an international system might be overcome. Various development strategies have been worked out, which differ in the extent to which they reject or try to work with international capitalism, but which are all deliberate efforts to improve living standards by changing the structural relationship with international capitalism. This generally means development is to be achieved through the actions of Third World states.

If we look at development from the perspective of the government of a Third World country, the implication of dependency thinking could be to advocate withdrawal from the international capitalist system, or at least strong local state controls on it. This might be in order to build up national capital or to institute some form of planned development, 'socialist' or otherwise. It might entail a kind of solo self-reliance, or could be in solidarity with other Third World countries.

However, such thinking hardly offers solutions. For one thing, it tends to be uncritical of Third World states, and assumes, contrary to the analysis of Section 6.2 above, that they are unified organizations with the power and the will to implement anti-capitalist policies. Again, without overseas involvement, where is the investment needed for economic development to come from? (Figure 6.9) The only possibility would seem to be from the savings of that poor country's own people; in other words, by squeezing surplus out of the same already poor population that may supply the state's political support base.

It's only some foreign aid mission members, sir. I told them we wanted to be self-reliant and didn't want to depend on any country and sent them away!

Figure 6.9

Neo-liberalism and structuralism – is there no other alternative?

So far, this chapter has stressed the differences between neo-liberal and structuralist views. It is worth also noting what they have in common.

There are two main areas of commonality. First, both view development mainly in terms of broad historical social change. As theories, though they have their normative aspects, they do not offer detailed prescriptions for what development agencies should do, beyond the idea of ensuring the positive features of the development process are enhanced as far as possible. Thus, neo-liberals stress removing obstacles to market regulation; Marxists favour the organization of socialized labour; dependency thinkers look to the actions of Third World states.

Second, both look favourably on industrialization as the only realistic way to the economic growth required to achieve massive improvements in Third World living standards. This is seen as a prerequisite for any other aspect of development.

What might be regarded as a third area of commonality is agreement on what is the main area of disagreement and conflict. Thus, the neo-liberal insistence on the materialist motivations of individuals and the market as a regulator is opposed to the structuralist view of the importance of social solidarity, class and collective forms of action. Since the only examples of large-scale collective action for development have occurred through the state, this opposition tends to be represented as *market versus state* or *profit versus planning*.

Is there really no other alternative?

Let us look briefly at two ways of looking at development which give pride of place to human voluntary effort (voluntarist views of development). This chapter has used the labels *interventionism* and *populism* for these two approaches, though once again they each cover a large range of views.

Obviously, individuals, particularly if they are charismatic and highly motivated, can achieve great things even against the general trend of history; even more so can governments, if united behind strongly held values. However, there is a general objection to elevating discussion about what individuals, governments and others might do to achieve development to a theory of development, in that it does not really describe any mechanism by which social change is continued over a long period. Nevertheless, without being general theories of social change, these two approaches to development are of great importance as potential sources of alternatives to the two main sets of views described above.

Interventionism

Here industrial capitalism is viewed positively but at the same time a need is perceived for non-market regulation through state intervention. The structural inequalities and contradictions inherent in capitalism are to some extent admitted. This view has a lot in common with structuralism, but, instead of hoping to replace the market, interventionism may be said to *combine* state and market. It is worth noting that interventionism and structuralism were the dominant views of economics and of development up to the 1970s; neo-liberalism is to a great extent a reaction to this dominance.

We can look at four important arguments for state intervention. First, in the *Keynesian* view, developed in the context of the great depression of the 1930s, periodic booms and slumps are inherent problems of capitalism, which has no inbuilt mechanism for ensuring a balance between supply and demand as economies grow. John Maynard Keynes proposed state spending to create employment and increase incomes, thus stimulating demand and restoring business confidence.

Second, there is the view first propounded by the nineteenth-century German thinker List, who advanced the case for protecting the 'infant industries' of newly industrializing countries from competition by well-established industries elsewhere. List was thinking at the time of protecting German industries from British competition; but his arguments for *protectionism* have been taken up by contemporary Third World governments. Today's newly industrializing countries are trying to

do so in the face of competition in global markets from powerful industrial countries and transnational corporations whose resources and sales may be greater than the annual incomes of many Third World countries. This kind of state 'protection' generally operates through tax concessions and other incentives, as well as restricting imports through the use of tariffs and quotas to ensure less foreign competition in the national market.

A third kind of argument may be labelled *welfarism*. You have met one version of this already from Seers (1979). Although capitalist development may create the potential for development, policies are needed to ensure that this is used to ensure the potential of the human personality, particularly since not all human needs are mirrored well by market valuations. This would require development planning to link investment with the creation of jobs, the eradication of poverty and inequality, and the achievement of other conditions for development such as improved status for women and so on.

Finally, there is *global environmentalism*. Concern for the environment is often a rather separate motivation for state intervention – and intervention by means of agreements between states. It is arguable that *advanced* capitalism always requires international regulation and owes as much to agreements between the major industrial capitalist powers as to market regulation. However, concern for the global environment adds a new dimension to such arguments, since the problems of managing the global commons necessarily require concerted international action. The Brundtland Commission and similar bodies always propose more international agreements to regulate the excesses of global capitalism.

Famine, war, and environmental catastrophe are all potentially linked in this way of thinking as overriding dangers to humanity. There is wide agreement on the importance of trying to eliminate poverty, and that this at least is an area where both international and state intervention is required. Galbraith (1990), for example, not only sees poverty as an inhumanity in itself but also as 'the source of oppression and conflict'. Thus he advocates economic assistance to achieve economic improvement in the poor countries, not only for the direct benefits of material progress but also to lessen the dangers of war and of violent repression of internal populations.

The most obvious problem with this approach is that there is no international state to implement any policies that may be suggested. The only possibility is to work through international agreements, but the question remains of how such agreements are to be policed.

Populism

Finally there is a set of ideas grouped together under the label 'populism'. By this is meant an emphasis on people themselves as agents of development, solving their own problems individually or through local organizations and networks. In reaction against large-scale and 'alienating' industrialization, this is a current of thought that favours small-scale individual and co-operative enterprise both in industry and agriculture. (There is another, quite different, usage of the term 'populism'; in a Latin American context, in particular, it has been used to mean a political style of leadership that legitimates authoritarian rule by appeal direct to the mass of 'the people'. This is not the usage in this text.)

In the previous chapters you will already have met concepts and ideas at different points that mesh with this notion of populism. These include the emphasis on local participation in primary health care in Chapter 2; on women's empowerment in Chapter 4; on development to sustain communities in Chapter 5. The attraction of the populist idea of direct action to meet one's own needs is clear and simple. There is a ready correspondence between many people's dreams and a vision of:

> "A world of 'humanized' production, based on a small scale but modern and scientific technology, a world of co-operation in villages and small towns, a world of enriched social relationships growing out of a process of production and exchange that is under human control rather than 'alienated'..."
>
> (Kitching, 1982, p.179)

It has to be admitted that there is little if any theory as to how such dreams could be replicated on a large scale, and how the kind of social change could be brought about that would safeguard them for the future. Kitching (1982) argues that despite the attractiveness of the vision, populism makes a 'very unsound and misleading basis' for economic theories of development and that 'agricultural and rural development can only occur...within the context of a sustained industrialization', so that the choice is still between state-led or market-driven industrialization.

However, Kitching also comments that 'economics is by no means the whole of life' and there are good reasons why populism has 'engaged the sympathy of both theorists of development and the peoples of underdeveloped countries'. Indeed, as we move into the 1990s there are some signs of consensus on the need to look more closely at the potential for local groups and individuals to be involved as their own development agents, if only because of the manifest failure of the main theoretical perspectives on development to deliver major improvements in living conditions to the world's poorest individuals and communities. Thus, even the World Bank is taking on some of the terminology of populism – its publication *Putting People First* (Cernea, 1985) echoes the title of perhaps the best known of the latest wave of writing in favour of self-sufficient, grassroots, participative development programmes: Robert Chambers' *Rural Development: putting the last first* (1983). It remains to be seen whether what has been termed the 'new orthodoxy' for development (Poulton & Harris, 1988) can really take over from the 'old orthodoxy' (Kitching, 1982) of the necessity for industrialization.

6.4 A final note

It could be objected that not all visions of 'the good life' require anything approximating to development in the senses introduced in this chapter. What about, for example, seeking personal salvation or transcendence? Or simply retiring to cultivate one's garden?

Moral values are clearly of central importance to the debate on development. By admitting the normative nature of the concept, one may appear to be suggesting that any view of its meaning is equally valid. This book, however, does have a clear bias: that is, whatever else may or may not be included in the meaning of development, it stresses the absolute human obligation to tackle chronic hunger and poverty and the equal right of all to economic prosperity.

Summary and conclusions

1   Development has a range of meanings, from basic economic well-being (measured more or less badly by GNP per capita) to broad notions of economic development incorporating wholesale societal change, modernization and industrialization, and comprehensive definitions comprising lists of human needs to be satisfied that go way beyond the material to include social and political aspects.

2   Development has to be viewed in the context of global capitalism, which is based around five elements: private ownership, regulation through the market, market distribution, enterprise management for profit, and the use of a combination of such ideas to legitimate actions.

3   Development occurs also in the context of national politics: it involves choice and steering, material interests and valued prefer-ences, political agents and institutions like political parties and bureaucracies, and state power. The state may be a primary agent for development, an enabling structure for development, or a structural obstacle to development. It is impossible to understand any pattern of development or how a more desirable development programme to benefit the poor might be set in motion without taking into account the patterns of interests and values and power which shape the develop-ment taking place.

4   There are four main sets of contending views on development and how it occurs: neo-liberal, structuralist, interventionist and populist. It can appear that the main area of disagreement is between the first two, i.e. a simple choice of how to achieve development: *market versus state* or *profit versus planning*. A key issue for development in the 1990s is to what extent interventionism, populism or some other approach represents a realistic alternative to the opposing views of neo-liberalism and structuralism.

THE MAKING OF THE THIRD WORLD

7

A THIRD WORLD IN THE MAKING: DIVERSITY IN PRE-CAPITALIST SOCIETIES

JANET BUJRA

Figure 7.1   A new world: Columbus' voyage to America is depicted in an Italian woodcut printed in 1493.

Europe claimed to have 'discovered' the non-European world of Africa, Asia and America (Figure 7.1). It would perhaps be more accurate to say that Europe discovered itself as the centre of that world – as the hub of world commerce and of an emerging and expanding capitalist system which, in the course of four centuries, was to annihilate or remould all other modes of livelihood. This process was by no means a purely economic one, either for Europe itself or for those regions which it dominated.

Q What impression can we gain of that pre-capitalist world which was undermined by European colonialism?

If we are to understand the Third World today we need to know something of its past.

7.1 A Third World in the making

The Third World as we know it today was a long time in the making. When we remember that the first European trading stations were established on the west coast of Africa in the fifteenth century, and that the European metropolitan powers were still squabbling about the boundaries of their colonial empires more than four centuries later, it becomes clear that in considering the Third World today we are analysing but one moment in a long historical process. European capitalism underwent many transformations during this period (see Chapter 8) and the regions which Europe brought under its domination were diverse in character.

In the period of European mercantilism (the accumulation of capital via trade) Spain, Portugal, England, Holland and France vied with each other to dominate the trade to and from particular areas – a trade in precious metals, luxury goods and slaves. In more fully fledged phases of European capitalist development (when the accumulation of capital began to derive from the harnessing of working-class labour power to new techniques of production), a whole variety of demands were made of less economically developed regions of the world: to supply the essential raw materials without which Europe's (or more particularly, in the first instance, Britain's) industry could not operate; to buy the finished goods which were the products of its manufacturing sector (Figure 7.2); to become investment areas for the wealth accumulated through European industrial expansion, so that more wealth could be created to feed back into that same process; to absorb Europe's

surplus population; and so on. Different areas might play different roles, and the significance of their contribution to metropolitan expansion might change over time.

European demands, then, were one side of the equation, but the other side depended on what the colonized regions had to offer in the first place, and how their inhabitants could be persuaded to part with what was theirs; in other words, it depended on their level of material development and forms of social organization. It is often assumed that these dominated regions were by definition 'less advanced' than Europe, and hence that the latter had some acceptable rationale for intervention. Such rationales certainly existed as part of the ideology that justified imperial expansion. But in many cases it is open to question as to whether this was the real state of affairs.

When Bernier visited the Mogul empire of India in the seventeenth century he was impressed by the sophistication and skill of its craftsmen, by the range of its products, and by the way in which manufacture was organized and controlled by the state (Bernier, trans. 1916). At that period Europe could not compete with India in the production of fine muslins and silks; indeed, it was the import of such luxury goods into Europe that was the mainstay of the British East India Company.

This high level of material culture and social organization had in fact to be *destroyed*, so that European capitalism might arise and transform the world. At the end of this process (it took some two centuries to complete) the East India Company was to boast in 1840 that, 'encouraged and assisted by our [own] great manufacturing ingenuity and skill, [we have] succeeded in converting India from a manufacturing country into a country exporting raw produce' – in particular cotton, to supply the burgeoning mills of Manchester (quoted in Davey, 1975, p.46).

It is worth comparing the respect with which European commentators regarded India in the seventeenth century with the lofty attitudes of travellers in Africa in a later phase of imperialist expansion. Thus, for instance, when Speke visited the East African kingdom of Buganda in the mid-nineteenth century he was astonished at the scale of the king's palace, consisting of 'gigantic grass huts, thatched as neatly as so many heads dressed by a London barber' (quoted in Perham & Simmons, 1948, p.157); Stanley commented on the King's navy of near 20 000 men in canoes; and Lugard praised the 'superior' skill of Baganda craft production of barkcloth, soap and pottery.

By now, however, such technical virtuosity was no longer comparable to that of contemporary capitalist Europe. As no demand was anticipated in Europe for Baganda barkcloth or pottery, the British paused briefly to admire and then proceeded to undermine this indigenous economic development. Within a decade or so after colonial rule had created the state of Uganda (1894), the traditional craft industries had almost disappeared, internal demand was being satisfied by imports from Britain, and the Baganda peasants were set to producing cotton and coffee for export to Europe.

In these two examples we have seen that the societies brought under European domination were not always markedly inferior in economic development to contemporary Europe (though it is clear that on the whole they were), and that as capitalist transformation revolutionized European production, this relative economic inferiority became increasingly marked. It is also clear that if we compare Mogul India with Buganda, at the point when they came under European domination, we find that they were quite different from each other. If this is true of levels of *material* development it is even more the case when we consider other aspects of cultural and social organization.

Within this Third World in the making there were peoples whose mode of life was a constant struggle for survival against nature, and there were others whose grand cities, manufactures and fine arts excited the admiration of European society. There were societies in which everyone was a farmer, others in which the farmer supported a nobility and a merchant class. There

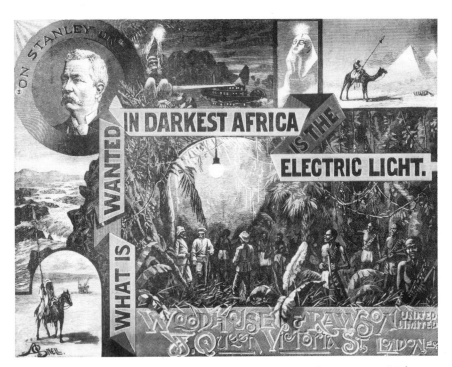

Figure 7.2   Lighting up Africa: an advertisement from the 1890s in which cultural assumptions are as evident as the desire to make a profit.

were peoples with written languages and a literature of their own, whereas others had only the spoken word. There were societies whose gods existed for them alone and others who recognized that the same world religion united them with other peoples in diverse areas of the globe.

When we speak of the 'Third World', then, we must recognize that it was not constituted as such by virtue of any original cultural uniformity. It *became* a 'Third World' as the result of a long historical process of subordination to the expansion of European capitalism in its various forms. By and large, the Third World countries did not experience the social and economic upheavals of industrialization, but remained agrarian societies, producing raw materials for industry elsewhere.

But we cannot hope to understand the Third World today if we see it simply as a one-sided product of European capitalist expansion. The manner in which this expansion took place varied considerably, depending on the pre-existing and co-existing forms of social organization of the dominated peoples and the differences between the colonial powers.

7.2 Subsistence producers

The world in which Europe discovered itself included pre-capitalist states and empires as well as communities of people living on the margins of subsistence. Many of the pre-capitalist states had swallowed up pre-existing communities of independent producers, while in other areas such communities still maintained a viable existence. It is vital that we look at both of these phenomena.

This pre-capitalist world is not easily recoverable. Its own voice has been muffled by centuries of domination, especially in areas where the spoken language had no written form. Our understanding is often dependent on the perceptions of European colonizers, missionaries and scholars who belonged firmly to their own, Euro-

pean, culture. Or it may be based on twentieth-century studies of those groups of hunters and gatherers, pastoral nomads or shifting cultivators which have survived – and none has survived in original form. Pre-capitalist states and empires have long succumbed to superior European force, and what we know of them comes from historical accounts, often written by the victors and flawed by their intrusive arrogance.

We look first at what are often termed 'subsistence' producers. Such groups seem to have generally formed the base on which states were built and, even though they were rarely independent, in some places small communities producing mainly for subsistence succeeded in maintaining their integrity right through to colonial times and beyond. Communities previously unknown to Europeans were being discovered as late as the 1930s in New Guinea (Connolly & Anderson, 1988). A closer view of how such a way of living might have been organized is important for three reasons:

1   A global historical process can be discerned, whereby self-sufficient communities producing little more than their own subsistence were overwhelmed in the expansion of centrally organized states (pre-capitalist as well as capitalist) founded on the appropriation of surplus.

2   When such communities were swallowed up by more powerful states, they resisted in a variety of ways which the superior power had to accommodate. They might retreat into relatively inaccessible areas or refuse to yield up tribute.

3   Many of the ways in which people in the Third World live and work today owe much to their grounding in such small rural communities.

One group which survived Spanish intervention in Latin America, and some of whose descendants still lived in the 1950s as hunter–gatherer–cultivators in the Brazilian Amazon, were the Mundurucu 'Indians' (Murphy & Murphy, 1974). Although an account of the Mundurucu at that point can only be suggestive as far as earlier times and places are concerned, the study by the Murphys is a first-hand account.

The Mundurucu

In the 1950s, the Mundurucu were divided between those who lived a relatively independent existence in the savannah contiguous with the Amazon forest (Figure 7.3), and those who had given up this life to become little more than outworkers, collectors of wild rubber for the handful of traders established on a tributary of the Amazon river. It is the former category, reduced in the 1950s to a mere 350 people living in a series of tiny villages, that we shall consider.

Kinship and the sexual division of labour

In capitalist society, we are accustomed to think of family affairs as belonging to the realm of

Figure 7.3 Some 'Indians' of the Brazilian Amazon have tried to maintain older ways of life in the face of the forest's destruction: a man carving a bird for a fertility festival.

private life, and private life, while often of central importance to individuals, as peripheral to public affairs.

Amongst the Mundurucu things were different. There was no sharp dichotomy between the public and private: family relations were also work relations and exchange relations, and they determined the form and pattern which learning took. Kinship – the network of relationships stemming from 'blood' and from marriage – was central in Mundurucu social life. It was 'the anvil on which social relationships are forged and the language in which they are conducted' (Murphy & Murphy, 1974, p.68).

The central social relations in any society are those which organize production and reproduction. The Mundurucu lived by a combination of exploiting what nature provided in the immediate neighbourhood and of actively intervening in natural processes through cultivation. The major item in their diet was manioc, made into flour, which they cultivated on patches cleared in the forest and fertilized by burning the undergrowth ('slash and burn' cultivation). A variety of fruits and vegetables was also grown, and their diet was supplemented by hunting, fishing and the gathering of wild foodstuffs.

It was kinship which structured the organization of work in these Amazonian villages, but whereas women tended to work with immediate family relatives (a mother and her daughters, her sisters and her sisters' daughters), men were forced to co-operate with more distant kin. We shall see presently why this was so.

The sexual division of labour was very marked in this society. Women did the major and most tedious work of cultivation: planting, weeding and harvesting. It was they who shouldered the time-consuming task of processing the manioc flour; they also gathered wild nuts and fruits, did a little net fishing, and took on the major share of household labour – cooking and the care of small infants. Men were the hunters, they caught most of the fish and did the heavy work of clearing the ground and firing undergrowth in preparation for cultivation. Nevertheless, it was on women's work that the major subsistence depended, and it was they who controlled food distribution, even where that food was produced by the men.

> "The distribution of [manioc flour]...is handled completely by the women. It is a women's product, and women control its disposition. But even the distribution of game eventually falls under female control. The man brings his kill to his wife, or his closest female relative if he is unmarried, and she and her housemates butcher it. They send pieces to other houses, but they determine who gets which parts. And if the take has been small, the food may be shared with only one household – the choice is the woman's and she generally opts for the one housing her closest relatives."
>
> (Murphy & Murphy, 1974, p.131)

This sexual division of labour was reflected in living arrangements. The houses of a village contained clusters of closely related women and their children, 20 or more per house; the men lived and ate in a separate men's house, visiting their wives discreetly at night. The women of different households in the village were generally related to each other, so that the co-operation of women beyond the household could also be validated in terms of kinship. And indeed, many of the tasks women performed (in particular the production of manioc flour) required a wider grouping than that of the household. Men's tasks were also communally organized, even though most of the men living in the men's house were not closely related. These patterns of work and residence resulted from the fact that the preferred pattern of work and residence after marriage was matrilocal: men moved to live in their wife's villages and women stayed with their mothers. 'How would a girl manage without her mother to help her?' asked the Mundurucu, thus explaining marital residence patterns in terms of the organization of household labour (Murphy & Murphy, 1974, p.122).

In terms of village co-operation, then, Mundurucu men related to each other effectively through their wives, and this is illustrated neatly by the distribution of the kill. But the people also related to each other in terms of patrilineal descent, which meant that children belonged to their father's line. The Mundurucu were divided into two categories or 'moieties', the 'White' and the 'Red', and each moiety was subdivided into clans. Both clan and moiety were groups claiming a common ancestry, although this could not be precisely traced. These kinship divisions structured marriage choices, for a White had to marry a Red; this meant that whereas it was unusual to find a father and son, or two brothers, sharing the same men's house, men could relate to each other by more distant associations of clan or moiety, and could discover such links wherever they might go in search of a wife. Ultimately all Mundurucu believed themselves to be related.

Political organization

Although citizenship of a modern nation state may be spoken of in the idiom of kinship ('the mother/fatherland'), we do not generally believe ourselves to have a single ancestry as did the Mundurucu. We are more likely to define ourselves as members of a single political unit, within which ethnic diversity is the norm. It is clear that our kind of society depends for its maintenance on centrally organized political institutions and a bureaucratic structure, backed up by the law and ultimately by control of means of coercion. The Mundurucu had none of this, and indeed in no real sense did all the Mundurucu villages constitute one political unit: each village organized its own affairs. Political effectiveness, as opposed to vaguer feelings of cultural identity, was restricted to the organization of relationships essential to production.

Each Mundurucu village had a chief, but he had no authority, only influence.

> "He does not make decisions on his own, nor does he give orders to others. Instead, decisions affecting the entire community (hunting, raiding, trading) are made in the course of conversations in the men's house with most of the adult males present, and the older and more prestigious men exercising the most weight. The chief acts as

the manipulator of the consensus, guiding the discussion and seizing upon the moment when compromise is possible."

(Murphy & Murphy, 1974, p.79)

The chief's influence was enhanced by a subtle twist of the normal rules of marriage. When the chief married, his wife had to come and live in his village rather than he removing to hers; similarly his sons stayed with their father after marriage, and in addition he attracted sons-in-law for his daughters. This already gave him a core of probable supporters, but he remained a 'first amongst equals': he had to earn respect, rather than automatically commanding it.

Real power requires control over vital aspects of people's lives. The Lord of the Manor in feudal Europe had power over his serfs because he held the land on which they depended in order to live, and he used his power to make the serfs labour for him as well as for themselves. Among peoples like the Mundurucu no one controlled resources on which others depended. All had free access to what nature could provide, but they depended on each other to exploit their natural environment more effectively. People did not co-operate here as a consequence of coercion by higher authorities; they co-operated partly because it was in their interests to do so as a contribution to the general well-being, partly because of sanctions from others if they did not. Such sanctions were not in the form of prison sentences, whippings or fines, but expressed in public scorn, derision and ostracism, or the fear that the co-operation of others might be withdrawn in retribution.

The supernatural

Each Mundurucu men's house contained a set of sacred instruments, the *Karökö*, which were believed to be the repositories of the ancestral spirits of the patrilineal clans. The *Karökö* were long horns, hidden away in an inner room so that women might not see them, although they often heard them being played. When men hunted, a portion of the meat always had to be laid before each instrument: 'the ancestral spirits are pleased by the playing of the *Karökö*, just as they are by the presentation to them of meat after a

successful hunt' (Murphy & Murphy, 1974, p.93).

The beliefs and rites surrounding the *Karökö* served to reinforce certain basic features of Mundurucu society: patrilineal descent and male forms of co-operation. The interrelatedness of all Mundurucu seemed to be more important for men than for women – presumably because women's daily activities were carried on with other closely related women, with whom patterns of co-operation had become familiar since childhood, whereas men had to co-operate with others to whom they were not close kin.

The *Karökö* also carried other symbolic references, however: they selectively emphasized one aspect of the Mundurucu mode of livelihood – hunting. The Murphys point out that although women's labour accounted for the bulk of the Mundurucu diet, meat was valued above vegetables and 'it is the skilful hunter who is honoured, not the industrious tiller of the soil' (ibid., p.62). One of the major Mundurucu myths may give you pause for thought. It recounts how it was *women* who found the sacred *Karökö* and how they forced men to do 'women's work' and to submit to women's sexual desires. 'The men could not refuse, just as the women today cannot refuse the desires of the men' (ibid., p.89). But women were unable to maintain their hold over the sacred instruments because they could not present them with offerings of meat; women did not hunt and therefore they lost their ascendancy in society.

'Science'

When Mundurucu women engaged in the complex process of producing manioc flour in such a way that the prussic acid which it contains was eliminated, could one understand them to be applying science to survival? Similarly, what of the Mundurucu practice of fishing by introducing a poison into the streams – enough to kill the fish but not to poison those who ate them? The differences between these practices and the image we commonly have of science is that the principles and logic behind such effective action could not be stated abstractly by the Mundurucu – they knew only that it worked, and that they had 'always' done it so.

Another craft which is often a repository of practical wisdom is that of healing. The Mundurucu knew of, and used, drugs such as ginger root and quinine bark which are used in modern medicine. Efforts are now being made in many parts of the world to record the knowledge of 'traditional doctors', as it is now recognized that in many cases their expertise has a sound basis, and efforts have been made to incorporate them into primary health care programmes.

Some general comments

Encountering peoples like the Mundurucu, the Sioux Indians of North America, the !Kung of the Kalahari, the Maori of New Zealand, or the Masai of eastern Africa, Europe confronted societies unlike its own. Although culturally different from each other, such peoples shared certain characteristics of socio-economic organization.

1 These societies were not organized to produce a surplus above comfortable subsistence requirements. Given favourable conditions, 'subsistence' might allow for the occasional feast, or for the expenditure of energy on non-utilitarian activities such as house decoration, rituals or celebrations. Certainly it allowed for periods of rest between bouts of work – there was no pressure to *maximize* productive activity above what could be consumed in everyday life.

2 Such societies were relatively egalitarian, since no surplus-appropriating class had arisen. However, this egalitarianism did not always include women. Although women were almost always the major producers of daily food, they were often excluded from political life and it was men's activities that were more publicly valued. Some feminists have seen this as evidence of the universality of male domination in human society: what do you think?

3 Everywhere kinship provided the framework for society as a whole, and in particular it validated the forms of co-operation which ensured people's livelihoods. There were many variations upon this theme: some peoples emphasizing descent through the father, others organized into networks of co-operation through the female line, others drawing equally on both.

These systems of kinship were linked to rules of marriage, residence and patterns of inheritance.

4 Since co-operation was essential in producing the basic subsistence, various devices existed to enhance the unity of the co-operating group (myth and ritual, political leadership, etc.) and to punish those who failed to conform (ridicule, sickness caused by ancestral ghosts, etc.). These were not communities held together by political coercion. Their lack of authoritarian structures made them pathetically vulnerable to the more technologically advanced and predatory state-organized societies.

7.3 Pre-capitalist states

The kingdom of the Bakongo

In 1483 the Portuguese discovered the estuary of the River Congo, and some time later, 150 miles inland, the capital of the kingdom of the Bakongo people, in what is now Angola. The king of the Bakongo maintained an 'empire' which was estimated to number two and a half million people and covered an area at least as big as England. The Portuguese received gifts of carved ivory and cloth made from palm raffia, and on their return highly embroidered accounts of the grandeur of the Bakongo king's court were published.

At the base of this empire were communities of subsistence agriculturalists probably like those I have described above. In this case they cultivated millet and sorghum by slash-and-burn techniques, and supplemented their diet by gathering wild fruits and by hunting. The king was able to maintain a hold over this territory by a system of provincial chiefs, who rendered tribute to the capital in the form of raffia cloth and other commodities. The founders of the kingdom were said to have been smiths who could work both iron and copper and who thereby became skilled as hunters and warriors. The Bakongo, therefore, could produce weapons, and knives and hoes for cultivation, in addition to cloth and pottery. It would appear that such items were bought and sold within the country. Exchange was facilitated by the existence of a currency in the form of cowrie shells, which came from the island of Luanda, controlled by the king. This empire was extended by conquest, although the loyalty of its outlying provinces could not be guaranteed; indeed, several had asserted their autonomy by the time the Portuguese appeared on the scene. In extending the kingdom, the Bakongo took captives; it is said these became slaves of a kind, though we do not know what labour they did.

The Portuguese were able to gain a foothold here, as in so many other instances of imperial expansion, by exploiting internal dissension, in this case over succession to the throne. When a pro-European candidate emerged, they worked out a mutually advantageous agreement. It required the Bakongo to capture slaves from their neighbours to be shipped off by the Portuguese to the sugar estates of São Tomé off the West African coast, and, later and more importantly, to Brazil. By the 1520s around 5000 slaves a year were leaving the port of Mpinda. In return, the Bakongo received beads, cloth, tobacco, metal goods, and new crops such as cassava and ground-nuts; but also the services of artisans (masons and blacksmiths) and the concern of missionaries for their spiritual well-being. The Bakongo king had been converted to Christianity as early as 1491.

Imperial expansion and pre-capitalist states

In many ways the colonists, merchants and missionaries found it easier and more profitable to deal with societies which had already developed centralized political institutions (Figure 7.4). To begin with, the very existence of such states spelled the presence of a surplus to be exploited, and sometimes of great riches to be plundered. Secondly, where territorial acquisition was the goal, a ready-made structure of administration and a people used to political control were tempting prizes.

Some pre-capitalist states were more vulnerable than others to European intervention. The rise of European mercantile power coincided with the decline of an earlier Islamic mercantile expansion, but the declining Islamic empires of the Middle East, North Africa and India were still equal to Europe in technical and maritime skills as well as in military capacity (see Section 7.4). It was a different matter in the rest of Africa, or in the Americas, where resistance to conquest pitted spears, clubs and knives, even bows and arrows, against guns and cannonry. An example of the ease with which Europeans exerted domination over native kingdoms is shown by the conquest of Peru.

The Incas of Peru

The Inca peoples had created a highly organized social and political system by the time the Spanish *conquistadores* arrived in the early sixteenth

Figure 7.4 King Alvaro II of Congo receiving Dutch ambassadors (from an eighteenth-century text).

century. At its foundation were small communities of cultivators, apparently co-residing patrilineages or patriclans, each with its senior elder. Over the considerable region that now includes Peru, part of Ecuador and part of Chile, such communities were welded together by a remarkable state apparatus which allowed the common people a subsistence living but was also able to support a superstructure of people withdrawn from directly productive labour. At the apex of the state was the Inca king and his queen, who was always his own sister. Believed to be descendants of the sun and moon (Inca religion was focused on great sun festivals) they had always practised sibling marriage in order, as they said, to preserve the purity of the royal blood line.

The Incas could work gold and silver and other precious metals; they had philosophers and poets, a developed knowledge of medicinal herbs, purgatives and blood letting, and a complex mathematical and architectural ability. Their language was, however, unwritten and the technology applied to agricultural production extremely simple. Cultivation was carried out by hand-plough, and commerce occurred only in the form of barter, as there was no currency.

This sounds like a bundle of cultural contradictions: how could such a state operate? The major factor seems to have been the centrally organized and to some extent coercive state apparatus, which played an active role in the organization of labour and the creation of an infrastructure for productive activity. State artisans were employed to construct and maintain irrigation works, and to carry out the terracing of slopes so that productivity could be increased (Figure 7.5). At harvest time each community had to pay its tribute to the king, and since the Incas had a good knowledge of storage procedures, these levies of grain and agricultural produce could be used to feed the nobility, the intelligentsia, the administrative functionaries and the artisans. Not only did the ordinary people have to provide

Figure 7.5 The terrace, housing and stone construction of the Inca ruins at Machu Picchu.

tribute in the form of food: they also had to produce clothing, shoes and weaponry for the Inca army, which was composed of enlisted men. The army was engaged mainly in extending the Inca empire, and to each new area the Incas extended their state-organized techniques of production.

A system of courts ensured that Inca law was obeyed and miscreants punished. Control was also exerted over the populace by a system of accounting, remembered here in the early seventeenth century by Garcilaso de la Vega, the half-Inca son of a Spanish *conquistador*:

> "Accounts…were kept by means of the knots tied in a number of cords of different thicknesses and colours… occasionally other, thinner cords of the same colour could be seen among one of these series, as though they represented an exception to the rule; thus for instance, among the figures that concerned the men of such and such an age, all of whom were considered to be married, the thinner cords indicated the number of widowers of the same age, for the year in question; as I explained before, population figures, together with those of all the other resources of the empire, were brought up to date every year."

(quoted in Gheerbrant, 1961, p.198)

The Spanish destroyed this ancient empire with a minimal show of force, by trading on Inca beliefs (a former king was said to have foretold the coming of bearded men, 'superior in every way', who would conquer the empire), and by exploiting rivalry between two heirs to the throne.

Elsewhere in Latin America, the Aztec civilization of Mexico had been subdued some few years earlier, though after rather more resistance. These highly organized and stratified societies, rich in skills and in treasure but pathetically vulnerable, were annihilated almost without trace under Spanish rule. Other colonial powers (e.g. Britain in India and Africa) exploited the existence of such societies by preserving the structures of authority, but subjugating them to new ends (see Chapter 10).

Some general characteristics of pre-capitalist states

Pre-capitalist states often seem to have arisen as a result of the extension of control by one people over another, rather than as an outcome of internal differentiation within communities of subsistence producers.

In the pre-colonial kingdom of Rwanda, for example, a complex stratification system had emerged, apparently from successive waves of immigration. At its apex were the pastoral Tutsi, who ruled over agriculturalists calling themselves Hutu, while at its base were the Twa, pygmy hunters and gatherers. The occupational and ethnic differences between the strata in this society led Maquet, a Belgian anthropologist, to liken it to the caste system of India.

In contrast with the relative egalitarianism of subsistence communities, these were states based on the 'premise of inequality' (to use Maquet's term). The labour of the many supported an élite who were no longer engaged in productive work. In many cases people continued to produce by the same methods as they had always done, but in addition to feeding their own families they now had to labour longer and harder to produce a surplus as tribute to their overlords.

Such communities could remain relatively unaffected in other ways by the overriding state apparatus. They might continue to define their community and its forms of co-operation in kinship terms, but this was rarely the bonding force for society as a whole. The relatively egalitarian notion of 'we of one blood' was out of place in a divided and unequal society. Bonds of a more overtly political nature had to be forged, and the genealogical tree was replaced by the administrative hierarchy.

Local beliefs regarding the supernatural might also persist: the fear of witchcraft or of witchcraft accusations was a potent force for social control in local communities. The Azande commoners in east-central Africa consulted 'oracles' to discover the witch responsible for a person's sickness or death. This was done mainly by administering a poison to fowls, some of which died while others survived. In this small-scale kingdom, however, there was a hierarchy of oracles; courts administered justice according to their verdicts , with the king's oracle as the final arbiter. Sometimes, however, we find a ruling group so effectively organized that all are drawn into the practices of a state religion – as with the cult of the sun promoted by the Inca kings.

Where the state was simply parasitic on the labour of simple farmers, hunters and cattle herders who continued to control the means to, and conditions of, their own production, it was always vulnerable. Where power was maintained simply by coercion and there was no reciprocal dependence of the populace on the state, secessions were a frequent occurrence. Occasionally, however, as among the Incas, there was *state intervention* to promote and improve production. State works such as irrigation projects have long been recognized not only as a means of raising productivity so that a surplus can be produced, but also as the basis of effective, strongly centralized and bureaucratized states.

In many of these early states most production was still for consumption, even if the levels of consumption were progressively greater as one moved up the social hierarchy. Foodstuffs particularly, but even weapons, clothing, ornaments or pottery had not necessarily become commodities for sale: the payment of tribute is not yet commerce. For the Baganda of East Africa, the aim of taxation:

> "was not the accumulation of wealth, for commerce seems to have been relatively undeveloped. Rather...[it was] to provide the Kabaka [King] with surpluses with which he could reward his favourites. The circulation must have been quite rapid for the buildings shown on [a] chart of the palace as constituting the treasury...would hold only a small part."

> (Fallers, ed. 1964, p.109)

Thus, surpluses were usually redistributed in such a way as to retain the political loyalties of powerful allies or subordinates, whereas in subsistence communities, although inequalities existed, resources were more likely to be distributed fairly among producers.

Not all the pre-capitalist states which were brought under the sway of European capital expansion were of this *pre-mercantile* character. In West Africa, ancient inland kingdoms such as Songhai or Mali flourished through their hold over trade routes spanning the Sahara (see Section 7.4). The Portuguese early traded with such kingdoms, though at one remove – through middlemen at the coast. In India, European commerce confronted an even more economically and militarily powerful society: the Mogul empire, contemporaneous with early Spanish and Portuguese exploration overseas.

The Mogul empire (1526–1761)

Long before the Moguls (sometimes spelt Mughals) established their sixteenth-century empire, a surplus-producing economy existed in north-west India – the ancient civilization of Harappa. City ruins and archaeological remains show that already, 4500 years ago, Harappan agriculture was sufficiently productive to support a sizeable city population engaged in non-agricultural production. Among the relics are depictions of agricultural technologies still widely used in India – wooden-wheeled bullock carts and bullock-drawn ploughs, for example. Other remains show that production here was not simply for use; that *production for exchange* had become important. Amulets and beads from the Indus Valley have been found in Mesopotamian cities and, conversely, luxury products of Sumeria and Mesopotamia found their way to Harappa. Clearly a *merchant* class had emerged, whose transactions served the demands of a leisured élite as well as allowing for their own enrichment. These features, of surplus production and of production for exchange, were further developed by the Moguls.

Over the centuries the Indian subcontinent has been subject to many waves of invaders and settlers from the north. The Mogul era began as one such invasion by Muslim northerners who,

Figure 7.6   The Taj Mahal, a mausoleum built by the Mogul Emperor for his wife, as depicted by Thomas Longcraft, 1786.

in the course of their 200-year rule, welded together the pre-existing petty kingdoms of north and central India into an administratively united country, making a qualitative break with the past and a significant step towards the modern nation state.

At the base of this empire was the village community, a more or less self-sufficient unit of peasant producers and artisans. Local chiefs or *zamindars* took a share of the products of the peasantry; more significantly, villagers also had to pay taxes to the central government. On this basis was erected an economic system characterized by a complex division of labour, with a good proportion of the population producing the necessities of life but specializing in various branches of manufacture. It was a system marked by cruel discrepancies in the conditions of life: at one extreme were the hard-pressed and illiterate peasantry, while at the other there was the Emperor with his peacock throne of precious stones, gold and pearls (Figure 7.6). There was also a flowering of the arts: literature, painting, philosophy, theology and music.

How did the Mogul empire extend to dominate almost the whole of northern and central India? A facilitating factor here was that Mogul India had a monetized economy. Habib (1969) tells us that, by the early seventeenth century, taxes on the peasantry accounted for approximately half the value of their production, and were predominantly in the form of cash rather than kind. Members of a village sold their produce in nearby markets in order to pay their taxes, or the produce was collected by the authorities for sale and the proceeds forwarded to the imperial coffers. Artisans within the village community were remunerated in kind by peasant cultivators, but in the urban areas they produced for cash.

The monetization of the economy eased the technical problems of storing and redistributing agricultural surplus. It also allowed for the rise of a class of merchants and money-lenders dedicated to the accumulation of money capital. In the city of Swat alone, Europeans found immensely rich merchants possessing between them 50 ships, some as much as 800 tons, trading with overseas countries (Persia, central Asia and Arabia). In addition, this cash economy greatly facilitated the division of labour, by allowing for an impersonal medium in which relations of exchange between strangers could take place (Figure 7.7).

More than the cash nexus was required to integrate such a vast empire, however. There was no unifying religion in Mogul India for (as we shall see in Section 7.4) the ruling élite were Muslims while the mass of the people remained Hindus. Nor could the bonds of kinship provide a framework here. Even at the village level ordinary people were divided into *castes* – hereditary ranked categories, the members of which did not intermarry, and whose relations were governed by a host of ritual regulations.

Caste is an institution whose origins are lost in antiquity, though it is probably the outcome of successive waves of invaders into India, each one exerting dominance over its predecessors. Its form was well established by the Mogul period. As a system, caste affected all aspects of life, from kinship to ritual, from consumption to production. Caste can be seen as a form taken by the *division of labour*, since each caste group was associated with a customary occupation: cultivating, weaving, soldiering, etc. More fundamentally still, the system was based 'on the exchange of labour and services for food between castes who had land and those who had little or none' (Moore, 1967, pp.333–4). It is clear that caste has always been concerned with access to the means of production.

The Mogul ruling class found that its interests could easily be accommodated by such a system. In each area the land was controlled by a dominant caste, always high in the caste hierarchy. Caste ideology provided a religious validation for economic inequality, presenting it as part of the sacred order and thereby unalterable. By means of their control over the local chiefs or *zamindars*, the Mogul rulers were able to capitalize on caste, both as a system within which a surplus was extracted from the direct producers and as an ideological reinforcement of the social order. A divided society was thus effectively subjugated.

The integrity of the Mogul empire also owed a good deal to its political structure. A large and powerful ruling class emerged, whose varied economic interests complemented each other and provided a joint interest in perpetuating the *status quo*. This dominant and largely urbanized category was composed of the nobility, the bureaucracy, merchants, and an intelligentsia dependent on patronage. Although members of the nobility were rewarded for their loyalty by the rights to taxes from particular blocks of land, they were never allowed to put down local roots. The rights to a share in revenue were assigned at the pleasure of the ruler and frequently ceased at the death of the assignee, when his wealth reverted to the Mogul treasury – a practice which encouraged the consumption rather than the accumulation of wealth. If the nobility were thus held in check, tendencies to fragmentation in the empire were also inhibited by the existence of a vast standing army, said to number over 200 000 in 1647, and employing matchlocks and cannon.

Although as a political entity the Mogul empire was highly successful, in terms of economic development it had severe limitations. High levels of aristocratic consumption left little surplus available to transform methods of production, and the peasantry had no incentive whatsoever to increase their level of production. The more the peasant produced, the more went in taxes to the imperial treasury.

In the end it was these harsh exactions which led to the decline of the empire. Increasing pressure on the peasantry sparked off revolts, helped on by minor chiefs in the countryside. New and smaller states, such as Maratha, were established as the Mogul empire fragmented (Bayly, 1983). At the same time inroads were being made from without, as a consequence of the rise of European mercantilism. English, French and Dutch merchants competed to gain footholds on the margins of the empire, aiding and abetting as they did so various ambitious and disaffected elements within.

It has been suggested that, given time and left to itself, Mogul India could have developed a

Figure 7.7 Workmen building the royal city of Fatehpur Sikri (an illustration from an Akbar Nama manuscript, c.1590).

capitalist economy. Habib discusses this possibility, and Basil Davidson (1978, ch.5) has asked the same question of certain of the African mercantile states. Other writers dispute this suggestion. Can you see any reasons why such a development might have been inhibited?

7.4 Cultural movements: commerce, politics and the rise of Islam

Pre-capitalist societies did not exist in isolation one from another. Patterns of interaction had always existed, even apart from outright conquest: barter, commerce, raiding, safe passage, intermarriage, tribute and so on. The expansion of Islam flowed along existing routes of this kind, as well as ideologically reinforcing the creation of new ones.

It is important to consider the rise of Islam for two reasons. First, because it is not only a religious system but was also the codified practice of new and expanding empires and mercantilist ventures. It carried with it not only literacy but also a set of rules governing the whole of social life. Its extended influence testifies to the vitality of merchant classes and circuits of exchange linking many parts of the pre-capitalist world.

Secondly, Islam had a powerful effect on European cultures, extending back to the mid-seventh century. Although Islamic influence had largely been excised from Europe by medieval times, it continued to be a potent and unifying force in many of the regions later to come under the political and economic domination of Europe. From West Africa to Indonesia in colonial times, Islam offered an alternative to that other world religion, Christianity, introduced largely by European settlement.

World commerce and the origins of Islam

Samir Amin (1976), in his account of the pre-colonial Arab world, notes that Islam was associated with the rise of 'rich and specifically

urban civilizations'. He then asks an important question: from where did the wealth come, on which these civilizations were built? Answering his own question, Amin contends that the Arab world was:

"a turntable between the main areas of civilization of the Old World. This semi-arid zone separates three areas whose civilizations were essentially agrarian: Europe...Africa, [and] Tropical Asia. The Arab zone fulfilled commercial functions, bringing together agrarian worlds which otherwise had little contact with each other... the crucial surplus which was the life blood of its great cities at their height did not come mainly from the exploitation of its [own] rural world."

(Amin, 1976, p.12)

The logical conclusion to such reasoning is that the surplus which created the Arab world came ultimately 'from the peasantries of other countries'.

This is a thought-provoking idea for, if we accept the argument, it distinguishes the major Islamic empires from those surplus-producing societies that we have considered so far which flourished on a peasant base (peasants being defined in this case as subsistence producers forced into producing a surplus for the state). Of course, once Islam spread beyond this Arabian epicentre it was no longer simply a 'trading formation'. We have seen this in the case of the Islamic Mogul empire which, while it was strongly mercantile in character, depended vitally on exactions from the peasantry.

Let us consider the extent and character of the commercial nexus within which Islam arose, looking first at the role of Mecca, the birthplace of the Prophet Muhammad. The Arabian peninsular is very largely desert, still occupied by semi-nomadic Bedouin whose livelihood depends mainly on their camels, sheep and goats. Scattered throughout the desert there are some small and large towns, generally situated at oases, whose inhabitants are settled townspeople. Mecca was one of the most important urban centres in Arabia during Muhammad's life (c.572–632). Its importance was due to its situation as a link in a network of trading routes connecting the Byzantine empire and Persia with southern Arabia, India and Africa.

As Shaban notes: 'It is impossible to think of Mecca in terms other than trade; its only *raison d'être* was trade' (1971, p.3). Muhammad himself was a member of the dominant Quraish tribe, which operated in this region as a trading community, and whose merchants collectively and regularly despatched great caravans (the guiding and safe passage of which depended on the Bedouin) to north and south. Muhammad himself was probably a merchant at one time, before he began to preach a form of monotheism which was at first unacceptable to his fellow Quraishi in Mecca. He found a base in another oasis town 200 miles to the south (later to be called Medina). Since there was no overall authority in this area, and since the profits of trade had always been subject to rivalry between the merchants of different towns and their Bedouin allies, Muhammad and his followers were merely exploiting pre-existing conflicts in making this move. From Medina he later mounted a successful attack on Mecca, after undermining its trade position.

The expansion of Islam

This first success was the forerunner of an amazing expansion of Islam after Muhammad's death in 632: by the end of the following decade his successors had conquered Syria, Egypt and Iraq. The initial impetus for this expansion was a response to the disruption of trade (and hence prosperity) in the region caused by Muhammad's activities and the wars which followed his death. In the process, energies were unleashed which led, as Shaban puts it, to the Arabs 'unintentionally acquiring an empire' (1971, p.14).

At first the military engagements against the crumbling Persian empire were merely for booty. The Byzantine empire, then in control of the area around the important merchant towns of Damascus and Jerusalem, was more vigorous, and greater organization was required to make inroads upon it. These initial conquests were

made not as the result of a centrally organized plan of action but as the outcome of decisions made by field commanders, faced with the necessity of feeding their troops and eager to advance the cause of Islam. The Arab tribal armies which had conquered this empire also settled and governed it – partly by taking advantage of existing systems of revenue collection and partly by taxing trade. Each area had its governor, at first either an army commander or his appointee. The Arabs did not, in the initial stages, integrate with the populations they had subdued: they remained segregated in garrison towns, each entitled to a share of the revenues. Not surprisingly, such towns gradually became rich and expanding urban centres, the locus of trading operations and production by artisans.

As the empire extended it became less and less possible to exert control over its far-flung provinces. At first a fixed percentage of the booty taken was sent back to Medina, and attempts were made to impose Medina appointees as governors of the provinces; but, given the communications of the period, it is not surprising that success was limited. Twice the capital of the empire was moved (from Medina to Damascus and later to Baghdad) in an unsuccessful attempt to exert centralized control. This was an empire held together by bonds of trade and religion more than by administrative control.

By the early Middle Ages Muslim influence extended from Spain in the west to Indonesia in the east; control of Spain lasted until the end of the thirteenth century, and in Granada it persisted a further 200 years. Such an empire was inevitably prone to fragmentation. When Mongols captured Baghdad in 1258 they destroyed the Caliphate (central leadership), and thereafter the empire became merely a collection of autonomous states. It did not, however, lose its impetus: these same Mongols became Muslims and went on to found the Mogul empire in 1526.

In the fourteenth century the traveller Ibn Batuta set off from his home in Tangier to make the pilgrimage to Mecca, and from there he was able to travel for a further 25 years in Muslim lands before returning home. He crossed Persia, Asia

Minor and the Crimea to Constantinople, then went overland to India (parts of which were under Muslim rule) and on to China. On his return he visited Sumatra, Arabia, and East and West Africa.

Though China was not part of the Islamic empire, Ibn Batuta found that 'in every Chinese city there is a quarter for Muslims, in which they live by themselves, and in which they have mosques' (McNeill & Waldman, 1973, p.293). Here Ibn Batuta met again a Muslim religious doctor, originally from Ceuta in North Africa, whose family he knew well and whom he had assisted earlier in India. This incident illustrates that Muslims from one part of the empire could feel equally at home in others, and that the pursuit of wealth by trade scattered rich families throughout its length and breadth. It also illustrates the intimate links which existed in Islamic countries between trade and religious learning: Ibn Batuta himself combined the two vocations quite profitably.

Islam: religion and society

What kind of religion, then, was Islam, and how did it come to play such a powerful ideological role in mercantile expansion and empire-building? One respect in which it made a break with the past was in its emphasis on *community* rather than kinship – the moral community made up of believers, not simply those who happened to co-reside. Prior to Muhammad's teaching, each Arab clan had its totemic deity, but Muhammad had the totemic 'idols' destroyed and demanded allegiance to a single God – Allah.

Islam did not repudiate Judaism and Christianity, except in certain important details, but claimed to build on them. Muhammad saw himself as the last in a long line of prophets which had begun with Abraham and which included Jesus Christ. Muhammad did not himself claim divine birth, which is why it is incorrect to refer to Muslims as 'Muhammadans'.

Allah is believed to be stern in his punishment of wrong-doers, but also compassionate and merciful towards the weak: Muslims are enjoined to help those who cannot help themselves, in particular the poor, orphans and widows. While fiercely egalitarian in some respects (at least as regards men), it does not disallow the accumulation of riches by trade. The merchant's occupation is an honourable one, though he must not profit by dishonesty, cheating or usury, and he must always do his duty towards the poor. As with certain versions of puritanical Christianity, abstemious accumulation is commended, together with a life of self-discipline (e.g. abstention from alcohol, fasting during Ramadhan). Political rulers are to be simply the first among equals in this moral community, and to rule by consultation, not fiat. Islamic legal codes cover the whole of social life.

Many attempts have been made to interpret Islam sociologically. There are those who have seen in it a reflection only of the harsh realities of desert life, of 'tent and tribe'. Others have fastened on it as a political manifesto rather than religious inspiration, and there are those to whom its essential significance is its tolerance of other religions. Elements of all these positions are present in Islam, but perhaps the essential point is that in its stern simplicity Islam equipped its believers with guidelines to which they could fully subscribe while at the same time interpreting them in relation to their own particular environment. The surface appearances of unity were always there, in the forms of worship and the social codes relating to vital human relationships. In addition there was the language of the Holy Book, Arabic: 'mastering the Arabic language to grasp the Qur'an's [Koran's] meaning, and learning it by heart became the necessary first step into the nascent civilization of Islam' (McNeill & Waldman, 1973, p.29). Arabic became the language not only of religion, of trade and of administration, but also of education and culture.

What I want to consider next is the impact that Islam had on the kinds of society we have been considering so far. I shall look first at some of the ancient African kingdoms of the southern Sahara; then return again to the Mogul empire.

The impact of Islam on ancient African kingdoms

Kingdoms such as Mali, ancient Ghana, Songhai and Kanem were in existence *before* the spread of Islamic influence and Arab empire in North Africa. We owe our first written account of ancient Ghana (situated roughly in the west of modern Mali) to an eighth-century Muslim traveller, Al Bakri of Cordoba (Spain), who made it clear that this was a pre-Islamic kingdom: 'the religion of the people of Ghana is paganism and the worship of idols.' He described how its kings were buried in huge mounds, accompanied by human sacrifices and 'offerings of intoxicating drinks' (quoted in Oliver & Fage, 1962, p.46). Regarding the ancient kingdom of Kanem (in the east of what is now Niger), a tenth-century Muslim traveller, Al Muhallabi, wrote: 'their King...they respect and worship to the neglect of Allah the most High... they believe that it is [their kings] who bring life and death and sickness and health' (ibid., p.47).

The prosperity of such kingdoms lay in their control of the trans-Saharan trade routes. Such routes had perhaps always existed, but they were given impetus by the extension of the power of Rome to North Africa (third century), to be followed by the Byzantine and Islamic empires. This trade came to consist of items such as gold, salt, copper and to a lesser extent slaves, ivory, ostrich feathers and hides (ibid., p.64). Al Bakri notes that the Ghanaian king levied tolls on this trade, and it would appear that certain mercantile devices such as credit were already being used by traders along the route. Following the establishment and extension of Muslim power in the north, an attempt was made to subdue such kingdoms, but this turned out to be not so easy as had been supposed – Ghanaian defences were well organized and it took 14 years before resistance could be overcome. The capital (Kumbi Saleh) was finally captured and sacked in 1076.

But it turned out to be an empty victory, since trade had already been disrupted as a result of the political upheavals in the north. The agricultural base of the Ghanaian kingdom was undermined by the process of conquest, and its constituent communities reverted to their previous mode of subsistence production, while the merchants moved further east, to Mali and Kanem. The kings of these states were converts to Islam; the best-known king of Mali, Mansa Musa, made the pilgrimage to Mecca in the fourteenth century.

There were at least two advantages which conversion to Islam gave these kingdoms. The first was in the field of trade: once they were Muslims, the ruling class and its merchants were treated as allies rather than infidels (Islam had its own ideology of cultural inferiority), and the good will of Muslim states to the north was clearly vital for the prosperity of trade. In turn the itinerant merchants helped to spread Islam to the south.

The second advantage, here as everywhere, was that Islam acted as a symbolic unifying force, its significance extending over and above the divisive loyalties of kinship and locality. To some extent the state religions of Ghana and Kanem had already provided such an ideology of nationhood, but its force had become progressively weaker as the kingdoms extended their control. Islam, by contrast, offered a developed and uniform system of formal education which, once instituted, could ensure the reproduction not only of religious adherence but also of a formally educated élite.

This process was strongly entrenched in Islam, so that the act of conversion, in theory at least, entailed an obligation on the part of the converter to instruct the convertee in religious knowledge. As we have seen, the repository of knowledge in Islam was in the Qur'an, a *written* document, so a basic literacy was demanded of all true converts.

The net effect of these processes was the production, in these ancient African kingdoms, of a ruling class of educated and literate merchants and officials whose 'interests were closely tied to the maintenance of the imperial administration and peace' (Oliver & Fage, 1962, p.88). Moreover, the ideological form taken by this process of class reproduction now linked Africa with the rest of the Islamic world. By the sixteenth century the Songhai empire could boast a university, situated at Timbuktu, where theology, Muslim law, rhetoric, grammar and literature were taught by lecturers from Cairo and Fez as well as local scholars.

By the seventeenth century these empires were falling into decay. Attempts by Morocco to dominate the trade routes more effectively led to the downfall of the Songhai empire; when attacked, it fragmented into petty kingdoms, and the trans-Saharan trade collapsed. It picked up again from new centres in the smaller Hausa states of what is today northern Nigeria (Figure 7.8), but by the time this adjustment was being made, European traders had already begun to turn their attention to the West African coastline as a source of gold, ivory and slaves. The focus of trade shifted from north to south and the new religion of prosperity became Christianity.

The Mogul empire: Islam and Hinduism

Islamic influence in India dates back to the thirteenth century, though at first it was restricted to the north-western corner of the country. As we have seen, the Mogul empire was founded in 1526 by invaders from the north, Turko-Mongols and Afghans, and survived until the eve of European penetration. In India, Islam faced a religion so deeply embedded into national life that it is difficult to define without describing Indian social structure. What was the impact of Islam on Hinduism, and vice versa? Before we can answer this question we need to look at the character of Hinduism itself.

Unlike Islam, Hinduism is polytheistic – a religion of many gods. In some of its versions there is an emphasis on a single personal God, while in others the conceptions of 'God' are so abstract as to be completely impersonal. There is no central theology in Hindu belief, nor any centrally organized priesthood. It was not founded at any precise period nor by any particular prophet; rather it unfolds as a set of diverse traditions which allows the individual believer a great deal of latitude.

One of its central ideas is *dharma* – the sacred order of things – and one aspect of that sacred order is the caste system. In Section 7.3 we looked at caste as a form of division of labour, and this is indeed one of its central features. However, it is not simply an economic phenomenon, but involves a structuring of the whole of social life. Not surprisingly, then, the caste divisions are given religious validation by Hindus.

Figure 7.8 Artist's impression of the city of Kano, capital of a Hausa state in northern Nigeria, 1877.

The basic idea comes from the Vedic scriptures where four categories in society are said to spring from different aspects of the Cosmic Being. The concept of *varna* (colours) is associated with these different categories: white for the Brahman (religious functionaries), red for the Kshatriya (warriors), yellow for the Vaishiya (cultivators and herdsmen), black for the Sudra ('one occupation only the Lord prescribed to the Sudra, to serve meekly the other three castes') (Manu I, 91, quoted in Burridge, 1969, p.89). Outside the *varna* altogether were the Untouchables.

Although social divisions in real life (the local caste grouping or *jati*) corresponded only roughly to this sacred model, caste was a clear determinant of life chances. One was born into and died within a particular caste group, and one's marriage choices were restricted within it. A host of ritual injunctions, many concerned with matters such as food preparation and bodily contact, regulated the relationships between members of different castes. It will now be clear that Hinduism is not so much a religion as a way of life. People do not become Hindus: they are born Hindus, and proselytization is rarely practised; indeed it is frowned upon by the orthodox.

After several centuries of Islamic influence the caste system remained intact, even though Islam promoted an overtly egalitarian message; indeed, India's Muslims were absorbed *within* the caste system. How did this happen?

In the era of Muslim expansion there was forcible conversion of Hindus in some areas. Later, however, there were two categories who most commonly became Muslims. On the one hand there were individuals in the highest ranks of Indian society who found it politically expedient to espouse Islam, the religion of the ruling power. Most conversions, however, took place at much lower social levels: for the socially disadvantaged in Indian society the attractions of conversion were clear enough. At this social level, however, conversion was rarely an individual matter; more commonly, a whole *jati* would convert *en masse*. Apart from their new ritual practices, the mode of life of such categories did

not change, so in certain areas Muslims have a monopoly of particular occupations: weavers, butchers and tailors are common examples.

Most significantly, the wealthiest groups in Indian society – landowners and merchants – remained loyal to Hinduism. India's external trade during this period was in the hands of Arabs and Persians, but Hindus monopolized trade within the country. The Muslim rulers were careful in their treatment of such powerful social groups. Indeed, since they did not transform the economy of India but merely centralized its political organization, they depended on these groups to organize the production of an economic surplus.

Two processes were set in motion by the interaction of Islam with Hinduism. On the one hand the Muslims took on many traits that were Hindu in origin, while on the other there were philosophical initiatives from Hindus to come to terms with Islamic concepts.

Writers have described the first process as the Indianization of Islam. Akbar, one of the most famous of the Mogul emperors, was accused by a Muslim faction of favouring Hindus, even of having Hindu sympathies. The Mogul rulers were not only tolerant of Hindu sages, they also adopted much in the way of Indian culture – food, dress, art and architecture. The *lingua franca* of the empire, Urdu, evolved out of a combination of Hindi syntax with Persian–Arabic vocabulary. The majority of Muslims were not foreigners but ex-Hindus, and they naturally reinterpreted Islam in ways that seemed appropriate to their situation.

Although most Hindus ignored the message of Islam, there were others who attempted a synthesis between Hinduism and the new religion. One such attempt resulted in the religion of the Sikhs, where the emphasis is on the oneness of God, the equation of Hindu and Muslim concepts of God, and on the promotion of a society that does not differentiate between castes.

We can make some general observations arising from our consideration of Islam in Mogul India. The first is that world religions must adapt to the particular environments into which they are transplanted – Islam in Mogul India was not the same as Islam in Mali. But Islam has a core of universalistic tenets of faith held to be applicable everywhere; Hinduism, on the other hand, is embedded in particular cultural forms that are not easily transplanted. Ask yourself: is Hinduism like this because the Hindus never created an empire, or did they refrain from creating an empire because their religious faith, being inwardly and specifically focused, contains no impetus for such action?

Summary

Throughout this account of Islam I have emphasized a certain 'fit' between ideology and economic practices – in particular, in this case, between Islam and international trade. It would be imprecise to see this 'fit' as deriving from Islam's religious message; it had more to do with the common language and code of social practice that Islam carried with it, and which undoubtedly facilitated economic transactions between people in far-flung corners of the globe.

Now it is clear that many other ideologies have promoted commerce, for international trade preceded the rise of Islam and persisted, indeed expanded, after the decline of its empire. Islam is but one 'appropriate' ideology among others. It is significant to us here, first because of its non-European origins, and secondly because of its success. The extent of that success is often unknown to those of us brought up on the glorious history of Western empires and European 'civilization'.

Pre-capitalist empires, as we have seen, were marked by their relative fragility. The common code of Islam, and the mercantile prosperity which it brought in its train, were factors making for a remarkable degree of cohesion, given the lack of any transformation in the organization of peasant production on which mercantile accumulation ultimately rested. When the empire dissolved into its constituent parts, Islamic culture persisted, providing a focus for resistance and identity in the face of European domination.

7.5 Conclusion

In this account of a 'Third World' in the making I have emphasized above all the diversity of the elements which went into its creation. The commercial and territorial expansion of Europe incorporated peoples whose modes of life and work were, at one extreme, similar in form to those of contemporary Europe, at the other completely different.

I have argued that in terms of material development in the seventeenth century, Mogul India was equal if not superior to contemporary Europe. If we consider its social organization we find a class-divided society held together by a strongly centralized state. At its base were the peasants, whose labour had to support not only themselves but also a hierarchy of officials, a military establishment and the nobility. A monetary economy and the existence of a wealthy merchant class distinguished Mogul India from, say, Inca Peru. In Mogul India we have a society comparable in many ways to the societies of feudal Europe. There are no exact parallels, however, for social phenomena such as the caste system or Hinduism.

Other societies which Europe confronted were quite unlike its own, smaller in scale, and more egalitarian. Communities such as these were beginning to be incorporated into more powerful and hierarchically organized states long before the arrival of European merchants and colonists. Subsistence producers were forced to sustain others more powerful than themselves. This did not necessarily mean that they began to produce in different *ways* – simply that they had to produce *more*; technology and forms of cooperation often remained unchanged. What was lost was the *control* of the producers over the intensity of their work and over the fruits of their labour. Adoption of the rulers' language and customs (what we might call 'cultural incorporation') was sometimes total, but often the conquered peoples maintained a degree of cultural identity – as the Hindus retained their distinctiveness from their Muslim rulers in Mogul India.

In many respects, the empires which Europe created in Africa, Asia and Latin America were merely the last (to date) in a long line of imperial ventures. There were few areas in the world which had not previously been subject to overlordship by external forces, sometimes to the suzerainty of successive powers. We still need to ask why (for example) the Islamic empire fragmented and lost its impetus, while European imperialism transformed the world economy. There were, in fact, two vital differences between the new empires and the old.

1 Whereas in previous empires merchants had merely served the demands of wealthy minorities for luxury goods (and in the process accumulated hoards of wealth themselves), now these stocks of wealth began to go *directly into the transformation of productive processes in Europe* (the Industrial Revolution) rather than into consumption. This transformation did not occur everywhere – the empires of Spain and Portugal (like that of the Arabs before them) were largely of the ancient kind, and their decline when mercantilism had been exhausted is evident.

2 Territorial acquisitions in the old empires were generally a means to ensure the appropriation of a surplus, *whose manner of production was usually of little concern to the colonizers,* whether it was in the form of agricultural or artisan products or slaves. (This generalization is not of course applicable to areas which were settled by large numbers of Europeans – in particular North and South America, where indigenous peoples were either wiped out or their cultures largely destroyed). In the new phase of empire the growth of capitalism in Europe led to the need for a closer control over the type and extent of production in areas under imperial domination. The sixteenth and seventeenth century trade in silk and spices, plus the plunder of the treasures of other nations, began to give way to the more direct organization of the production of specific items, vital as raw materials or as intermediate inputs to European industry.

European imperialism was thus the harbinger of a new world economic order. The harnessing of the labour of many millions to the creation and recreation of *industrial capital* in Europe led to the development of capitalist societies in the West, but it transformed a large part of the rest of the world into a backwater, supplying raw materials for European industry and consuming European manufactured goods. It played a major part in creating an economically dependent 'Third World' whose capacity to develop autonomously was – at least initially – stifled.

CAPITALISM AND THE EXPANSION OF EUROPE

HENRY BERNSTEIN, TOM HEWITT AND ALAN THOMAS

Understanding the Third World today in all its diversity involves looking back to yesterday. More than that, it means delving into European history. The development of capitalism and the 'expansion' of Europe from the sixteenth century onwards was instrumental in creating a world market and an international division of labour, the legacy of which is still very evident (though transformed) today.

The commodities that sustained this global economy were, of course, produced by the labour of those inhabiting what we now call the Third World, whether as slaves, tribute labourers, indentured workers, peasants or proletarians, as well as by the labour of workers and farmers in Europe itself and in other parts of the world. But the conditions and uses of the labour of these millions of people were structured by the interests and conflicts of the ruling classes of Europe as it underwent its long transition from feudal to capitalist society, and its subsequent industrial revolutions.

Q How exactly did Europe come to dominate the entire world, and what are the consequences of the way this occurred for the make up of today's Third World?

The main issue addressed here is the relationship between the expansion of Europe and the development of capitalism with its special characteristics. Within the sweeping heading of 'the

expansion of Europe', and the processes it encompassed over the last five centuries – from plunder, pillage and slavery to 'legitimate trade' and investment – special attention will be given to **colonialism** or direct rule of non-European areas and peoples. The boundaries of those areas typically established the political territories we know as Third World countries today.

> **Colonialism:** The direct political control of a people by a foreign state; control of a non-European people by a European state or the USA (see also Box 8.1).

8.1 Capitalism and colonialism

At a general level (and at the risk of being oversimplistic), capitalism is a system of production of goods and services for exchange on the market (rather than consumption by the producers) in order to make a profit (see Chapter 6).

The production of commodities (i.e. goods for exchange), the existence of markets for them, and the use of money as a medium of exchange, are not exclusive to capitalism. Pre-colonial societies generally produced a surplus and often had well-developed markets and trade networks, considerable specialization in the social division of labour, and classes of rich merchants and money-lenders – all, however, without having undergone a transition to capitalism as a distinctive mode of *production*. The word 'production' is stressed here because capitalism is distinguished by the emergence and central importance of *productive capital* – capital invested in production. Productive capital invests in means of production (land, tools, machines, etc.) and labour power, which it then organizes in a production process, making new commodities and creating new value as the necessary step towards realizing a profit. By contrast, *mercantile* capital is invested in the circulation of commodities (by wholesalers, chain stores) and *finance* capital in the provision of finance and credit (by banks). These contemporary merchants and money-lenders play an important role in a capitalist society in which commodity production is generalized, but it is the activities and needs of *productive* capital that give that society its special characteristics.

Only productive capital presupposes that labour power and the means of production are available as commodities. As most pre-capitalist societies were predominantly agrarian (hence the common synonym 'pre-industrial'), a crucial step in the transition to capitalism was that land should become a commodity, to be freely sold or rented without restriction by customary laws, the rights of monarchs, feudal lords, peasant communities, or whatever. In England such restrictions on the commercialization of land, imposed by the class relations of feudalism (one type of pre-capitalist society), were undermined far earlier than anywhere else in Europe. Historians have stressed the importance of the capitalist 'agricultural revolution' in England that preceded, and undoubtedly contributed to, the more celebrated industrial revolution.

Productive capital invested in means of production can do nothing, however, without labour power to use those means. Just as land and other means of production had become commodities, there had come into existence a class of people possessing no other commodity than their labour power. The related emergence of productive capital and a working class – basic

conditions of capitalist production – are part of the process called *primitive accumulation* that resulted from particular processes of change and disintegration in pre-capitalist societies. This meaning of 'accumulation' is broader than the usual notion of amassing wealth or capital, since it includes the historical formation of a class of people whose labour power is necessary to the production of wealth and capital.

The process of transition to capitalism in north-western Europe took place over a long historical period, mainly the sixteenth to the nineteenth centuries, when the industrial revolutions took off. This period of transition was one of continuous (albeit uneven) expansion of commodity production and exchange, facilitated by a range of social, political and cultural changes.

The process of primitive accumulation was helped by the 'expansion of Europe' in the same period, as a result of which vast amounts of wealth flowed into Europe from the plunder, conquest and colonization of many of the pre-capitalist societies of Latin America, Asia and Africa. In itself, this flow of wealth was not different in character from the riches amassed through other great imperial ventures in history, such as those of the Ottomans, the Moguls and the succession of Chinese dynasties. It would not have led to capitalism if it had not been able to feed into changes already taking place in Europe. For example, much of the treasure extracted from their colonies by Spain and Portugal went to buy commodities from north-western Europe, where the transition to capitalist production in manufacturing as well as agriculture was taking place. The relatively slow transformation of feudal relations in Iberian society resulted in the declining wealth and power of Spain and Portugal compared with those countries that were pioneering capitalism.

The development of capitalism had a global dimension from the beginning, therefore, which was experienced by the pre-capitalist societies of the Third World through their incorporation in an emerging world market and an international division of labour, typically initiated during a period of European colonial rule. In this sense, capitalism came to these societies from the 'outside' rather than resulting from their internal dynamics.

One is struck by the long period (about three centuries) between the beginning of the break-down of feudal society and the onset of the industrial revolution, which provided the emerging capitalist society of Britain with its distinctive type of production process – large-scale machine production. Once capitalist industrial production was firmly established and had begun to develop elsewhere in Europe, in the USA and in Japan, the striking feature by contrast was the 'acceleration' of history, caused by the tendency of capitalism constantly to revolutionize technology and methods of production and to accumulate capital on an ever larger scale.

This framework suggests key themes in the relationship between capitalism and colonialism, and also significant variations in the colonial experience. Such variations arose from:

1 different stages in the emergence of capitalism, and its uneven development between colonizing powers and within the areas they colonized;

2 different types of colonial state and the interests they represented;

3 the diversity of the pre-colonial societies on which European domination was imposed.

With respect to the first point, for example, Spain and Portugal colonized Latin America while they were still feudal societies, and did so at an early stage of the transition to capitalism in north-western Europe. At that time, the demands of the emerging international market focused on precious metals (gold and silver) and on tropical products for 'luxury' consumption by the wealthy classes of Europe (e.g. sugar, coffee, spices, precious woods and fabrics). But by the time Britain, France and Germany were competing for colonies in Africa in the last quarter of the nineteenth century, they were already industrialized or rapidly industrializing capitalist countries. The international market had changed with the industrial revolution to produce an enormous demand for raw materials for manufacturing (minerals and agricultural products like cotton, jute, rubber and sisal) and for mass consumption by new and large urban populations (e.g. tea, sugar, vegetable oils). It should also be remembered that periods of colonial rule in different areas of the Third World cut across those stages in the development of capitalism. For example, most of Latin America consisted of independent states, created from struggles against the Spanish and Portuguese crowns, before most of sub-Saharan Africa was incorporated into the colonial empires of European powers.

This connects with the second and third points, which entail consideration of the duration of colonial rule as well as when it was initially imposed. Most of Latin America, for example, experienced at least three centuries of colonialism, while in parts of Africa the period of colonial rule lasted less than the lifetime of some individuals. Again, for Latin America and the Caribbean colonialism was a brutal first introduction to the emerging world economy of the sixteenth century, and existing ways of life were shattered. In many parts of West Africa, on the other hand, the development of an agrarian commodity economy involved in international trade – 'the major revolution in the lives of the peasants' (Crowder, 1968, p.7) – had begun long before the beginning of the colonial era in the late nineteenth century, although it was certainly restructured and intensified under colonialism (Box 8.1).

8.2 Stages in European colonialism

The profound changes in the lives of millions of people brought about by colonialism and capitalism, and particularly in the conditions of their economic activity, are discussed in Chapter 9. This section elaborates the elements of the periodization of the expansion of Europe sketched above, and summarized in Figure 8.1. The final section concludes the chapter with some ideas about the formation and contradictions of colonial states which are developed further in Chapter 10.

Box 8.1 'Colonization' and 'colonialism'

Both these terms derive from the Greek notion of a colony as the permanent settlement of people who have moved from their original home territory to another. Some writers attempt to distinguish *colonization* (meaning setting up permanent settlements in this way) from *colonialism* (meaning direct political control of a people by a foreign state). Thus the settlements of Portuguese, Spanish, English, French and other Europeans from the late fifteenth century in the Americas would be examples of 'colonization', where the settlers intended to establish societies in many respects similar to those they left behind. By contrast, the administration of territories in Africa by British, French, German and other European states in the late nineteenth and twentieth centuries would be examples of 'colonialism', implying exploitation by the ruling power through perhaps a relatively small number of local agents. Thus Britain on the whole had a positive relationship with its North

American settler colonies, very different from the antagonistic relationship with its subject colonies in Africa.

In practice, however, all cases of foreign political control imply a degree of exploitation and opposed interests between the ruling foreign state and the local people, and also some links, both economic and cultural, between at least some sections of the colonized population and the colonial power. In this book, the word *colonialism* is used throughout for control of a non-European people by a European state or the USA, irrespective of the degree to which settlers are present (though this has an important bearing on the type of colonial state formed – see Section 8.3). Other cases of occupation and foreign control may or may not be regarded as 'colonial' – the term has negative connotations and there is generally a dispute as to whether an area is 'occupied territory' or simply forms part of the territory of the ruling state.

Periodization of the 'stages' of European colonialism and their relation to the development of capitalism provides an analytical framework for the historical processes of the making of Third World economies, as indeed of First World economies. This is not the same as, nor a substitute for, the detailed investigation of the economic histories of particular countries, of which many illuminating accounts have been written using this kind of framework.

The crisis of feudalism and the first stage of expansion

It can be suggested, following the work of Michael Barratt Brown (1963), that the motivations, forms and cumulative intensity of the expansion of Europe in the sixteenth century were closely linked to the crisis of feudalism there. One aspect of crisis in the old order was a transition from one kind of commodity economy, controlled and constrained by the power of landowning aristocracies, to another that was being initiated by

increasingly independent groups of merchants based in the towns. They encouraged the development of urban production (crafts, simple manufacturing) and exploited the weakening control of feudal lords over the agrarian economy and its peasant producers.

Late feudalism was marked by dynastic wars for sovereignty within and between existing political territories, and the emergence from them of new states confronting the effects of the massive costs of continuous expeditions, the disruption of the agrarian economy, and a series of peasant uprisings. The need of these states for further sources of revenue stimulated the search for, and seizure of, the wealth of other societies.

"The movement to the New World, the establishment of forts and trading posts along the coasts of Africa, the entry into the Indian Ocean and the China Seas, and the spread of the fur trade through the boreal forests of America and Asia all represent

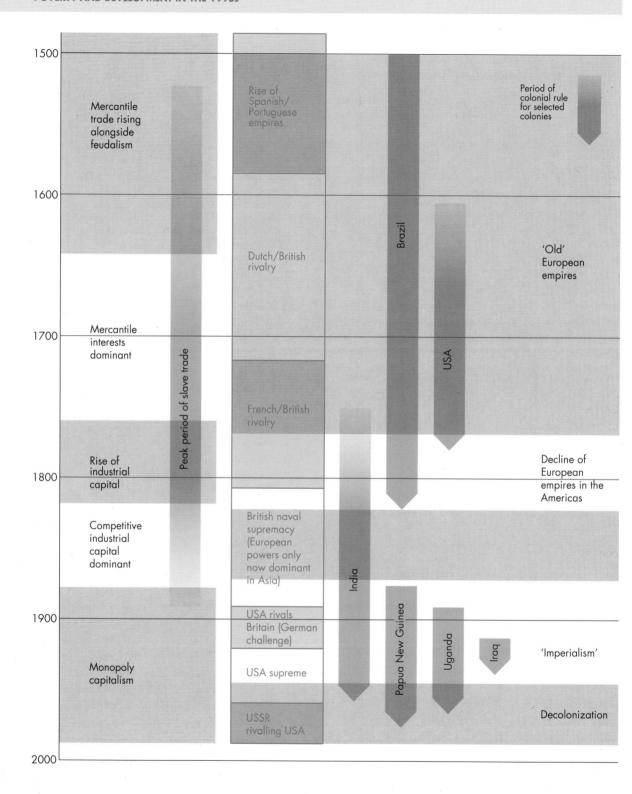

Figure 8.1 *Periodization of European colonialism in relation to the development of capitalism.*

ways in which these goals were sought and fulfilled. New goods entered the circuits of exchange: tobacco, cacao, potatoes, tulips. African gold and American silver, as Braudel has said, enabled Europe to live beyond its means."

(Wolf, 1982, p.109)

The agents of this first wave of expansion were explorers, mercenaries and merchant adventurers. From their forts and trading posts, these gangster entrepreneurs collected from local societies the luxury goods valued by the wealthy classes of Europe, whether by plunder, trickery, or establishing commercial monopolies. Portuguese traders, among others, used their forts and especially their fleets for 'not merely protecting their own trade but also selling "protection services" to others, forcing Asian merchants to pay for the privilege of sailing the seas in peace' (Curtin, 1984, p.137).

In the sixteenth century, systematic colonial rule was imposed only in the Caribbean and Latin America (Figure 8.2), where the aftermath as well as the immediate methods of conquest had devastating effects. The quest for treasure that had first spurred exploration of a western route to the Indies led to the opening of the great silver mines of Mexico and Peru. It is estimated that the 'silver mountain' of Potosi absorbed the forced labour of about 15% of the male population of Peru in the second half of the century. From 1503 to 1660, shipments from Spanish America to Castile tripled the amount of silver in Europe.

At the same time, although American silver sustained the feudal regime of Spain, it did so at the expense of that regime over the longer term. The domestic economies and overseas trade of Spain and Portugal were to face increasing competition from England and Holland in particular – small countries on the periphery of Europe that were moving much more rapidly towards capitalism.

Merchants, slaves and plantations

In the course of the seventeenth century, a different kind of European expansion was added

Figure 8.2   Spanish monarchs took their place in the sequence of Inca emperors.

to the Spanish pursuit of treasure by plunder and mining in the west, and to merchant–adventurer trade in luxury items from the east. Alongside these 'feudal' types of colonization and commerce, and ultimately displacing them, new forms of settlement and trade, exemplified by British interests in North America and British and Dutch activity in the Caribbean, linked more directly with the development of manufacturing and the transition to capitalism in Europe.

An example of this new type of colony, which combined large numbers of settlers with a plantation economy based on slaves and indentured labour, was the Virginia colony in British North America. In the first two decades of the seventeenth century, Sir Edwin Sandys, the treasurer of the Virginia Company, succeeded in reforming the Company (previously the preserve of the Crown and the merchant élite of the City of

London) to satisfy the needs of manufacturers and smaller independent merchants. On one hand, this was an important victory over royal and feudal-type monopolies and privileges in (and restrictions on) overseas trade. On the other hand:

> "the result of Sandys' efforts was to establish the slave owning plantation culture of the south, but at the same time to ensure supplies of Virginia tobacco and cotton for British manufacturers. These were to become far more important than all the spices and silks of the east. The American colonies later became, in fact, the main markets outside England for the products of England's new manufactories."
>
> (Barratt Brown, 1963, p.37)

In short, British colonization of North America and the Caribbean initiated a new kind of international trade linking the systematic large-scale production of raw materials for manufacturing in Europe, the development of markets for European goods in the colonies, and also, for several centuries, the procurement from Africa of slave labour for plantation production (Figure 8.3).

The first recorded slaves were imported into the New World from West Africa in 1518. Until the mid-seventeenth century their principal destination was the sugar plantations of coastal Brazil, which imported about 4400 slaves a year.

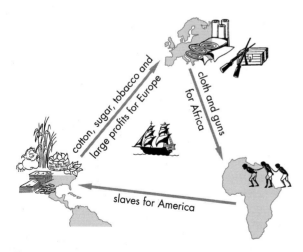

Figure 8.3   The triangular trade.

cotton, sugar, tobacco and large profits for Europe

cloth and guns for Africa

slaves for America

The Dutch then played a leading role in the spread of slave production to the mainland coasts and islands of the Caribbean, to meet the demand by merchants and sugar refiners in Holland, while the British developed the slave plantation system of what is now the southern USA.

Despite these important moments in the transition to capitalism, the latter half of the seventeenth century experienced a relative decline in international trade and the fortunes of European merchant companies. This was connected with turbulent events in Europe, including dynastic wars and, significantly, a new type of mercantilist trade war conducted principally at sea by armed fleets. The eighteenth century saw a revival and further intensification of European expansion, both reflecting and contributing to the resumed pace of the transition to capitalism. This was manifested in the growth of the Atlantic slave trade to meet the increased demand for tropical commodities. It is estimated that between 1701 and 1810, 6 million slaves left Africa, of whom 2.7 million were destined for the British and French Caribbean, 1.9 million for Brazil, and the rest for the Dutch Caribbean, Spanish America and British North America and the USA (Curtin, 1969). Recent studies regard even this as an underestimate.

It was merchants who financed and organized the slave trade and the shipment of tropical commodities and European goods, to their own benefit and that of the emerging industrialists of north-western Europe (Figure 8.4). During this period, adventurers and merchants also extended their exploration, pillage and pursuit of commercial advantage along the coasts of Africa and within Asia. These activities continued and developed the forms of European expansion that had begun in the sixteenth century, and were marked by armed conflict between Europeans (as well as between them and the people of the areas on which they sought to impose their domination) – for example, between the Portuguese and the Dutch in the Spice Islands (now Indonesia, Malaysia, and the Philippines), and between the French and the British in India where the power of the Mogul emperors was in decline.

Figure 8.4 Carved ivory tusk depicting an African view of mercantilism as a hierarchy: slaves at the bottom, African producers in the centre and a European merchant at the top.

The Dutch wrested control of the Spice Islands from the Portuguese after a long struggle. The Dutch East Indies Company profited considerably from remittances, dividends and exporting spices between 1650 and 1780, and then from the establishment of systematic plantation production in Java. In India, Clive's bloody victory in Bengal in 1757 put paid to French hopes and hastened the downfall of the Mogul order. While Clive's activities, in their purpose and methods, had much in common with the original phase of European expansion, and were accompanied by massive plunder organized through the British East India Company, their effect was soon to establish the systematic integration of India and its mostly peasant producers in the developing international division of labour, shaped by the

emergence of industrial capitalism in Europe. As British manufacturers succeeded in banning the import of Indian textiles, which had been the major trade item of the East India Company, the latter promoted tea production and the tea trade with China, for which opium produced in Bengal (and forced on the Chinese) was the principal commodity exchanged to finance the trade.

This extremely schematic outline has suggested that, in the course of the seventeenth and eighteenth centuries, the expansion of Europe intensified in ways connected with its accelerated transition to capitalism and the international division of labour that was emerging from it. At the same time, most colonization in this period was undertaken by merchant companies rather than by European states themselves (however much these states assisted their merchants through political, diplomatic, and military – above all naval – measures).

The consolidation of more systematic colonial rule including state formation (see below and Chapter 10) during the nineteenth century, as well as the last great wave of colonial expansion towards the end of the century, involved a more direct role for European states in an international context structured by the effects of industrial revolution. Again, it is highly suggestive that the original 'feudal' colonialisms of Spain and Portugal were losing their American possessions at a time when capitalist colonialism was about to embark on its most significant period of domination, from the mid-nineteenth to the mid-twentieth centuries.

France and Britain also lost large parts of their original overseas possessions in the second half of the eighteenth century. Britain took over much of what is now Canada from France at the end of the Seven Years' War in 1763; then by 1783 the American War of Independence had ended with Britain in turn losing most of its North American colonies. However, France and Britain were both able to increase their importance as colonial powers in the context of the dominance of industrial capital.

The types and volumes of raw materials needed by a rapidly industrializing Europe, new market outlets for its factory-produced commodities, the character of overseas investment, new types of shipping and of communications more generally (railway, telegraph), together with their strategic implications, all made the capitalist colonialism of the nineteenth and twentieth centuries very different from its sixteenth-century antecedent in Latin America.

Colonialism and imperialism

In India, the rule of the East India Company was replaced by that of the British state after the 'mutiny' (uprising) of 1857–58. In subsequent decades, colonial rule was also imposed and/or consolidated by the British in Burma, Sarawak and the Malay States, and by the French in Indo-China. The most rapid and dramatic wave of European expansion in this period, however, was 'the scramble for Africa'. In 1876, European powers ruled about 10% of Africa. By 1900, they had extended their domination to 90% of the continent, which was thus the last great 'frontier' of colonial capitalism. Africa was carved up principally between Britain and France, with substantial areas also seized by Belgium, Germany and Portugal.

> **Imperialism:** Whereas *colonialism* means direct rule of a people by a foreign state, *imperialism* refers to a general system of domination by a state (or states) of other states, regions or the whole world. Thus political subjugation through colonialism is only one form this domination might take; imperialism also encompasses different kinds of indirect control. Also, whereas *colonialism* may be used as a purely descriptive term, *imperialism* is almost always used in an ideological way, usually as part of a particular theoretical view of the causes, nature and effects of such domination, such as Lenin's view of imperialism as 'the highest stage of capitalism'.

The causes of the partition of Africa in the late nineteenth century are fiercely debated by historians. Colonialism was a controversial issue among leading European capitalists and politicians of the time, not least in Britain where some preferred the 'imperialism of free trade' to that of direct political rule, with what they considered its unnecessary costs (Figure 8.5).

JUDY, OR THE LONDON SERIO-COMIC JOURNAL.—Jan. 28, 1885.

DIGNITY AND IMPUDENCE.

Figure 8.5   Rivalry between European nations was exemplified by Britain's occupation of Egypt despite French claims.

"Britain's industries were reared behind protective walls, nourished on imperial tribute and encouraged by the destruction of all competition from the east. But, once established, they needed protection, plunder and protected markets no more... the factory product could undersell the work of handicraftsmen in any country in the world. All the industrialists asked was the freedom to trade – to obtain food and raw materials wherever they were most cheaply produced and to open up the whole of the world as markets for their wares."

(Barratt Brown, 1963, p.52)

Even the advocates of such international free trade, however, had to recognize the strategic nature of their trade routes (and the sources of their raw materials) and, hence, the need to guard them. This meant, at the least, an effective network of naval bases, and often the political and military capacity to guarantee communications and the flow of commodities across great land masses.

The scramble for Africa occurred during the great depression of late nineteenth century Europe (1873–96), which was the first major manifestation of the cycles (boom followed by slump and crisis) of the new world economy of industrial capitalism. The connection between these two processes was made by Lenin in his pamphlet *Imperialism: the highest stage of capitalism*, written in 1916 with two immediate and related objectives: explaining the causes of the First World War, and winning the workers of Europe away from mutual slaughter in the interests of 'their' ruling classes (Lenin, trans. 1939).

For Lenin, the great depression of the late nineteenth century marked a critical turning point in capitalism, from an earlier 'competitive' stage to what he termed **monopoly capitalism**. This does not mean that competition ceased to exist, but rather that it took more extreme and dangerous forms (leading, in 1914, to war) which were generated by the massive **concentration and centralization of capital** among rival capitalist countries. At that time, the most

potent conflict was between Britain and the new industrial power of Germany.

Monopoly capitalism: A stage in the development of capitalism dominated by giant corporations, each of which controls a relatively high proportion of the local or world markets for its products. This means that instead of simple price competition between small independent producers, there is greater importance for finance and investment. Competition between large corporations each with monopoly control in different areas takes the form of competition for finance, for sources of raw materials and for profitable investment opportunities.

Concentration of capital: The combination of small capitals into large corporations (by takeover, merger or joint venture).

Centralization of capital: The tendency for certain geographical centres to become dominant, combining the administrative centres of finance and industrial capital with political centres.

Thus, local competition between small firms is replaced under monopoly capitalism by rivalries between centres – and nations – as well as between corporations.

Lenin also argued that the expansion of colonialism in this period was due to the need to find new outlets for the export of capital for two reasons. The first reason was competition for overseas sources of raw materials and markets for European manufactured goods (both in ever-increasing volumes). The second was the search for investment opportunities that would be more profitable than those available in Europe itself.

Britain at this time provided the best example of this 'thesis of capital export' of Lenin's. British capital exports accelerated rapidly in the later nineteenth century and early twentieth century, and from the 1880s about 40% of British overseas investment was directed to railways, plantations, factories, government stocks and finance in the empire.

It was Germany, on the other hand, that best exemplified the concentration and centralization of capital in the form of giant industrial corporations closely linked with banks. Lenin termed this particular combination of industry and banking 'finance capital', which he saw as the distinctive and dominant form of capital in the period of imperialism or monopoly capitalism.

Lenin's account has been criticized as factually inaccurate (among others, by Barratt Brown, 1963, and Warren, 1980). His analysis may be considered 'exaggerated' in that two of the principal characteristics of imperialism he identified (capital export and the formation of modern finance capital) were respectively exhibited by two countries with quite different paths of capitalist development (Britain and Germany) rather than being combined. However, the trends in the internationalization of capitalist production and finance, which Lenin may have 'exaggerated', have become much more powerful and evident since then, notably in the operation of transnational corporations and banks. For this reason, analysis in terms of imperialism retains its force towards the end of the twentieth century, and may even have increased its importance.

A second reason that Lenin's analysis retains its significance is that it highlighted a striking feature of capitalism in its imperialist phase, namely the gap between the continuously increasing internationalization of capitalist production and finance and the persisting political organization of capitalist societies through national states. Lenin and another leading Bolshevik, Bukharin (and, indeed, other contemporary socialists), were very alert to the dangers of instability and war implied by this widening gap.

While Lenin was able to connect the great depression of late nineteenth-century Europe, the emergence of imperialism, and the last great wave of capitalist colonization in Africa, a third reason for the continuing relevance of his analysis is its insistence that imperialism, as 'the highest stage of capitalism', does not necessarily depend on colonies (Figure 8.6).

"Since we are speaking of colonial policy in the epoch of capitalist imperialism, it must be observed that finance capital and its foreign policy, which is the struggle of the great powers for the economic and political division of the world, give rise to a number of transitional forms of state dependence. Not only are there two main groups of countries, those owning colonies, and the colonies themselves, but also the diverse forms of dependent countries which, politically, are formally independent, but in fact are enmeshed in the net of financial and diplomatic dependence typical of this epoch. We have already referred to one form of dependence – the semi-colony. An example of another is provided by Argentina...

It is not difficult to imagine what strong connections British finance capital (and its faithful 'friend', diplomacy)...acquires with the Argentine bourgeoisie, with the circles that control the whole of that country's economic and political life.

A somewhat different form of financial and diplomatic dependence, accompanied by political independence, is presented by Portugal... Great Britain has protected Portugal and her colonies in order to fortify her own positions in the fight against her rivals, Spain and France. In return Great Britain has received commercial privileges, preferential conditions for importing goods and especially capital into Portugal and the Portuguese colonies, the right to use the ports and islands of Portugal, her telegraph cables, etc. Relations of this kind have always existed between big and little states, but in the epoch of capitalist imperialism they become a general system, they form part of the sum total of 'divide the world' relations and become links in the chain of operations of world finance capital."

(Lenin, trans. 1939)

Figure 8.6 European domination without direct political control. China remained independent throughout the period of European colonialism, but, in the late nineteenth century at least, European powers were able to suppress anti Western movements. (left) A French magazine reports massacres of Christians and missionaries in China in 1891; (below) Punch's view of Western action in China against the 'Boxer' movement of 1900.

THE AVENGER!

As the last sentence quoted here indicates, imperialism as the distinctive form of modern capitalism has a different and more precise meaning from imperialism in the colloquial usage of 'empire'. The latter tends to include the British Empire, for example, as simply one of a line of great empires in history. Lenin suggests that the British Empire was different from those that preceded it, and that British imperialism could survive the end of its formal empire and decolonization (on the analogy of Argentina and Portugal). In this respect, too, it is worth noticing that the First World War, caused by the rivalry of industrial capitalist powers, also resulted in the final demise (after long decline) of the remaining pre-capitalist empires of Eurasia: those of the Hapsburgs (Austro-Hungary), the Romanovs (Russia), and the Ottomans (Turkey and its possessions).

In fact, the inter-war years saw an increase in the number of British (and French) dependencies overseas. The vehicle of a League of Nations 'mandate' was used to allow what were effectively new areas of colonial rule in the Middle East, where various parts of the old Ottoman empire were divided between Britain and France. Although Iraq, for example, was only ruled by Britain for 16 years, during that period that rule was enforced as strongly as anywhere in the British Empire, and new weapons technology in the form of air raids and mustard gas was used in policing parts of the rural population in order to enforce payment of taxes.

The Middle East also gave rise to new examples of imperialism continuing beyond a period of direct political control. New boundaries were drawn and states granted independence at different times in such a way that what was becoming a vital natural resource, namely oil, was divided geographically between several states so that the capitalist states of Europe and the USA were able to control supplies through the various giant oil corporations.

League of Nations mandates granted Germany's West and East African colonies to France and Britain (plus what is now Namibia to South Africa, and Papua New Guinea to Australia). Apart from Namibia, which remained occupied until 1990, these were treated like other colonies. Japan also consolidated the area of its overseas control in this period; and the Soviet Union was established as a non-capitalist rival to the imperialist powers.

In the decades following the Second World War, the capitalist colonial empires were dismantled. Decolonization occurred relatively quickly in the Caribbean, Asia and Africa, compared with the period during which European domination had been established over these areas. The end of empire was primarily the result of anti-colonial and anti-imperialist struggles pursued by the peoples of the Third World, but post-war decolonization was also supported by both the USSR and the USA, though for different reasons.

The USA was to prove the dominant international capitalist power after 1945, hence the dominant imperialist power in Lenin's sense. It had little in the way of formal colonies; the Philippines which had been taken over by the USA from Spain in 1898 (having been Spain's principal Asian colony for nearly four centuries) became politically independent in 1946. The expansion through which the economic power of US capitalism emerged had taken place mostly through its own internal 'frontier', at the expense of indigenous Americans and of Mexico to the south. Before the First World War, however, US capitalism had actively expressed its imperialist character in the countries of Central America and the Caribbean (as well as in the Philippines). Following the decolonization of Asia and Africa, its economic, political and military activity extended to other areas and intensified, confirming the role of the USA as the leading power in an imperialism now without colonies (Figure 8.7).

Figure 8.7 Lenin was not the last to locate the causes of modern war in the crises of capitalist economies.

8.3 Colonial state formation

The formation of colonial states was central to establishing some of the conditions of generalized commodity production, including various labour regimes geared to export production (see Chapter 9). However, this does not mean that the politics of colonialism (as any other politics) are simply reducible to economics, to the economic interests and projects of particular classes.

For one thing, political conflicts have a complex relationship to specific differences of interest that exist between and within dominant economic groups, whether divided on lines of national rivalry (as emphasized in Lenin's *Imperialism* of 1916) or by their particular economic base (e.g. metropolitan manufacturers, colonial settler farmers and planters, imperial traders, mining capital, banking capital). More fundamentally, the political systems of colonial rule, and the economic aims they were typically called on to promote, were resisted by those on whom they were imposed. Whether or not methods and means of resistance were overt, direct or organized, the actions of the colonized inevitably frustrated, in a variety of ways, the blueprints of colonial 'good government' drawn up in the capitals of Europe and those of the colonies themselves.

Common themes in colonial state formation

Despite the enormous variations in time and place, colonial states were generally maintained through the use of force, administrative structures, and various means for attempting to legitimate colonial rule. However, there was a fundamental contradiction between changes required for colonies to serve their rulers' interests and the need to maintain stability. These major common themes will be elaborated on in Chapter 10, but are worth a brief discussion here.

First, colonial territories were established as political entities typically through conquest: the actual or implied *use of force*. Once colonial states were created, their domination over their subjects rested ultimately, like that of all states, on their ability to mobilize and deploy military and other forms of coercive power. This could present problems in colonies where Europeans were generally a tiny proportion of the population, and which had distant lines of communication to the imperial country.

Secondly, together with military and other coercive apparatuses, colonial states established structures and procedures of *administration* to carry out the daily tasks of government. As noted, the latter included important economic objectives and had important (if often unintended) economic effects. Colonial states were expected minimally to develop a revenue base to be self-supporting financially, and usually also to promote the provision of commodities, markets, investment opportunities and profits for the traders, capitalists and exchequer of the imperial power.

Thirdly, in order to establish and maintain political stability, administrative efficiency and economic profitability, the *legitimation* of colonial rule was desirable. Colonial states often tried to secure consent to their subordination, or even active participation in it, on the part of colonized peoples, or at least key groups among them. This project could employ both material means (limited opportunities for selected groups in relation to land, trade, Western education, jobs in the colonial administration, etc.) and ideological means, above all through conversion to Christianity and/or access to Western education (the latter again for a select few).

However, according to the specific circumstances of time and place, viability depended to a greater or lesser extent on introducing potentially wide-ranging changes while attempting to control and contain their potentially disruptive effects. These changes could be political (administration, law, taxation) and cultural (religious conversion, education, new consumption needs, medical practices) as well as economic (labour tribute, commercialization of production, monetization of exchange). To try to achieve desired changes while maintaining social and political stability, colonial states operated selectively both 'traditional' forms of social organization and legitimation ('traditions' that were often invented by the colonizers), and those associated with European states and a (developing) capitalist economy. Again, their success was limited by the fundamental contradiction noted above, and also by the frequent ignorance of colonial rulers about indigenous societies and cultures which provided the colonized with a variety of means of hidden and open resistance (including the invention or reconstruction of their own 'traditions').

Variations in colonial states

These common themes, problems, contradictions and struggles in the politics of colonialism were, of course, manifested in ways as complex and varied as the long history of colonialism itself.

One source of such variation, linked to the economics of European expansion and its periodization, concerns *when* colonial rule was established and by which European country. The stage in the (international) development of capitalism and the position of a given European country within that stage determined the type of colonial state that was formed.

Another determinant was the nature of the indigenous society or societies occupying a territory on which colonial rule was imposed. One broad difference between pre-colonial societies (noted in Chapter 7) was the extent to which any

surplus production was used as a basis for more or less elaborate state systems of their own. A further dimension of complexity and variation thus follows recognition of the interaction of different *types of colonizing countries* in different periods with different *kinds of pre-colonial economies and political structures*.

Also relevant in this context are what have been called the technologies of colonial conquest, including armaments and means of transport and communications, and their forms of military and political organization. Colonial powers drew on different technologies of conquest and rule in the various stages of European expansion. These technologies became increasingly powerful over time as Europe industrialized, so that the gap between them and the military technology and organization of pre-colonial societies also tended to get bigger.

The extraordinary account of the conquest of Mexico by Bernal Diaz, one of Cortés' officers, shows how similar in many respects the military methods and politics of the Aztecs and the Spanish were in the early sixteenth century (Diaz, 1963 edition) (Figure 8.8). Both represented pre-capitalist cultures of a hierarchical kind, in which state religion, the accumulation of treasure, and intense personal factionalism all

Figure 8.8 The battles of Cortés in Mexico; a sixteenth-century drawing showing the Tlaxcalans as allies of the Spaniards.

featured. Despite their horses and guns, the Spanish were only able to defeat the empire of Montezuma through alliances with other Mexican groups and after protracted and bloody campaigning. By the 1890s and the expansion of the British into the eastern Cape (South Africa), the colonial confrontation had a very different character: 'Rhodes mowed down a mealie field with machine guns before the eyes of the paramount of eastern Pondoland and his councillors and explained that their fate would be similar if they did not submit' (quoted in Low, 1973).

Finally, a source of variation in the nature of colonial states, and colonial society more generally, was whether colonies had substantial European settler populations or not. Colonial settlers were important in North America, Spanish America and the Caribbean, and in Africa in Algeria, Kenya, southern Rhodesia (now Zimbabwe), Angola, Mozambique and South Africa, but were largely absent in West Africa and India. Where there were large settler populations, they sought to exercise control over the colonial state as an instrument of their own particular interests, often coming into conflict with the imperial power (not least when the latter presented itself as the guardian of the rights and interests of 'the natives'). The different fates of settlers in the Americas and Africa are instructive.

The USA and the countries of Latin America were established as independent countries through revolts and wars of independence against the colonial powers of Britain, Spain and Portugal in the late eighteenth and nineteenth centuries. Two aspects of their experience deserve emphasis. First, these anti-colonial movements achieved their success before the capitalist colonialism of a rapidly industrializing Europe had entered its major period of global domination. Secondly, while struggles for independence mobilized and combined the energies of different classes and groups in colonial society, they were led by landed aristocracies of settler descent whose project of an independent state excluded any reforms detrimental to their own position. The colonial aristocracies of both North and South America derived their wealth and power

from a history of plunder, enslavement and oppression.

In the African colonies mentioned above, settler populations were established during the peak period of capitalist colonialism. Generally, their fate was sealed with decolonization and 'the end of empire' brought about on the one hand by the struggles of the colonized peoples and, on the other hand, by changes in the global development of capitalism which no longer required direct colonial rule (see above). In most cases, political independence in the former settler colonies of Africa was accompanied by a mass departure of their resident European populations. South Africa is an exception both in its relatively longer history of European settlement, and in its survival to date as a white minority state reproduced through the systematic racial oppression that was virtually definitive of colonial rule. The term 'internal colonialism' is often used to characterize the social and political structure of South Africa, and sometimes to refer to the oppression of 'Third World' populations in the Americas, e.g. indigenous Americans ('Indians') and blacks descended from African slaves.

Just as the politics of colonial rule and the nature of colonial states differed according to the presence or absence of large settler groups, so did the politics of decolonization. In general, since 1945, it was politically easier for European countries to relinquish those colonies without substantial settler minorities. Independence with majority rule in former settler colonies was typically achieved only after periods of bitter armed struggle against the forces of the colonial power and/or settler society.

In conclusion, it can be suggested that different 'models' of colonial rule were more the results than the causes of different practices of domination: that is, they were reflections on, and justifications of, specific methods and results of colonial state formation. While the latter had certain common conditions and features, these were necessarily manifested in historically complex variations from sixteenth-century Mexico to nineteenth-century India and twentieth-century Nigeria.

Summary

1   European colonialism from the sixteenth to the mid-twentieth centuries was marked out from other examples of occupations and empire. The expansion of the European colonial empires occurred along with the development of industrial capitalism in Europe.

2   In the early period, when merchant capital was dominant (mercantilism), European expansion was not so different from other empires where local surpluses were extracted through plunder, tribute and trade to enrich foreign aristocracies and merchant classes.

3   Later, however, particularly where slave-based plantation production added to these surpluses, they were able to feed into the process of primitive accumulation in (north-west) Europe, which led to the industrial revolutions there from the late eighteenth century.

4   The peak period of European colonialism, from the mid-nineteenth into the twentieth century, accompanied the dominance of productive capital in Europe. The colonies provided both raw materials and markets for manufacturing production by European capitalists.

5   It is suggested that the twentieth century has seen the rise of monopoly capitalism in the form of giant corporations, and that this corresponds to imperialism, a general domination by the major capitalist states, which is able to continue beyond the formal ceding of political control to newly independent states.

6   Colonial states, and thus their independent successors, have certain historical factors in common. They also differ enormously, according to the nature of the indigenous society or societies on which colonial rule was imposed, the degree of importance of settlers, and the stage in the development of capitalism when each colonial state was established.

LABOUR REGIMES AND SOCIAL CHANGE UNDER COLONIALISM

HENRY BERNSTEIN, HAZEL JOHNSON AND ALAN THOMAS

This chapter continues the discussion of the historical processes involved in the formation of the Third World by examining the relationship between colonialism and the organization of productive labour in the colonies.

Q How did the colonial powers control the organization of labour in the colonies?

Q How did the reorganization of local labour affect the development of what was to become the Third World in the longer term?

9.1 Colonial political economy

Whether the initial reason for the colonization of a territory was strategic or economic, few metropolitan governments were prepared to bear the financial cost of colonial administration for long. It was therefore necessary to organize the productive capacity of the area concerned so as to generate sufficient income to sustain the administrative and military presence that maintained European control.

This was seen as a minimal requirement, though it was one that some colonial territories were barely able to satisfy. As well as this requirement of being self-financing, colonies were expected to contribute to the economies of their metropolitan rulers.

Colonies therefore had to be integrated in an international economy, initially formed by the expansion of Europe in its period of transition from feudalism to capitalism (see Chapter 8), and subsequently shaped and reshaped by the dynamics of capital development on a global scale. As noted in Chapter 5, the making of colonial economies within the international division of labour occurred through the production of commodities for export, above all from extractive industries and tropical agriculture. The production of these commodities took different forms. Mining and larger scale agriculture (whether organized by plantation companies or by individual colonial settlers granted large areas of land for this purpose) required some initial capital and a sufficiently large (and cheap) labour force. Alternatively, peasants were 'encouraged' to grow particular crops for sale and export, by various means ranging from direct coercion to more indirect pressures, including the need for a money income to pay taxes and to purchase the new kinds of goods and services introduced with colonialism.

The kinds of commodities produced in the colonies, and how they were produced, varied in time and place according to the interests represented in different colonialisms. Large-scale trade dominated by metropolitan companies and colonial entrepreneurs was a consistent interest throughout the history of colonialism, although (as noted in Chapter 8) the composition and scale of that trade changed as industrial capitalism developed. In the sixteenth and seventeenth centuries, the mining of gold and silver in Spanish America and their transport to Europe provided the profits of colonial entrepreneurs, shipping and mercantile interests, and also met the needs of the Spanish Crown for revenue to finance its dynastic ventures in Europe. But as the growth of industrial capitalism in Western Europe and the USA accelerated during the nineteenth century, it required a more diverse range of products for processing and manufacturing, and in ever larger quantities: minerals like copper, and industrial crops like cotton, rubber, sisal and jute. The rapid urbanization that accompanied industrialization in the Western countries also resulted in a new market demand for tropical products that became items of mass consumption (sugar, tea, coffee, palm oils) and their production in the colonies was consequently expanded.

Changes in the international economy also led to changes in forms of colonial exploitation. For example, the massive development of the mining industry in South Africa in the late nineteenth century was a very different matter from the earlier Spanish adventurers' colonization of Latin America in search of 'treasure'. In the late nineteenth century, gold was needed to support the Gold Standard, on which the stability of the vastly expanding international trade and international monetary transactions was held to rest. Diamonds were needed for new industrial processes, as well as continuing to be an item of luxury consumption. The historian Colin Bundy (1979) has argued that prior to the discovery of diamonds and gold, a thriving African commercial agriculture had emerged in certain areas of South Africa as part of a previous phase of capitalist development. This was undermined by state policies to restrict African agriculture and access to land and independent incomes, at a time when the spread of mining (and the stimulus to settler or white commercial agriculture it provided) required a plentiful and continuous supply of cheap labour power.

The variation and complexity of the economies created by European colonialism was thus a result partly of different stages in the formation of a capitalist world economy, and also of different forms of colonial incorporation and exploitation. The making of colonial economies required the 'breaking' of pre-existing types of economy and the social relations within them. In different cases, the rupture could be more or less abrupt, more or less brutal, and effected by more or less direct means, as the rest of this chapter illustrates.

9.2 Colonial labour regimes

The term 'labour regime' is used here to refer to different methods of mobilizing labour and organizing it in production. The essential mechanisms of four broad types of labour regime are described and illustrated briefly below, namely forced labour, semi-proletarianization, petty commodity production and proletarianization. Connections between those labour regimes and the development of capitalism are then suggested.

Forced labour

Of the regimes of **forced labour**, *slavery* in the Caribbean and the Americas is probably the most widely known because of its scale, its duration over more than three centuries, and the intense violence of the slave trade and of the conditions of plantation production.

Forced labour: The mobilization and organization of workers based on extra-economic coercion. Workers do not enter the arrangement by their own volition or by selling their labour power in the market. Examples of forced labour are *slavery*, *tribute labour* (labour services or payments in kind) and *indentured labour*. In some circumstances, forced labourers may well own or have access to their own means of production, from whose produce they may make forced payments in kind as well as, or instead of, providing labour service.

There were two main historical factors that contributed to the development of the slave trade.

1 The demand for tropical products (sugar, cotton, tobacco) increased with the expansion of production, trade and incomes in Europe, which was associated with the development of capitalism, itself stimulated by the in-flows of precious metals and treasure acquired from colonial conquest in Spanish America, and subsequently from the plundering of large areas of Asia.

2 The indigenous people of the colonized areas of the New World were too few to provide sufficient labour to produce these commodities, or were resistant to enslavement, or were destroyed by European arms and diseases, or some combination of these factors.

Significantly, the slave trade and plantation production reached a peak in the eighteenth century, as north-western Europe was completing its long transition to industrial capitalism. Slavery was profitable as long as a plentiful and cheap supply of slaves could be assured. This might be met by the reproduction of the existing slave population, although reliance on this placed limits on the intensity with which slaves could be exploited: the slave population could not be replaced at the desired rate if they were literally worked to death after a few years or less (Figure 9.1). Alternatively, plantation owners had to rely on continuing shipments of slaves from Africa at prices that suited them. This strategy was undermined by the abolition of the slave trade by Britain in 1807.

Two other factors in the eventual decline of slavery are worth noting. The first is that in the course of the nineteenth century, new and superior technologies made available by the industrial revolutions of Europe and the USA were increasing the productivity of labour in agriculture as well as in manufacturing, thereby rendering slave production less competitive. The nature of slave production on plantations, including brutal forms of control and slaves' resistance to them, meant that it was very difficult to operate new and more sophisticated techniques of production with coerced and antagonistic workers. Secondly, the social and political costs of maintaining control over slave populations grew as they themselves increased in number, both absolutely and as a proportion of the population in plantation colonies. There were numerous strike waves and slave revolts in

Figure 9.1   Slaves on a treadwheel in Jamaica.

the sugar regions of Central America and the Caribbean throughout the eighteenth and early nineteenth centuries, and the ratio of slaves to others was ten to one in Jamaica by the time of the abolition of slavery in the British Empire in 1833.

The major effects of slavery over this long period illustrate the spatial dimensions of the processes contributing to the formation of a capitalist world economy.

1 In West Africa, slave trading brought about massive social disruption and depopulation. The raiding and warfare necessary for the provision of slaves was mostly carried out by indigenous groups (who consequently increased their own wealth and power by, for example, acquiring European firearms), in collaboration with the European traders on the coast.

2 In those societies it created (in the Caribbean, Brazil, the southern USA), the experience of slavery had profound consequences for social differentiation and cultural patterns that are still felt today.

3 For Europe, where the often vast profits of slave traders and shippers and plantation owners were directed, slavery contributed to the accumulation of wealth and facilitated the transition to industrial capitalism.

More or less contemporaneous with the long history of slavery was a variety of forced labour regimes in the Spanish colonies of the Caribbean and Central and South America. Originally adapted from the feudal institutions and practices of Spain, these allowed the Spanish settlers to establish themselves as a colonial aristocracy, in relation to the subjugated indigenous populations on one hand, and to the Spanish Crown on the other.

In the first half of the sixteenth century, individual colonists, churches and agents of the Spanish Crown were given rights to exact tribute from the labour of indigenous 'Indian' communities, which they could extract as *labour service* (for agricultural production, porterage, construction, personal services, etc.) and/or *tribute in kind* (agricultural and craft products).

This was the *encomienda* system. Technically it did not bestow rights to Indian land, although individual grants of land could be made by the Crown independently of the *encomienda*.

With the massive decline of the indigenous population (in Mexico, for example, from about 11 million in 1519 to 6.5 million in 1540 and 4.5 million in 1565), as the direct and indirect result of conquest and early colonization, and with the arrival of new colonists demanding their grants of Indian labour, there were conflicts over access to a diminishing labour supply. These were intensified by the discovery of massive silver deposits in Mexico and Peru in the mid-sixteenth century. The Crown introduced new mechanisms of forced labour service (known in Mexico as *repartimiento* or *cuatequil* and in Peru as *mita*), through which colonists had to apply to the state for 'Indian' tribute labour (Box 9.1). In this sense, the allocation of labour was centralized and bureaucratized as an action of the colonial state against colonial settlers (who accumulated large numbers of Indians through *encomienda* and passed them on to their heirs). At the same time, the aim of the new systems of tribute labour was to rationalize the supply of labour from wider areas for the concentrated demand brought about by the new mining boom. This resulted in long journeys undertaken by large convoys of Indians to work out their labour service at distant mines – from which many of them never returned (Figure 9.2).

The real practices of forced labour regimes in Spanish America differed considerably from the legal theory, and both changed over the centuries with changes in patterns of economic activity and with social and political struggles between settlers and 'Indians' and settlers and the Crown. Without going into the intricacies, it should be noted that when the new republics were established in the first half of the nineteenth century following wars of independence against Spain, their constitutions granted Indians equal citizenship and abolished forced labour. However, during the course of the eighteenth century many of the formal mechanisms of labour coercion had already given way to

Box 9.1   Women and forced labour in Peru

From the beginning of the Spanish Conquest, women were brutally exploited by colonial administrators and *encomenderos* (recipients of royal *encomienda* grants) who needed women's labour to produce goods (particularly cloth) destined for the colonial and European markets. The first forms of industrial labour draft emerged...when the *encomenderos* established a tribute in cloth. Many *encomenderos* introduced the practice of locking women in rooms and forcing them to weave and spin; these women were so exploited that in 1549 a royal decree was published prohibiting the continuation of the practice... this decree, like many others issued to alleviate the burden of the peasantry, was effectively ignored.

One hundred and fifty years later, the judge responsible for indigenous affairs in Cuzco was imprisoned because he had a private jail in his house where he forced Indian women to weave...

Colonial magistrates (*corregidores*), who primarily saw their stay in the colonies as a way to make a fast buck, forced women to weave clothing for them for less than half the free market rate... Spanish tribute demands and taxes were so high that women saw themselves by necessity having to weave...in their homes in exchange for grossly depressed wages, while their husbands and male kin were away working in the *mita* service... In addition, the wages paid to a *mitayo* (a man performing *mita* service) in the mines were equivalent to approximately one-sixth of the money needed to cover his subsistence requirements. Since *mitayos* often were accompanied by their wives and children...one way in which the difference may have been made up was for the labourer's wife and children to work also...but at wages which were certainly lower than the already depressed wages of their husbands.

(Silverblatt, 1988, pp.167–8)

Figure 9.2   Mining at Potosi, the Peruvian 'silver mountain'.

other labour regimes based on debt bondage (see below), to secure labour for the estates of what had become, in effect, a *landed* aristocracy.

> "In the process of debt bondage, the employers gave an advance on wages, or paid their labourers' tribute debts, after which the debtor was obliged to work for the man who had loaned him money until such time as he was able to settle the debt. However, since the debt slaves always received very low wages, and since the employers provided them with essential commodities and possibly also tools at a very high price, the workers became more and more involved in debt the longer they worked – so that the system really in effect boiled down to life-long compulsory labour."
>
> (Kloosterboer, 1960)

Forced labour regimes were also features of other colonies at other times, particularly during the early stages of colonization. Throughout sub-Saharan Africa in the late nineteenth century, and in a number of Asian colonies, tribute labour was directed to the construction of railways and roads and to work on European plantations. The control over labour and the conditions experienced by workers were little different from those of slavery.

Another, distinctive type of forced labour regime was that of *indenture*. Indentured labour is a practice whereby people contract themselves to work for an agreed number of years for a particular employer. This was an important device in the early settlement of British colonies in the Caribbean and the southern USA, those who were indentured as workers and servants coming from the poorest sections of the British population. Probably some of the small settlers in the Caribbean, who were displaced by the spread of sugar plantations in the seventeenth and eighteenth centuries, had first gone there as indentured workers and had become small farmers when the period of their indenture had been worked out.

In the nineteenth and early twentieth centuries, indentured labour occurred on a far larger scale,

drawing particularly on those masses of people in India and China whose poverty and destitution resulted from European domination (even though China was not formally colonized). Most of them were peasants driven from the land by crippling debt or hunger produced by intensified commercialization and exploitation, and craft workers like spinners and weavers, whose livelihoods were destroyed by competition from the cheap textiles of Britain's new factories.

Indian and Chinese indentured workers went to the plantations of the Caribbean, Mauritius in the Indian Ocean and Fiji in the South Pacific; to the rubber plantations of Malaya; to British East Africa, where the economically strategic railway from the port of Mombasa to Lake Victoria was built by Indian indentured workers in the first decade of the twentieth century; and to South Africa, as workers in agriculture and mining.

In principle, indenture was a contract freely entered into by workers for a limited period, and in this sense it differed from slavery, where the body and person of the slave was exchanged and used as a commodity. However, given the circumstances of destitution that drove people into indenture, the tricks and coercion often employed by licensed labour recruiters to get them to sign the indenture contract, and the power of the plantation owners and other employers (backed up by the colonial state) in the countries where they went to work, the experiences of indentured workers were often similar to those of the slaves of earlier generations (whom they replaced in the Caribbean). This has been amply documented by Hugh Tinker (1974) who termed indenture 'a new system of slavery'.

Semi-proletarianization

Indentured workers who completed their contracts often stayed in the colonies they had been shipped to, some of them subsequently becoming semi- or fully proletarianized. Full proletarianization refers to a generalized process of wage labour employment. When Karl Marx analysed capitalism, he called wage labour 'free' because workers own no means of production and are 'free' to sell their labour power as a commodity in the market without any form of coercion beyond the economic necessity of earning a living.

The meaning of **semi-proletarianization** is often elusive, but two somewhat different uses of it can be suggested. The first refers to conditions where producers are unable to pay back debts and are required to carry out labour services or make payments in kind to creditors (usually landlords). This is called *debt bondage*. This kind of situation was (and still is) found among the poorest strata of peasants and rural semi-proletarians, caught in a permanent cycle of debt to their landlords and others with a claim on their labour. It was indicated above that in the eighteenth century, with the consolidation of the colonial aristocracy of Spanish America as a *landed* class, many of the former forced labour regimes gave way to debt bondage as a means through which landowners secured a 'captive' (and resident) labour force for their estates.

> **Semi-proletarianization:** A process where people who have inadequate access to means of production, or have been dispossessed from them, have to provide labour for others. One mechanism of semi-proletarianization is *debt bondage* in which producers provide labour because they have fallen in debt with their creditors over land rents, cash loans or other resources. Another type of semi-proletarianization occurs through *periodic labour migration*. Historically, semi-proletarianization has involved a dimension of extra-economic coercion as well as economic compulsion. Current forms of semi-proletarianization may mirror characteristics of colonial forms but are generally regarded as being based on economic compulsion. Additionally, contemporary semi-proletarianization often combines production using own means of production with wage labour for local farms and industrial enterprises.

Significantly, the introduction of new and profitable commercial crops into particular areas was often accompanied by new types of debt bondage, or the intensification of existing ones. The term 'semi-proletarianization' seems appropriate here to the extent that debt bondage is a way of securing labour for commodity production within capitalism, but labour that is clearly not 'free' in the full sense suggested by Marx. In most Latin American countries, legislation to abolish debt bondage was passed between 1915 and 1920, but debt bondage as one type of labour regime established within capitalism (and not only colonial capitalism) remains widespread in many parts of the Third World today.

Those who have to supply their labour because of debt bondage may also have some land of their own or other resources which contribute part of their livelihood through subsistence or small-scale commodity production. This is the major characteristic of the second form of semi-proletarianization in which periodic labour migration is combined with other economic activity (and especially subsistence agriculture) to provide a means of livelihood and household reproduction. Cyclical or periodic labour migration regimes were a major feature of many colonial economies, notably in sub-Saharan Africa where semi-proletarianized migrants supplied much of the labour for mining and commercial agriculture – both large-scale (particularly in southern Africa) and small-scale (particularly in West Africa) – and continue to do so. Like debt bondage, then, the regime of semi-proletarianized migrant labour is reproduced, or even recreated, more generally within capitalism beyond the specifically colonial origins it had in many cases.

Petty commodity production

Petty commodity production is widespread today within an international division of labour (see Chapters 3 and 5). Under colonialism its conditions were typically established by the need for a money income to pay taxes to the new states, and subsequently also to purchase new means of production and consumption that the extension of the capitalist market made available (and often necessary). In some cases, colonial states ordered particular cash crops to be grown and attempted to regulate their methods of cultivation; in other cases, peasants seized or created opportunities to pioneer new cash crops and ways of farming.

> **Petty commodity production:** The production of commodities for sale based on economic necessity and using own means of production and household labour. Petty commodity production is therefore small scale but is based on a high level of integration in product markets, frequently leading to integration in credit and input markets. The use of family or household labour is an important characteristic, although temporary or seasonal wage labour may also be employed.

Here is an example from late nineteenth century Guinea:

> "A considerable poll tax was imposed on a population which had little or no contact with a cash economy and thus quite literally had no money to pay with. So Africans were forced to gather rubber to sell at derisory prices to the companies in order to get the money with which to pay the tax. Indeed, in the early days the tax was itself payable in rubber. The Fonta and Savannah areas were systematically defoliated of rubber plants as every year the African peasants moved out in ever wider circles from their villages to gather rubber. Thus the Administrator's fiscal policy was aimed at coercing Africans into the cash economy for the greater profit of the trading companies, and coincidentally to help balance the Administration's budgetary books."
>
> (Johnson, 1972, p.235)

Generally, rural populations preferred to meet the new needs for cash imposed on them through petty commodity production, in which they could exercise some control over the uses of their labour, rather than periodic wage labour for others in the harsh conditions of plantations, settler estates and mines. There is a parallel here with *encomienda*, in which the payment of tribute in kind was experienced as relatively less oppressive than labour service.

In some colonial economies the preference of rural people to undertake petty commodity production, and their success in doing so, confronted capitalists requiring large numbers of workers at low rates of pay. An example was given earlier, in the context of the mining boom in late nineteenth century South Africa, which can be repeated for many other areas where the interests of powerful types of capital demanded a plentiful and 'cheap' supply of labour rather than commodities produced by peasants. In colonial eastern, central and southern Africa (as in South Africa today), the solution to the problem of competition over labour between (European) capitalist and (African) peasant production was to undermine the ability of the latter to generate an adequate income through growing cash crops. This was done by restricting African farming and rural residence to limited and usually agriculturally marginal areas (the 'Native Reserve' system), and discriminating against peasant commodity production in terms of prices, transport charges, access to credit, etc. These various measures, directly imposed or facilitated by colonial states, institutionalized some of the conditions of semi-proletarianization.

Proletarianization

Full **proletarianization** in colonial economies occurred when impoverished peasants and craft producers (including previously indentured workers) either lost access to land and other means of production, or were driven by debt or hunger to try to secure a living through selling their labour power. From the side of capital, there was sometimes a demand for a more stable and skilled workforce than the labour regime of semi-proletarianization could provide. This applied to some jobs in mining, in manufacturing (although this was very limited in most colonial economies), and in such branches as railways, ports and road transport, which played a strategic role in the circulation of commodities and in the administration of the colonial state.

While the emergence of a stable working class in colonial economies was usually limited relative to the numbers of those proletarianized, that working class was able to develop trade unions and other forms of political action (both legal and illegal) which played an important role in the movements for independence from colonial rule.

Colonial labour regimes and capitalism

The different types of colonial labour regimes discussed here are summarized in Table 9.1. The table does not represent a rigorous classification of mutually exclusive categories, which in any case could be misleading. In particular, if we view the global history of capitalism, we see that

Table 9.1 Colonial labour regimes

Labour regime	Separation of producers from means of production	Extra-economic coercion	'Free' wage labour	Examples
1 Forced labour				
Slavery	Complete	Yes	No	Caribbean, Brazil, southern USA, 16th–19th centuries
Tribute, tax in kind	No	Yes	No	Spanish America, 16th–17th centuries; Africa, 19th to early 20th centuries
Labour service	Partial	Yes	No	Spanish America, 16th–18th centuries; Africa, Asia, 19th to early 20th centuries
Indenture	Complete	Partial	'Transitional'	Caribbean, East Africa, Malaysia, Mauritius, Fiji, 19th–20th centuries
2 Semi-proletarian labour				
Debt bondage	Partial or complete	No	'Transitional'	Spanish America, 18th–20th centuries; Asia 19th–20th centuries
Periodic labour migration	Partial	No	'Transitional'	Africa, and Third World generally, 20th century
3 Petty commodity production	No	No	No	India and Africa, 19th century; throughout Third World, 20th century
4 Proletarianization	Complete	No	Yes	Some sectors of colonial economies: 18th century (Latin America), 19th century (India), 20th century (Africa)

Proletarianization: The process (and result) of generalized employment of wage labour in commodity production. Proletarian labour is one of the fundamental characteristics of capitalism in which workers are separated from their means of production, and sell their labour in the market to capitalists (owners of capital). The notion of 'generalized commodity production' (often used to describe capitalism) therefore suggests not only the generalized production of goods for sale but the employment of commoditized labour (i.e. wage labour) to do so. Proletarianization is based on economic compulsion.

it has absorbed, created and combined many diverse social forms in the course of its uneven and contradictory development. For example, capitalism is usually characterized by the employment of wage labour (or full proletarianization) but other labour regimes can and do co-exist under capitalism.

Directly coercive labour regimes were characteristic of the period of primary or 'primitive' accumulation (see Chapters 7 and 8) on a world scale, during the sixteenth to eighteenth centuries when Europe was undergoing its long transition from feudal to capitalist society. This does not mean that directly coerced labour then disappeared all at once. In Cuba, slavery was not fully abolished until 1889, and various forms of tribute labour were imposed on the people of the new colonies of Africa and Asia in the late nineteenth and early twentieth centuries (and, in the case of Portugal's African colonies, continued until the 1960s) (Box 9.2).

Nevertheless, from the turn of the nineteenth century, there were several clear changes. These led to the establishment of forms of production in colonial economies based on semi-proletarian and proletarian labour (capitalist production) and household labour (petty commodity production). Whereas the initial creation of these types of labour within capitalism (as distinct from, for

example, pre-capitalist and pre-colonial peasant production) often required direct and indirect forms of *extra*-economic coercion, the latter were replaced sooner or later by *economic* compulsion. That is, people came to depend on commodities for consumption (and in the case of petty commodity production, for means of production too) and therefore needed cash incomes to buy these commodities. Semi-proletarian, proletarian and household labour were all reproduced within capitalism as a result of economic compulsion, and consequently have persisted in Third World economies after the demise of colonialism.

This is the reason why these three types of labour are shown in Table 9.1 as not requiring extra-economic coercion as a condition of their reproduction. At the same time, one should not regard the 'freedom' of labour under capitalism too literally. Marx's reference to such 'freedom' was ironic: it consists precisely in economic compulsion rather than other types of compulsion. Nevertheless, capitalists do also use political, ideological and legal means of coercion to structure, or to augment, economic compulsion in ways that will deliver the kinds of labour they want on the terms they want (levels of pay, conditions of control and discipline, etc.). This is evident in labour regimes using debt bondage to try to secure a captive and compliant work force, but it also applies to class struggles more generally, including those circumstances in which labour power is 'freely' exchanged through the market.

A similar point applies to the separation of the producers from the means of production as a condition of 'free' wage labour. Semi-proletarian labour is generated within capitalism no less than full proletarian labour. But conditions which produce semi-proletarian labour are usually regarded as 'transitional'; i.e. they are considered to be only part way towards proletarianization in its full sense. In practice, semi-proletarian labour is not necessarily transitory or short-lived, and, along with fully proletarianized labour and petty commodity production, is a general feature of Third World economies today.

Box 9.2   Late examples of forced labour

(a)   Coffee production in colonial and independent Brazil (eighteenth and nineteenth centuries)

Slaves worked from 3 a.m. until 9 or 10 p.m. Even during the rainy season, slaves had to pick coffee at night. One coffee planter calculated upon using a slave no longer than a year, 'longer than which few could survive', but that he got enough work out of him not only to repay his initial investment, but even to show a good profit.

On isolated *fazendas* [estates], amid numerous slaves, planters perceived the precariousness of their situation. Many declared openly 'the slave is our uncompromising enemy' and the enemy had to be restrained and kept working on schedule through fear of punishment, by vigilance and discipline, by forcing him to sleep in locked quarters, by prohibiting communication with slaves of nearby *fazendas*, and by removing all arms from his possession.

Slaves were punished by the whip or *tronco*, a form of heavy iron stock common on plantations. Arms and legs were imprisoned together forcing the victim hunched forward with arms next to ankles, or to lie on one side.

(Stein, 1972, pp.121–2)

(b)   Railway construction in the Belgian Congo, 1920s

'I work on track repairs with a group of fellow villagers. Men work on one side of the track, women on the other. When a woman can't be sold or gets too old, she is made to do more work than a man. In scorching sunshine they carry large stones on their heads, level the ground and drag blocks of marble along, all to the sound of continual sad moaning. There is also a black overseer. The monotonous beating of a drum gives rhythm to the work, but when the music stops, the negro overseer brings down his whip on the shoulders of 50 or 100 male and female workers, passive, weakened and hungry. This is how we build the road to civilization. That's how progress goes. The engine's whistle blows where there was once the silence of the impenetrable forest. But the train runs on the bones of the thousands who died without even knowing what was this progress, in whose name they were made to work.'

(Congolese worker's letter to the International Trade Union Committee of Negro Workers, quoted in Nzula *et al.*, 1979 edition, p.85)

9.3 Colonialism and the `labour problem'

The labour regimes so briefly described here were the outcomes of processes of struggle to impose new structures of economic activity on untold millions of colonized peoples, against their resistance or their attempts to adapt them to their own survival and needs. The 'labour problem' was a preoccupation of colonial officials and interest groups in many different places at different times. Why should labour supply have been so general a 'problem' for colonialism? The answer to this question suggests a number of connected factors.

The first is that the conditions of work in plantations and mines and on large-scale public works like railway construction were extremely bad, and labour was intensively exploited. Secondly, the payment for such work was minimal, sometimes barely enough to enable workers to survive from one day to another (or not even that). Thirdly, this kind of labour withdrew people from their customary farming and other productive activities, although this usually remained an important source of their subsistence and reproduction. But it was not only their productive activities and capacities that were materially undermined: so also were the social and cultural relations that were an integral part of their productive activities. In short, their whole way of life was threatened – their ways of making and doing things, and the values associated with them.

This is particularly striking with respect to practices and ideas connected with *work*. While pre-colonial societies varied widely with respect to types of work (who did it, and how much of it), the development of colonial political economy almost invariably led to an *intensification* of work (see Chapter 3), as well as different (and frequently oppressive) kinds of work, for most of the people subjected to colonial incorporation. The intensification of labour sprang from the material interests of colonial states and settlers, but was also explicitly recommended in their ideologies of the 'civilizing mission' of European rule. In the ideological package of colonialism, hard work could conveniently be considered a moral (and specifically Christian) virtue as well as an imperative for economic progress (i.e. creating wealth for European colonists, capitalists and states).

A Spanish decree on *encomienda* in 1513 stipulated that 'Indians…were to work nine months a year for the Spaniards (i.e. without pay), and were to be compelled to work on their own lands or for the Spaniards for wages in the remaining three months'. The stated intention was '*to prevent them spending their time in indolence and to teach them to live as Christians*' (Kloosterboer, 1960, emphasis added). More than 400 years later, the 1922 *Annual Report* of the Governor-General of the Belgian Congo stated that 'under no circumstances whatsoever should it be permitted to occur that a peasant, who has paid his taxes and other legally required obligations, should be left with nothing to do. The moral authority of the administrator, persuasion, encouragement and other measures should be adopted *to make the native work*' (Nzula *et al.*, 1979, emphasis added).

As these quotations suggest, the other side of the colonial 'labour problem' was the equally widely expressed stereotype of the 'lazy native', which justified the use of coercion to get the natives to do more work as part of their apprenticeship to European 'civilization'. Certainly, people in pre-colonial societies had generally worked less than their colonial masters now wanted them to. First, less labour was required to satisfy their needs (including the needs of their dominant classes), and labour was spread unevenly over the year, according to seasons and modes of interaction with nature. Secondly, time not spent in directly productive work was used in social, ceremonial and creative activities that were an integral part of pre-colonial cultures. It is ironic too – and would hardly have escaped the attention of the 'natives' – that the virtues of intensive manual labour were advocated (and enforced) by groups of Europeans notable for their aversion to it, whether colonial officials, missionary 'princes' of the church, traders, landowners or the managers of mines and plantations.

The labour regimes of slavery, indentured labour and debt bondage presupposed the removal of alternative ways of earning a living. Slaves were forcibly taken from their homes in West Africa and shipped to the New World, where they were totally subordinated to their owners (as were also, of course, those born into slavery in the Americas). Indentured labour was recruited from the pauperized peasants and craftsmen of India and China, dispossessed of land or their traditional means of livelihood. Debt bondage presupposed the loss of (sufficient) land for subsistence production, and the indebtedness that ensued in an increasingly commercialized and monetized economy.

Other types of labour regime were based on the continuing existence of household production, though with varying degrees of viability. Where the surplus was appropriated in the form of a feudal-like tribute or taxes in kind, people were left in control of their land and of the organization of their labour to produce both their own subsistence and the surplus they had to deliver. This was also the case where peasant production of commodities was encouraged (as in India and West Africa), although pressure could be exerted to make peasants produce more for sale through imposing money taxation and paying low prices for their crops.

Those situations where labour on a large scale was demanded required some sort of balance between the amount of labour that could be compelled to work for the colonial state and settlers, and the amount of labour and other resources (e.g. land) necessary to reproduce the households of those performing labour service. The measures used to enforce the supply of labour ranged from straightforward legal compulsion by the colonial state (*repartimiento* in Spanish America, *chibalo* in Southern Africa) to economic compulsion – when cash was needed to pay tax and could not be obtained in any other way. Economic compulsion generally came to replace direct legal coercion as the need for a cash income extended beyond tax payments to include means of consumption which now had to be purchased as commodities (food, clothing, medicines, Western-style schooling) and new manufactures such as household utensils, farming implements, etc.

Colonial authorities could ensure that people had no means of buying goods and services (such as Western education) other than by working for wages. This was done by restricting them to overcrowded Native Reserves (colonial Kenya and the Rhodesias, South Africa's 'Bantustans' today), preventing them from growing cash crops and so on. Such policies effectively undermined what may have been a viable subsistence agriculture, and replaced it with a '*sub*-subsistence' agriculture that necessitated continuing periodic migration to make up for the shortfalls in subsistence. Moreover, as most labour migrants were men, their absence intensified the labour of the women left behind to try to maintain agricultural production.

It is hardly surprising, then, that there was a 'labour problem' under colonialism. This constant preoccupation and complaint of colonial rulers and settlers, and plantation and mining capitalists, further indicates that the making of colonial economies (with all their differences of time and place from sixteenth-century America to twentieth-century Africa) entailed a *common* process of breaking pre-existing ways of production and restructuring the uses of the labour

thus 'released'. The resistance of the colonized to tribute and wage labour was particularly strong, and generally, when it was possible, they preferred what Ranger (1985) has termed the 'peasant option': that is, the attempt to retain at least some control over the uses of their labour in petty commodity production, even with all the pressures to which this too could be subjected.

9.4 Labour and the 'land question'

Given that most people in pre-colonial societies gained their living from the land, the alienation of land to settlers and colonial companies by formal decree or other means (including outright land-grabbing), the restriction of indigenous people to agriculturally marginal (and sooner or later overcrowded) areas, and the competition for land as a commodity generated by commercialization, all contributed in major ways to the 'releasing' and the restructuring of the uses of labour in colonial economies. People were dispossessed of land, or their access to and modes of using land were radically changed. This has already been indicated in relation to debt bondage in Latin America, the recruitment of indentured labour, and processes of proletarianization. Some of the ways in which colonialism created a 'land question' with critical implications for labour, including peasant labour, can be briefly illustrated in the cases of India and Kenya.

India

In India, colonial land policy was seen initially as an aspect of revenue collection rather than as a method of (re)organizing agricultural production. There were no European settlers and peasant farming remained the main form of agricultural production, except in limited areas of north-east and south India and in Ceylon (now Sri Lanka), where tea plantations were introduced during the nineteenth century.

Under the Mogul emperors, the collection of taxes from agricultural producers (peasants) was conducted by an important class of intermediaries known as *zamindars*. They were not landowners, as the concept and practice of private property in land did not exist, but they had the right of control over 'estates', some of which encompassed as many as 800 villages. Each estate formed a hierarchy with the *zamindar* at its top and various levels of 'tenants' and 'subtenants' with different types of claims to the use of land. *Zamindars* collected an annual tax from their estates, based on a proportion of the harvested crop; this was subject to annual negotiation, according to weather and other conditions, and was often paid in kind.

Lord Cornwallis's Permanent Settlement of 1793 was designed to rationalize the collection of taxes via *zamindars* for the treasury of the East India Company, rather than for that of the Moguls. In retaining the role of the *zamindars*, the Permanent Settlement in Bengal conformed to the prevailing idea of non-interference with indigenous social arrangements. However, it contributed to the undermining of these arrangements by conferring on *zamindars* the status of 'land-holders' according to Western ideas of landed property, and by replacing the above flexible tax with an annual tax of a *fixed* amount of money.

These two measures had the most profound consequences, contributing to processes of commercialization of the rural economy. The legal establishment of individual property rights in land meant that, henceforth, it could be bought and sold: land became an exchangeable *commodity*. This allowed the development of a market in land as an object of speculation and profitable investment, yielding rent to landowners. The monetization of taxes (and of economic transactions more generally) pushed the peasantry into producing cash crops for export markets, as British trading and industrial interests wanted (cotton, jute and indigo, and also opium, introduced into Bengal by the East India Company to finance its tea trade with China). At the same time, the introduction of fixed annual taxes to be paid in money (rather than a share of the crop) led to peasant indebtedness in years of bad weather and poor harvests.

For peasants, the tyranny of fixed monetary obligations in relation to the variability of nature and of yields could lead to a cycle of debt-bondage through borrowing to pay taxes (and rents), and ultimately to eviction from the land they farmed. Bad harvests affected not only the peasants' ability to pay *zamindars*, but also the latter's ability to deliver the taxes demanded by the colonial administration. In this situation, the weaker *zamindars* were 'bought out' by the Indian merchants and financiers of Calcutta, while the stronger ones became absentee landowners living in towns, where they tended to merge with the mercantile and money-lending class. The effect of the Permanent Settlement was, therefore, to contribute simultaneously to the impoverishment and 'squeezing' of large numbers of peasants, and to the displacement of a local rural aristocracy in favour of a class of urban absentee landlords, whose principal interest in the countryside was the income they derived from rents and from buying and selling land, rather than investment to enhance agricultural production. This also encouraged the conversion of rents in kind to fixed-money rents, parallel to the introduction of fixed-money taxes by the colonial state.

The Permanent Settlement and its effects provide a good example of a colonial policy which, while preserving the existing *form* of rural society (*zamindars* and peasants), profoundly changed its *content* in terms of property relations, the intensification of peasant labour through commercialization and monetization of the rural economy, and the appropriation of a surplus through taxes and rents.

The same principle, of building administration on an existing indigenous social structure, led in south India to the establishment of the *ryotwari* system, so called because the basis of taxation was conferring land titles on *ryots*. According to Sir Thomas Munro, the architect of this system introduced in Madras in 1812, *ryots* were 'a crowd of men of small but independent property, who, when they are certain that they will themselves enjoy the benefits of every extraordinary exertion of labour, work with a spirit of activity which would in vain be expected from the tenants and servants of great land-holders' (quoted in Low, 1973, p.46).

In Munro's conception (and by contrast with the aristocratic *zamindar* 'landlords' of Bengal), *ryots* appeared as small and 'improving' yeoman farmers on the model of those found in England at the time. In fact, those who benefited from the land titles bestowed by Munro's policy were local leaders and notables, often from powerful castes. With the possession of property titles, many of them were now able to constitute themselves as a landowning class, similar to the *zamindars*, although on a more modest scale.

In effect, landlords used the opportunities presented by the *ryotwari* settlement in ways which had never been envisaged by Munro and the British administration in Madras. Because landlords were the major source of recruitment of local revenue officials, or were allied with locally influential persons, or even chose and paid for village officers, they had opportunities to shift many of their tax liabilities on to others. Madras revenue records provide hundreds of examples of land acquired by revenue officials or through family agents, and the use of torture and coercion in the collection of revenue was tacitly accepted by the administration (Stein, 1977, pp.72–3).

Thus, colonial legislation and administrative practice had profound effects quite unanticipated by the rulers, which often reflected their ignorance of important aspects of the indigenous societies they sought to shape. Together with other aspects of the commercialization of economic life brought about by integration with the international economy and the impact of a developing industrial capitalism in Europe, the colonial state in India contributed to new forms of exploitation of the peasantry, and the formation of new social classes (the absentee landlords of Bengal, the local landlords of Madras, the landless labourers resulting from the dispossession of the peasants).

Kenya

In colonies like South Africa, southern Rhodesia (now Zimbabwe) and Kenya, where large areas of land were alienated for use by European settlers, changes in existing patterns of landholding and land use in the process of the commercialization of agriculture resulted in other types of social differentiation. In Kenya, vast areas of the best land were appropriated by the colonial state and handed over to the white settler farmers and landowners, while Africans were confined to Native Reserves and forbidden to grow cash crops such as coffee or to compete in the same market with food crops grown on settler farms. The 'land question' became very important in Kenyan politics and was a major factor in the Mau Mau revolt of the 1950s.

The taking over of land by conquerors was not a new experience for the main group in Kenya, the Kikuyu. 'What was novel, and produced great bitterness, was that the new white conquerors appeared to suppose that the rights of use conferred on them by victory were exclusive' (Kitching, 1980, p.284). Traditionally, this concept of exclusivity did not exist. Land was abundant, with no fixed property rights in the Western sense; land settlement proceeded by the formation of new groups of cultivators based on kinship, and their movement into new areas. In some places two separate groups of people might use the same land for different purposes – Dorobo hunter/gatherers reaching a mutual agreement to share territory with Kikuyu cultivators, for example.

The alienation of land to Europeans transformed these former methods of land use by placing a rigid restriction on the amount of land available. By the time a Royal Commission on land was appointed in 1931, following decades of land alienation to European settlers, Africans were having to establish their claims to land in terms of Western concepts of 'ownership', 'purchase', 'sale' and 'tenancy'.

Kitching's account describes the way in which, from the mid-1920s onwards, household heads began to dispose of land by sale, or to rent it to tenants outside their own *mbari* (lineage). Colonial observers referred to this as the 'individualization' of tenure, but this again was based on a misconception, since household heads had always had considerable latitude in land use once they were actually in possession of an area of land. The change was not one from communal to individual landholding and use, but from the concept of 'redeemable' to 'irredeemable' sale. This distinction had been of minor importance in a situation of land abundance but became central once land had become a relatively scarce *commodity* due to property titles, land alienation to settlers, and the increasing commercialization of agricultural production.

Once the concept of irredeemable sale had replaced the traditional concept of redeemability, sales of land increased, and a significant number of African 'land accumulators' began to emerge. Most sellers of land sold out of distress, to meet essential subsistence and other expenses. The increasing inequality of land ownership, combined with population pressure in Native Reserves with fixed boundaries, placed small landowning households in a vulnerable position. There were two additional important factors. First, the transition from a livestock to an agricultural economy gave land accumulators the opportunity to buy land from others still oriented to pre-colonial notions of wealth in cattle, who used the proceeds of land sales to acquire livestock, which they could graze as squatters on European farms. Secondly, the 'chiefs' and headmen could use their positions to get land from others, by direct and indirect coercion.

Kitching's exhaustive historical study on Kenya brings out the way in which changes in land tenure and land use and the beginning of land accumulation were instrumental in the creation of new social classes. On one hand, there was an embryonic African bourgeoisie which even during the colonial period was able to use its links with the colonial system (through mission education, government employment, headmanships, etc.) to increase income, saving and investment in land and business. On the other hand, a poor landless class developed, unable to take advantage even of the opportunities presented by the process of land reallocation after independence in 1963, and which came to form the basis of both a rural and an urban proletariat.

9.5 Colonialism and social changes

It is important to remember that restructuring the uses of labour was only one element, albeit a fundamental one, in the interlinked processes of change under colonialism. Chapter 10 addresses mainly political dimensions of change; here four related aspects of social and cultural change are summarized.

First, the colonial experience involved resistance and adaptation (sometimes combined) by colonial peoples to the changes imposed on them. Two major instances in the sphere of cultural change concern the introduction of Western education and of the Christian religion (sometimes closely connected, as in most of Africa). Both illustrate contradictions of colonial rule, and the impossibility of ensuring its effective legitimacy over the colonized. Western

Figure 9.3

education was introduced to train people for the lower ranks of the colonial civil service (as clerks, medical assistants, teachers) but those who acquired literacy in Western languages were able to continue their education beyond the limits set by their colonial (often missionary) teachers. They were able to articulate their resistance to foreign domination through turning Western principles of democracy and justice (and sometimes the vocabulary of socialism) against their colonial masters.

Similarly, while Christianity was a central element of Western imperialism's ideology of its 'civilizing mission', and missionaries often functioned as informal agents of the colonial state, the meaning of Christianity could be assimilated and interpreted in different ways. It could facilitate the acceptance of colonial rule by preaching the virtue of hard work, sobriety and due deference to authority, both spiritual and temporal. On the other hand, its message of universal brotherhood and equality in the sight of God could be used to criticize the inherent racial oppression and inequality of colonial society.

Second, responses to colonial incorporation included initiatives and innovations by those who were colonized (often using their ability to draw on aspects of their culture and social organization of which colonial authorities were ignorant, or which they misunderstood). Within the (varying) constraints set by different forms of economic domination, some of the colonized became entrepreneurs and were able to accumulate through trade, land grabbing and renting, agriculture and transport. In the sphere of religion,

in Africa many 'native' churches developed in opposition to the European monopoly of Christianity (as also happened among the black populations of the Caribbean and the USA, themselves descended from African slaves). Whether overtly resistant to colonial rule in their teachings or not, these independent churches were necessarily subversive of the ideology that tried to justify European domination.

Third, colonial society was marked above all by the ethnic divisions of labour, of legal status, political influence and social standing, between colonizers and colonized, justified by ideologies of European racial superiority. This was a potent factor contributing to the unity of anti-colonial movements, overriding (at least temporarily) many of the differences emerging among the colonized people themselves. However, such unity could be fragile and subject to intense strains following independence from colonial rule (decolonization) (Figure 9.3).

The final point is that the fundamental racial differentiation of colonial society could obscure the developing social differentiation among the colonized. The examples of the land question in India and Kenya provided glimpses of class formation and differentiation. These processes were often abetted, whether intentionally or unintentionally, by the policies and practices of colonial states. These included strategies of 'divide and rule' (contributing to the potentially explosive combination of extreme regional economic and social inequality with distinct cultural identities, including language and religion), incorporating and reconstituting ruling groups from pre-colonial society within the

hierarchy of the colonial order (giving offices to chiefs and princes and educating them, granting land and tax offices), and conducting other experiments in 'planned' class formation. For example, the Tanganyika Agricultural Corporation was set up by the colonial government in 1953 to promote 'a healthy, prosperous yeoman farmer class, firmly established on the land, appreciative of its fruits, jealous of its inherent wealth, and dedicated to maintaining the family unit on it' (quoted in Cliffe & Cunningham, 1973, p.134). This was partly a response to events in Kenya (Mau Mau), and partly an attempt to create social groups which, it was hoped, would underwrite social stability and friendliness towards the former colonial power (and the West more generally) following the inevitable moment of political independence.

In 1961, President Nyerere expressed the hopes of that moment of independence: 'This day has dawned because the people of Tanganyika have worked together in unity... from now on we are fighting not man but nature.' In quoting these words at the end of his book on his history of Tanganyika, Iliffe (1979) noted 'but it was more complicated than that.' It proved to be more complicated precisely because the end of colonialism was not the end of capitalism. Not only did Third World countries still have to confront the unequal structures of the capitalist world market and international division of labour in efforts to achieve economic development, but the contradictory social relations and divisions of capitalism were now as much part of their societies as of those societies in which capitalism had its origins.

Summary

1   The organization of commodity production (usually for the world market) by colonial states, settlers and companies, entailed the 'breaking' of pre-existing patterns of economic and social life, through the introduction of new labour regimes (systems of recruiting and organizing labour).

2   All colonial labour regimes involved an intensification of labour compared with pre-colonial modes of organizing labour in production.

3   Particularly in the earlier stages of colonialism in Latin America and Africa, compulsory means were employed to obtain a supply of labour for plantations, mines and settler estates.

4   Direct coercion of labour, characteristic above all of 'primitive' accumulation, tended to give way to semi-proletarian and proletarian labour in the nineteenth and twentieth centuries, along with the continuation of household labour in petty commodity production.

5   Colonial legislation and administration concerning money taxation, property titles in land and the commercialization of agriculture linked to the international economy radically altered the nature of rural social relations, and caused the formation of new social classes among the indigenous population.

6   The kinds of commercialization and monetization of economic life introduced with colonialism survived the end of direct foreign rule, and have continued to develop in independent Third World countries.

7   The colonial experience involved adaptation and resistance to cultural as well as political and economic changes.

8   The fundamental type of social differentiation in colonial societies was between the colonizers and the colonized; at the same time, changes introduced under colonialism led to new forms of social differentiation among the colonized people, which became particularly important after the achievement of independence from colonial rule.

10

COLONIAL RULE

DAVID POTTER

One of the things that has conventionally distinguished the Third World from the rest of the world is that nearly all of it experienced European colonial rule. Some of the reasons for the Europeans going to the Americas, Asia and Africa have already been mentioned in Chapter 8. They include the economics of the expansion of Europe, the development of capitalism, and the creation of a world market and international division of labour. The colonialists, once established, were gradually drawn to control the area they had conquered or in which they had their investments. Eventually these political interventions resulted in the formation of what came to be called colonial states.

Colonial states gradually ripped apart pre-existing economic and social relations and created colonial economies, including various labour regimes as discussed in Chapter 9. These and other actions of such states were more or less resisted by those on whom they were imposed. The result was that colonial rule was never easily created and maintained. In the end, the problem of rule became insuperable and the European colonialists withdrew or were driven out.

In this chapter, as in the preceding chapters, colonialism refers to European colonialism between the sixteenth and twentieth centuries, which developed into the global system of imperialism discussed in Chapter 8.

To speak of colonial rule and the colonial state in the singular suggests that there was some standard form and practice in all European colonies. No such standard model ever existed. As Chapter 8 has indicated, the forms and practices of colonial rule varied from colony to colony because of (a) differences in the starting and ending points of any particular rule, (b) the particular experience of initially imposing that rule, (c) the great diversity of indigenous societies on which that particular rule rested, and (d) the extent to which colonial settlers were prevalent.

It is not my intention in this chapter to try to discuss colonial rule generally nor to attempt an overall assessment of its economic, political and social consequences. My main concern is with an essentially political question: how was colonial rule maintained?

Q How was it possible for so few Europeans to hold down so many colonial peoples?

I shall argue in this chapter that the beginnings of a useful answer to this question can be found through an understanding of seven political features that all colonial states shared to some extent, at least in Asia and Africa: (1) an international political dimension, (2) bureaucratic élitism and authoritarianism, (3) statism, (4) use of 'traditional' authority figures, (5) use of force, (6) technological advantage, (7) hegemonic ideology.

The subject matter of this chapter is, of course, enormous. To make it a little less unmanageable, I have focused the discussion primarily on colonial rule in India. The chapter opens with a section tracing the history of colonial rule there. This provides a basis for a somewhat more general discussion of the seven features of colonial states based on Indian and African evidence. Towards the end of the chapter I comment briefly on certain aspects of why European colonial rule ended when it did and remark on several legacies of colonial rule for post-colonial states.

10.1 Colonial rule in India: a thumbnail sketch

When representatives of the British East India Company first arrived in India at the beginning of the seventeenth century to commence trading, they sought permission from Indian rulers to establish 'factories' at ports like Surat, Calcutta and Madras to hold and process goods in readiness for the arrival of the next ship from Europe. These tiny enclaves eventually had presidents, councils, administrators and small military and police establishments. Their survival depended on the support of nearby Indian political and economic institutions which controlled the surrounding countryside.

The character of British power in India was gradually transformed as a result of two main phases of territorial expansion by force of arms: between 1740 and 1760, and between 1792 and 1818 (for a general explanation of this transformation, see Bayly, 1989). The Company acquired vast tracts of populous and productive land, and land revenue began to replace trade as one of the most important sources of the Company's income. At first, the Company's officials in the mid-eighteenth century relied on the administrative organizations they inherited to collect the revenue for them and maintain a semblance of order in the countryside. In this,

they were behaving roughly in accordance with Indian political tradition, where state power rarely implied direct territorial control. The normal mechanism of state formation in India in the past had usually been through the forging of alliances between a ruler and petty princes, priests, and other intermediaries with considerable power locally (Wink, 1986; Stein, 1990). Ceremonial gift-exchange and acts of conspicuous piety helped to cement such alliances (Dirks, 1988).

Company officials soon realized that they knew little about the agrarian system they had acquired and the shifting alliances and rivalries on which it was based. They were losing considerable revenue because of their ignorance, and indeed in many places they found it almost impossible to agree any tax settlement with their new 'allies'. Company officials began to press for revenue arrangements with which they were more familiar, so that more revenue could be obtained and closer control ensured. This began to happen first under Warren Hastings in Bengal in the 1770s; for example, European collectors were posted in each revenue district. Lord Cornwallis in Bengal went further in the 1790s with the 'Permanent Settlement' converting the *zamindars* there into something like English landed aristocrats (see Chapter 9) and at the same time altering the structure of administration and excluding Indians from senior positions in it. In parts of Madras, revenue settlements were made directly with *ryots*, the actual producers on the land. During the early decades of the nineteenth century, variations on the *ryotwari* revenue arrangements were established throughout Madras and elsewhere in British India. The Company began to press down hard on the agrarian order in its endeavour to extract enough revenue to satisfy the directors in London and to pay for the insatiable demands of the Company's army which was engaged in seizing an empire in India. The land revenue system was at the heart of Company rule: 40–50% of the Company's receipts came from this source.

On the more commercial side of its operations, the mercantilist character of the Company's mode of thinking about the economy led to the Company taking direct responsibility for the organization of production in many of the most valuable areas of commerce, such as coffee, indigo, opium, sisal.

The shift from colonial rule by commercial enterprise to rule by a far-flung colonial bureaucracy coincided with Britain's industrial revolution. The Company lost its trade monopoly with India in 1813, lost its trading connections altogether in 1833, and was replaced in 1858 by a Government of India directly responsible to the Crown (i.e. the British Cabinet) in London. As this was happening, India was increasingly perceived 'as a financial asset of enormous potential value to Britain's industrial economy – a captive market for its manufacturers, a source of foodstuffs as well as of cotton, jute and other industrial raw materials, and, in its railways, plantations and public utilities a dependable recipient and multiplier of capital investment' (Arnold, 1986, p.12). To provide a favourable 'law and order' context for this 'enormous financial asset' became the principal purpose of the colonial state. The army was too crude an instrument for this purpose, so new police forces were set up in the provinces of British India in the 1850s, modelled on the authoritarian formation of the British police in nineteenth-century Ireland.

There was no way that the British could control their vast empire in India in the latter part of the nineteenth century (Figure 10.1) by using only the police and the army. No government lasts long on the basis of force alone. The British had to have people on whom to rest their rule. They found that essential political support primarily amongst princes in the princely states (occupying about 40% of the area of India in the mid-nineteenth century) and the landlords in the provinces of British India.

To the extent that political legitimacy attached to any person in a princely state, it attached traditionally to the prince or *maharaja*. Since the prince controlled his people, and the British controlled and preserved the prince, the arrangement amounted to an important political asset for British rule. This was demonstrated

during the Indian mutiny or revolt of 1857–58, when most princes remained at least neutral. The British rewarded them afterwards by preserving these princely states (or most of them) until 1947 (Figure 10.2).

In the provinces of British India, the *raj* (rule) rested on landlords for the most part; for example, in the United Provinces (or 'North-western Provinces and Oudh') in North India in the mid-nineteenth century:

> "Government needed the support of those groups in the province that were powerful and therefore potentially dangerous. Over 80% of the provinces' population lived off the land, over 50% of its revenue came from the land; it was logical that government should seek the support of those with social and political power on the land: the landlords... the policy of the British in the U.P. looked to the landlords as the main prop of their rule.
>
> Government's technique was simple. It doled out patronage, disbursed contracts, supported educational projects, bestowed a myriad of honours, consulted allies and did its best to make them strong."
>
> (Robinson, 1971, pp.10–12)

The logic of the government's political strategy *vis-à-vis* the landlords was essentially the conservative one of preserving the agrarian order. This was coupled with the fear 'that if competitive capitalist relations were allowed freedom to take over the countryside, the resulting conflict would destroy the *raj*', which was 'paranoid' about 'the prospects of an active Indian capitalism, gnawing at the bases of the agrarian order' (Washbrook, 1981, pp.684–5, 690). The upshot was that the colonial state failed to direct the economic surplus being generated in the Indian countryside towards industrial growth. The consequences of this were to stifle the development of an urban bourgeoisie and to stunt industrial development in India.

One example of government policy which dampened the growth and development of Indian industry in the nineteenth century was the

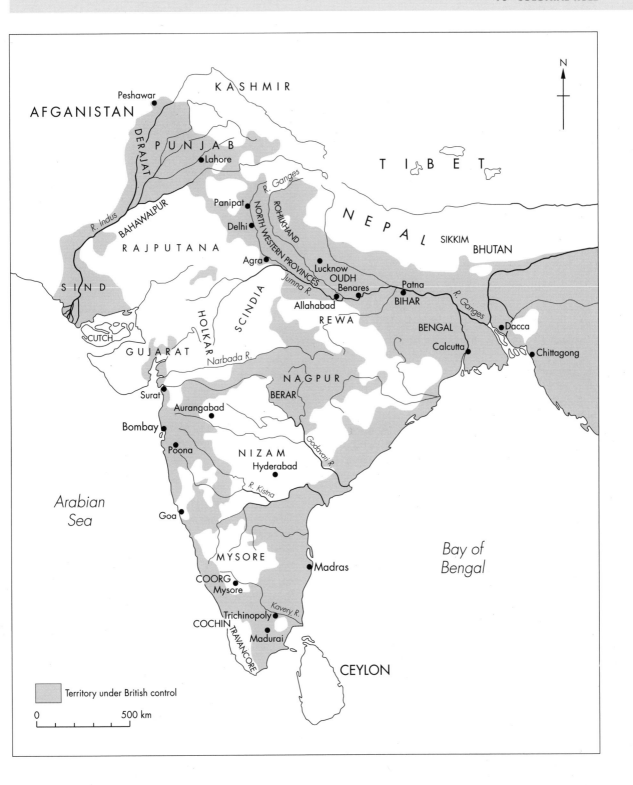

Figure 10.1 India in 1857, indicating territorial acquisition by the British at that time.

Figure 10.2   Preserving the prince. Durbar procession of Akbar II, Emperor of Delhi, accompanied by the real rulers in top hats. Delhi school, c.1815.

restrictions placed by the Secretary of State for India in the British Cabinet on government purchase of locally manufactured machinery, iron, steel, tools and plant. Such government support was important in getting industrialization off the ground in other countries, but was not available in India.

Another example of a government policy detrimental to India's industrialization was 'free trade'. During the period 1851–1900, 'other nations were able to develop their industries under the umbrella of protective tariffs, but the Government of India (or rather the Secretary of State for India) stubbornly maintained a policy of *laissez-faire*, prepared neither to impose protective tariffs nor actively to encourage domestic industries in other ways' (Rungta, 1970, p.69). Such a policy, however, was advantageous to capitalist interests and industrialization in Britain.

During the latter part of the nineteenth century, recurring financial crises propelled the *raj* to introduce local government institutions in municipalities and rural districts in order to increase revenue by increasing the tax base. This decision had important consequences. More taxation meant more representation. It also meant in this case the opening of political office to elections (although on a very limited franchise). The consequences of this were that additional avenues of political advancement were opened

up to aspiring social groups. Indian society got stirred up. Agitations occurred. Force was used to quell these (Figure 10.3), but the government also took more positive steps to counter such agitations. Increased consultation with Indian political leaders was the government's method. To provide that consultation, government created new political institutions through which such consultations could take place. More people, consequently, began to participate in political life. As political mobilization occurred, enhanced by the spread of education and the introduction of new means of communication, areas of political association widened. This widening process led eventually to the formation of the Indian National Congress in 1886. It was through this political organization that, subsequently, the main thrust of India's nationalist movement was organized.

During the final phase of colonial rule, the gradual extension of representative institutions and the franchise continued. This process reached its climax in the Government of India Act, 1935, which provided for popularly elected assemblies in the provinces of British India and other institutional reforms. Indian political leaders therefore had some experience of working in (semi) democratic institutions before the end of colonial rule.

The main opposition to colonial rule in the twentieth century was orchestrated through the

JUSTICE.

Figure 10.3 Agitation and force: Punch showing Britannia meting out 'justice' to Indian mutineers.

Indian National Congress. Initially an élite institution, the Congress later became a capacious, umbrella-like political organization representing a broad spectrum of social groups that extended from the Westernized and mildly radical intellectuals to sections of the business community and to politically active sections of the peasantry. To put the structural positions (too) simply, urban bourgeoisie and peasant leaders joined in opposition to landed upper classes and colonial bureaucracy. The great mass of the peasantry (including landless peasants) were not involved directly and consistently in the nationalist movement, but many did get caught up in it towards the end – a development that distinguished India from other colonial rules (and became important later with India's comparative success with democratic politics). The political programme that helped to forge this curious political connection between urban bourgeoisie and peasant interests was the work principally of M. K. Gandhi, the leading figure in the Indian National Congress in the 1920s and 1930s.

Key elements in the programme of the Congress were the political goal of independence *(swaraj)* from foreign rule and various economic and social programmes of peasant uplift. Gandhi's programme managed to hold out a promise to the peasantry while at the same time not threatening the interests of the urban bourgeoisie. Gandhi's insistence on the use of the political technique of *satyagraha* (truth-force), which emphasized the importance of non-violence in pursuit of economic and social objectives, also helped to keep the movement safe for the Indian bourgeoisie and others with property. Independence from foreign rule had in the 1930s become profoundly important to Indian business interests, which felt severely cramped by the imperial order. Since the Indian National Congress wanted to free India from the shackles of colonial rule, Indian business in due course began actively to support the Congress. Indeed, the Congress became to some extent dependent from the 1930s on the financial support provided by business houses.

What tends to get obscured in all the attention given to Gandhi and the nationalist movement

is that the *raj* held firm, despite nationalist pressures, and continued to rule and have an enormous impact on the politics of the day until the very end. By 1942, for example, at a time when Britain had its back to the wall in Europe and on India's border with Burma, the *raj* still managed to quell the Quit India movement, the most serious and violent disorder in India since the Mutiny or revolt of 1857–58. It is therefore an oversimplification to say that the Indian nationalist movements (Congress and Muslim League) forced the British out. The British finally left for several different reasons – the consequences of the nationalist struggle, the crumbling of support structures in India, and exhaustion from the Second World War and other causes at home in Britain.

10.2   Seven features of colonial rule

I want now to draw attention to the seven political features of colonial rule listed at the beginning of this chapter, making use of the preceding sketch about India but also referring to several other colonial rules in Africa.

International political dimension

First, all colonial rules involved political direction and control of a subject society ultimately by a foreign power, be it a Board of Directors (like the East India Company) or a government organization (like the British Cabinet). This international dimension has been frequently forgotten by students of colonial rule who have tended to be preoccupied with how rule was maintained locally. The thumbnail sketch of India illustrates how important such foreign direction and control could be. For example, economic policy in India in the latter part of the nineteenth century was shaped by directives from London, to ensure that local decisions and rules worked to the advantage of capitalist interests and industrialization in Britain. These directives included restrictions placed by London on purchase in India of locally manufactured goods and refusal to allow protective tariffs in India to assist the growth of Indian industry. More broadly, the economic surplus being generated in the Indian countryside was not directed toward industrial growth in India because of government policies determined in London.

Foreign rule also had the consequence of there being little resource to assist agricultural development. In the 1870s, departments of agriculture were set up in the provinces of British India, but they were always desperately short of money. Nearly all the land revenue was consumed by the high costs of maintaining the Government of India and the imperial system. The land revenue helped to meet the costs of the civil service, the army, the police, the Afghan wars, the magnificent public buildings, the pomp that attended the Viceroy and his staff, and so on. In addition to the actual costs of the government in India, there were various other expenses in Britain and elsewhere that were met from the Indian revenues. For example (Whitcombe, 1972), every government item in London that remotely related to India, from the cost of training Indian regiments in preparation for duty in India to the fees of the charwomen in the India Office at Whitehall, were charged to the Indian account. Indian peasants, as the original source of the government's land revenue, paid among other things for a lunatic asylum in Ealing, gifts to members of the Zanzibar mission, the consular and diplomatic establishments of Great Britain in China and Persia, part of the permanent expenses of the Mediterranean fleet and the entire cost of the telegraph line from England to India. When the Sultan of Turkey made a state visit to London in 1868, his official ball was held at the India Office and the bill charged to India. Little was left after all this for state support for agricultural improvement in Indian provinces.

The international dimension must not be conceived narrowly in terms only of direction from the British Cabinet in London. The Cabinet responded to what they conceived to be the interests of powerful classes in British society. Colonial rulers in India also were alive to the importance of these interests in shaping the character of their rule. For example, the Viceroy stated publicly in Calcutta in 1888 that he considered the prime duty of his Government to be to watch 'over the enormous commercial interests of the mother country', and that 'it would be criminal to ignore the responsibility of the Government towards those who have…invested their capital in [India]' (cited in Misra, 1976, p.90).

Bearing in mind this international dimension, it is useful to see colonial rule as operating on different levels: the levels of the metropolitan power (e.g. London), the colonial government (e.g. Delhi), the provincial government (where colonies had provinces) and the local district presided over by a District Officer. Complex relationships operated between levels. Subordinate levels could have considerable autonomy from time to time, and disagreements did occur between levels in this structure of rule. Normally, however, in important decisions to do with vital imperial interests the views of the metropolitan power prevailed.

All colonial rules had an international dimension of the sort indicated here, from Spanish and Portuguese rule in Latin America, to the various European rules in Africa and Asia, to American rule in the Philippines and Japanese colonial rule in Korea. The politics of the metropolitan power need to be constantly borne in mind when trying to understand the dynamics of any colonial rule.

Bureaucratic élitism and authoritarianism

Most of the people who did the work of the colonial state in India were Indians. Although many civil servants and policemen and soldiers were needed to govern and control a subcontinent the size of Europe, containing about 20% of the world's population, few were recruited in Britain. The rest had to be found cheaply in India. Pre-eminent among the administrators sent from Britain were the men (women were not allowed) of the Indian Civil Service (ICS), trusted agents of the British Government placed in posts specially reserved for them in the districts and secretariats in each province and the secretariat of the Government of India. In the secretariats the ICS was responsible for handling questions of policy arising in India in a manner loosely consistent with the economic and strategic interests of the British Government. In the districts, even while overseeing the routine work of Indian subordinates collecting revenue and maintaining law and order, ICS men were inevitably engaged in political work with local Indian collaborators and others, nursing support structures and moving for advantage in fluid situations while never losing sight of imperial aims and requirements (Potter, 1986).

The colonial state in India was bureaucratic, at least from the beginning of the twentieth century, in the sense that the state apparatus was made up of large-scale organizations characterized by hierarchy, continuity, impersonality (in principle at least) and expertise (in the sense that most colonial officials had to have some qualification and/or minimal training). It was bureaucratic **élitist** in that most of the key positions in the bureaucracy at all levels of the state were reserved for a miniscule group of political administrators in the ICS (there were only about 1000 ICS officers in a bureaucracy of several million); all political and administrative decisions of consequence were referred to the ICS (and if need be, back to London); young British (and a few Indian) men were recruited each year in Britain and sent out to India, and only young ICS officers had a clear career run in such key positions to the top of the colonial bureaucracy, commanding exceedingly handsome salaries most of the way. No wonder that British rule was referred to as a 'Rolls-Royce administration in a bullock-cart country' (Schiff, 1939, p.145). Political control by a bureaucratic élite in India under the broad rule of a foreign power is an instance of **authoritarianism** in that participation in government decision-making by those being governed was virtually non-existent or, towards the end of the *raj*, quite limited. This was so even though more popular participation in government was allowed in India than in most other colonies, especially under the provisions of the Government of India Act, 1935, when popularly elected ministries (elected on a restricted franchise) governed

during 1937–39 in the provinces of British India under the watchful eyes of the provincial Governors and the Government of India.

> **Elite:** A small group within the state or other organization which has disproportionate power over important decisions.
>
> **Authoritarianism:** Non-existent or very limited participation in government decision-making by those being governed.

Colonial rule was marked by élitism and authoritarianism (Figure 10.4). A few foreigners were sent out to control key positions in the colonial state; in African colonies, for example, they 'enjoyed very wide powers without brakes from below' (Crowder, 1987, p.15). The Belgians allowed hardly any African participation in the Congo, the Portuguese none at all in Angola, and 'even the more liberal systems such as those developed by the British in West Africa had allowed only limited participation in government, and this itself was confined to a small élite' (Gann & Duignan, 1967, p.331).

Statism

Colonial rule in India was **statist** in the sense that the *raj* intervened far more in economic life than the state did in Britain's political economy.

Figure 10.4   Elitist authoritarianism. Nineteenth-century wooden model of an Indian court, or cutchery, with a British official in the chair.

The thumbnail sketch indicated that the mercantilist East India Company had monopoly control over many important areas of commerce. Later, even when British rule backed away from such monopoly control, the state nevertheless had exclusive control of most of the economic infrastructure. It owned the railways, controlled external trade, favoured foreign firms, and more generally imposed an array of laws and regulations which had the consequence of stifling industrial development in India to the advantage of industrialization in Britain. Even towards the end of the *raj*, from 1900 to 1939, 'the emergence of Indian entrepreneurship in most parts of India was systematically discouraged by the political, administrative and financial arrangements maintained by the British rulers' (Bagchi, 1972, p.423). The *raj* was statist not because the bureaucracy was gigantic (it was actually quite small in relation to the population) but because its command over the economy was fairly comprehensive.

> **Statism:** Comprehensive (although not total) command over the economy by the state.

Colonial rules generally were more or less statist, although there were a few exceptions, e.g. Hong Kong in the early 1990s. African colonial states tended to be very statist. The usual colonial economic controls and state ownership obtained, but a special aspect of statism in Africa was the establishment of state monopoly control over the purchase and export of agricultural goods through various statutory marketing agencies or boards, which meant that the colonial state controlled the principal source of cash income in the economy (Bates, 1981).

Use of 'traditional' authority figures

In the early stages of colonial rule the Europeans were on the lookout for 'traditional' authority figures in society on whom they could rely for political support and who could bring to the new state the cloak of legitimacy. Where they could not find authority figures, they created them.

The thumbnail sketch shows that in India the British found these authority figures, so important to the success of their rule, among the *rajas* in the princely states and among the *zamindars* and other big landlords in what later became the provinces of British India. Similar alliances were forged in Africa, where colonial officials were constantly on the lookout for chiefs who were traditional authority figures there. Establishing and then maintaining such alliances was never easy. The politics of these relationships kept constantly changing. In Uganda, for example, when the British moved to establish authority over the African kingdoms there between 1890 and 1900, they made alliances with a number of younger African chiefs, who were in opposition to a generation of older chiefs. Forty years later, a new generation of colonial officials backed older chiefs against younger chiefs, with the consequence that the authority of colonial rule ran into some difficulty and rural police stations had to be placed throughout the colony for the first time, thereby affecting the structure of the colonial state (Low, 1973, pp.19–20). Similar stories involving chiefs, landlords and other traditional authority figures feature in the history of all colonial rules.

The main reason why it was so important to forge and then maintain alliances with traditional authority figures was that, as long as such political support was maintained, the colonial rules could be fairly sure of controlling the countryside. In India, for example, the landed groups' 'influence over the entire agrarian base was strong…through influence over access to land, through provision of employment and credit opportunities, through traditional ideologies of deference, and through often extensive connections of caste and kinship' (Washbrook, 1981, p.688). One example of the usefulness of local landlords to colonial rulers and of the power they could exercise within their domain must suffice here. When Mohammed Raza was the ICS head of Larkhana District in Sind Province in the early 1940s, he found that the presence of the principal *jagirdar* (landlord) there 'was worth six police stations' because 'whenever my police were unable to trace an absconder and reported that he had moved to the Ghaidero Jagir [the *jagirdar*'s estate], all that I had to do was to write a letter to the [*jagirdar*] and the absconder was handed over to us within 48 hours!' (Raza, undated). In due course, a knighthood was conferred on the *jagirdar* for his services in helping to maintain colonial rule.

Not all landlords supported the *raj* in this way. In Bombay Province, for example, a confidential *zamindars*' book was kept at each district headquarters and ICS District Collectors recorded for the next ICS Collector their opinions about *zamindars* and others in the area on whose support they could rely, and also the names of prominent people who were not helpful. Collectors could make life difficult for the latter. Raza gives an example of what is reputedly an entry for one *zamindar* who fell foul of successive Collectors in one district, as recorded in the *zamindars*' book:

"First Collector: 'This *zamindar* is a scoundrel. He needs to be crushed'.

Second Collector (5 years later): 'I have crushed him'.

Third Collector (5 years later): 'Found him crushed'."

(Raza, undated, p.113)

Crushing opponents and rewarding powerful supporters was at the heart of the political work done by colonial officials.

Use of force

The history of colonialism is one of extension of political control by force of arms and 'crushing' opposition. For example, British history is studded with the names of colonial wars – Maratha Wars, Sikh Wars, Burmese Wars, Afghan Wars, Zulu Wars, Ashanti Wars, Matabele Wars. And this list does not reflect the active resistance of countless groups whose names never reached the imperial history books. This resistance took many forms, ranging from hit-and-run tactics, kept up by the coastal forest people of Côte d'Ivoire for almost 30 years, to the sophisticated and well-armed war waged by Samori Touré against the French in West Africa (Markowitz, 1977, pp.49–50). In many cases hostility was sustained over a period of years, but ultimately it foundered against the military superiority of the Europeans (Figure 10.5).

Figure 10.5 Military superiority. H. M. Stanley with the Maxim Automatic Machine Gun: the Emin Pasha relief expedition, 1887.

Later on, the military, having performed the task of conquest, retired to the background, to be used only in emergencies. Colonial police establishments began to be formed to do most of the routine work of maintaining order. Troops were too expensive to be used for maintaining law and order on a regular basis and they were rather inflexible as regards the degree of coercion used. Troops were also not well suited to guarding property or detecting and preventing crime. The civilian police were more suited to this work, although they always retained an exceptional capacity to use force through the development of sizeable armed and paramilitary units within their organizations.

Two important features of colonial policing are brought out clearly in the standard work on the police in Madras Province (Arnold, 1986). First, force and violence were used extensively as a regular feature of police practice: 'a contempt for Indian lives (in marked contrast to the special protective role adopted towards Europeans) and a belief in the positive value of displaying and deploying armed force were basic and enduring elements in the psychology and practice of colonial control' and 'with repeated use, the frequent resort to high levels of state violence became sanctioned by departmental custom and entrenched in police procedures and mentality' (p.233). People in the towns on the receiving end of colonial policing were treated in a manner that would have been quite unacceptable in England. As for the villages, the Madras police (indeed all colonial police forces in India) were not large enough to provide a permanent police presence there, so when police intervention in a village came it tended to be 'abrupt, brutal and partisan' (p.114); invariably, it was landlords or other dominant villagers who called in the police, not the poor.

This last point relates to the second main feature of colonial policing in Madras Province – 'its close identification with the propertied classes…, both European and Indian' (Arnold, 1986, p.234). Such partisan policing served:

> "to frustrate or negate such coercive power as the rural and urban poor possessed and

to enable employers, traders and landlords to resist demands for higher wages and lower prices to which they might otherwise have been obliged to succumb. While police intervention shielded the propertied classes from direct attack and left them secure in their pursuit of profit and the exploitation of wage labour, it also contributed to the brutalization of emerging class conflict in India."

(Arnold, 1986, p.235)

Such partisan policing created a fund of bitterness against the police among ordinary people in urban and rural India.

Colonies may have differed in the extent to which the army or the police were used, but they were alike in not flinching from the widespread use of state violence and repression as a central characteristic of their rule. Throughout Africa, it has been said that 'the colonial state was conceived in violence rather than by negotiation', and 'it was maintained by the free use of it' (Crowder, 1987, p.11).

Technological advantage

The use of force usually involved 'tools' and effective organization. Headrick (1988, pp.5–6) claims that a major reason why the conquest of Asia and especially Africa was so swift was the technological advantage of the Europeans. This included the continual improvement being made in the organized use of firearms: 'muskets and machine guns gave small European-led units an overwhelming advantage over their African and Asian enemies'. Another advantage was the discovery and efficient use of quinine as a prophylactic against malaria: the extremely high death rate among Europeans due to malaria was reduced, especially in West Africa. Two other technological achievements of major importance were telegraph cables and railroads. These tools made it easier for Europeans 'to control their newly acquired colonies efficiently.' The telegraph could (among other things) be used by local district officers to alert superiors to incipient revolt, and the railroads (or riverboats in places like lowland Bengal) could then be used to bring troops and armed police swiftly to the local area (Figure 10.6).

Railroads and steamships were not only used to advantage by the Europeans to police colonies. They also enabled bulky commodities and new materials to be transported from the colonies to feed the factories and breakfast tables of Europe.

The role of technology and organization in the expansion of Europe was not a simple one of cause and effect, with technologically more advanced societies inevitably overcoming more backward (and by implication 'inferior') ones. The relationship was circular rather than linear, in that the technological advance of Western Europe to some extent required and depended on the subjection of other societies to European control. Nevertheless, both the technological supremacy and the ideology of superiority which it helped to engender were important assets in the struggle to maintain colonial rule.

Hegemonic ideology

British rule in India relied to some extent on the acquiescence by Indian people in a belief system or ideology about the nature of the colonial state. Two clusters of ideas were central to the ideology of the *raj*. One was that the British in India were benevolent, just, and gradually 'modernizing' or developing India economically, socially and culturally. The second cluster of ideas centred on the belief that the colonial rulers were invincible, it was futile to oppose them, and that Indians were too weak and disunited to oppose them successfully. These principal ideas about the benevolence and invincibility of British rule were carefully nurtured by the British, and to the extent that this ideology entered into the hearts and minds of Indian people it was an important asset for the British in their struggle to maintain their rule. Even during the heyday of the *raj* in the late nineteenth century, such ideas were never totally dominant, but they *were* the leading or hegemonic ideas carefully promulgated and reproduced by the state. As with all **hegemonic ideologies**, at least as important as the main interconnected ideas *in* the ideology were the ideas that were organized *out*. This is the context for a remark by an Indian historian that 'a major objective of the hegemonic colonial ideology was to hide the face of the real enemy – colonialism – that is, to hide the primary contradiction between the interests of the Indian people and colonialism' (Chandra, 1989, p.507).

Figure 10.6   Technological advantage. French soldiers on the River Congo, 1914.

Hegemonic ideology: The dominant or ruling set of ideas in a society which is reinforced regularly by the state as part of a process of legitimation supporting the continuation of the existing political regime.

The importance of ideology in the maintenance of the *raj* is perhaps underlined by the fact that Indian nationalist leaders saw ideological work as the most important element in their political strategy for ending British rule. The idea that British rule was benevolent was repeatedly undermined by nationalist politicians in public statements forcefully demonstrating the fundamental contradiction in the colonial situation between the interests of the *raj* and the interests of the Indian people. The idea that British rule was invincible was challenged by the law-breaking mass movements led by Gandhi and others – movements whose basic objective was to show that British rule could be challenged and to build up confidence and courage among the people so that they could develop the capacity to struggle successfully against that rule.

Nationalist leaders were almost certainly right to see that building up confidence and courage was extremely important for people who had been for generations subject to colonial rule, the consequences of which had led to what has been referred to as widespread cultural and psychological dependency. Classic studies have addressed this idea. Worsley (1967, p.29) pointed out that it was 'the internalization and acceptance of the total superiority of European culture, not force alone, that was to lead the non-European in lengthy psychological subordination'. Frantz Fanon's (1963) *The Wretched of the Earth* focused on 'the colonization of the personality', and Mannoni (1956) in *Prospero and Caliban* suggested that colonial rule led to a 'dependence complex' – meaning that 'the colonized transfers to his colonizers feelings of dependence, the prototype of which is to be found in the effective bond between father and son' (p.158). In the Indian context, Ashis Nandy (1983) analysed what he calls the 'loss and recovery of self under colonialism'. More generally, Dhaouadi (1988) and other Third World scholars have argued that the colonial relationship tended to underdevelop the self-possessed resources and capacities of the dominated parties, such underdevelopment having not only socio-economic consequences (which Western scholars – both Marxists and liberals – have concentrated on) but also adverse cultural and psychological consequences (which Western scholars have largely ignored). To the extent that such dependency syndromes did obtain in colonial contexts, they provided another prop to the maintenance of colonial rule and help to explain why so few Europeans were able to rule so many colonial peoples.

10.3 Why did colonial rule end when it did?

Being aware of the seven political features of colonial rule identified in this chapter can help one to think cogently about how to approach another large and important question: why did colonial rule end when it did? For example, in the case of India, it would be useful to examine at least the following set of factors in combination:

1 *The international political dimension.* There are several things here. For example, it was probably important that British rule in India ended in 1947, shortly after a Labour Government more sympathetic to Indian nationalist aspirations came to power in Britain. Secondly, the Second World War had just ended, a war which stretched to the limit the capacity of the British Government to continue to provide the colonial administrators and others needed to maintain the *raj*. Thirdly, reference can be made to the pressure of 'world public opinion' (more especially the US Government) at this time on the British Government to begin dismantling the empire (thereby opening up 'closed' colonial economies to US capitalist penetration).

2 *Use of 'traditional' authority figures.* The *raj* traditionally relied on the support of princes,

zamindars, jagirdars and other 'big' landlords. By the end of the 1930s, it was becoming clear that the power of this rural class was fading as political workers in the Indian National Congress mobilized a successor class of smaller but more numerous peasant proprietors who had power locally. Colonial rule had come by the 1940s to rest precariously on a dying class, and because of the Congress the colonial state was now blocked from forging another alliance in the countryside.

3 *Use of force*. British power in India depended on the support of subordinate Indian personnel in the military, the police and the civilian bureaucracy. By 1946–47, this support (which had previously been unquestioned) began to crack. The British officials realized that they could no longer rely entirely on the police and others to respond to orders to use force to maintain British rule. It is significant, for example, that in the days immediately preceding the announcement on 19 February 1946 by the British Cabinet in London that a Cabinet mission would be going to India to discuss the transfer of power with the Indian political parties, there were clear indications of widespread disaffection in the police in Madras Province and there was a revolt by a section of the Royal Indian Navy in Bombay.

4 *Hegemonic ideology*. I have mentioned the decades of work by political leaders in the Indian National Congress to undermine the ideology of the colonial state as benevolent and invincible. By the end of the Second World War, the colonial ideology was no longer hegemonic; too many people had been educated to perceive the colonial state and its consequences for India rather differently. Also, many people no longer believed that the power of the colonial state was invincible.

Consideration of these four aspects does not constitute a full answer to why the variety of colonial rules in Africa, Latin America and Asia ended when they did (Figure 10.7). Nevertheless, I suggest that in all the many unique cases of colonial rule ending, the most useful places to start looking for an answer to this large question would be in the areas of: (1) What was happening internationally? (2) Could 'traditional' authority figures in the colony still be relied on? (3) Did the colonial state still have the capacity to use force? (4) Was the ideology of the colonial state still hegemonic?

Figure 10.7 Another colony becomes independent. Why? The Union Jack is lowered in colonial Bechuanaland at midnight, 30 September 1966 to make way for the blue, white and black tricolour of independent Botswana. Taking the salute is Sir Hugh Norman-Walker, whose term of office as Queen's Commissioner ended with the striking of the flag.

10.4 Legacies of colonial rule for post-colonial states

Being aware of the seven political features of colonial rule discussed in this chapter helps us understand distinctive features of post-colonial states in the Third World. By way of illustration mention can be made of certain features in the context of Africa:

1 *Authoritarianism.* All European powers with colonies in Africa set an example of authoritarian government where little or no popular participation was permitted until towards the end of colonial rule. Most Africans, therefore, had little experience of democratic government until independence in many cases was rather suddenly thrust upon them. The anti-democratic character of post-independence politics in many African countries can be traced, at least in part, to the colonial legacy of authoritarianism.

2 *Use of force.* The widespread use of state violence and repression was characteristic of colonial rules in Africa, and it was 'colonial rulers [who] set the example of dealing with…opponents by jailing or exiling them, as not a few of those who eventually inherited power knew from personal experience' (Crowder, 1987, p.15). It has been argued that this legacy helped to foster the repressive character of many post-colonial political regimes in Africa.

3 *Statism.* The colonial state in Africa intervened profoundly in the economy for the benefit of the metropolitan economy in Europe, thereby discouraging the development of indigenous capitalist classes in Africa. As Diamond, Linz & Lipset suggest (1988, pp.7–8), 'both for its resonance with socialist and developmentalist ideologies and for its obvious utility in consolidating power and accumulating personal wealth, this legacy of statism was eagerly seized upon and rapidly enlarged by the emergent African political class after independence' and 'mushrooming state ownership and economic control, and the consequent emergence of the state as the primary basis of dominant class formations, was to have profound consequences' for the nature of African politics.

4 *International political dimension.* This feature of colonial rule, which was snapped at independence, left most African states acutely vulnerable to new forms of economic dependency on external forces. For example, changes in international commodity prices have had adverse effects on attempts at planned economic development.

Summary

Major political features of colonial rule include its (1) international dimension, (2) bureaucratic élitism and authoritarianism, (3) statism, (4) use of 'traditional' authority figures, (5) use of force, (6) technological advantage, (7) hegemonic ideology.

Being cognizant of these features equips us to get at least an initial bearing on the main question in this chapter: *how was colonial rule maintained?* Important elements in any answer to this question would include the following (I have grouped the seven features under five headings here):

1 *The élitism and authoritarianism of the colonial state.* Important decisions were always referred to a small group of colonial officials recruited by the imperial power. Colonial peoples had little or no control over this decision making power. Major decisions were referred to the metropolitan power in Europe (the international political dimension).

2 *The statism of the political economy.* The colonial state intervened extensively in economic life, partly in order to damp down economic developments that might threaten the metropolitan power and the interests it represented.

3 *The colonial state's capacity to use force* and its *technological advantage.* Colonies were acquired by force of arms, and colonial states, once established, were always quick to use force whenever 'disturbances' took place which threatened their rule.

4 *The colonial state's reliance on traditional authority figures.* Colonial states relied on powerful groups and classes (e.g. the landed upper classes) who in effect 'ruled', at least to some extent, on the state's behalf. Colonial officials constantly 'nursed' these colonial support structures.

5 *The colonial state's hegemonic ideology.* There was constant reinforcement of ideas in colonial society about the benevolence and invincibility of colonial rule. This 'official' ideology was regularly promulgated in the media and in other ways, and contrary ideas about the state were usually suppressed (e.g. through press censorship).

Each of these aspects of rule begs numerous other questions which I cannot pursue here, but they do provide the beginning of an explanation of how so few ruled so many. The same list of aspects of colonial rule can also be used to ask questions about why European rule ended when it did, and to help understand distinctive features of post-colonial states.

11

DEVELOPING COUNTRIES – 1945 TO 1990

TOM HEWITT

Chapters 8 to 10 have emphasized how the influences of colonialism have been felt at different times and in different ways. But it is not only colonial history which has shaped developing countries. The post Second World War era has seen many profound changes, apart from the repercussions of national liberation movements and independence from colonial powers.

By the 1960s, much of the Third World was politically independent from colonial rule, frequently after prolonged conflict between colonizers and colonized. The British did not leave India until intense and sustained opposition to their presence forced them out. Neither did they give up Kenya or Aden without taking military action against independence movements. Similarly, the French generated bitter confrontation in Algeria and Vietnam, and the Portuguese brutally suppressed opposition in Mozambique and Angola. Militarization in many former colonies has been a lasting legacy of colonialism.

Political sovereignty did not necessarily bring with it economic independence. This led some to question whether formal independence was indeed nothing more than a formality. Certainly the colonial experience (even in Latin America) resulted in ties with a world economic order which long outlived independence. Neo-colonialism, as these inequitable ties became

known, cannot be discounted, but there is more to it than that.

The post-war years to 1990 (the period under scrutiny in this chapter) have seen economic booms and slumps, wars and alliances, social transformation and political struggles, economic development and severe recession. Most of these have been mediated in some way by changes in the global economy.

Each country, region, community and individual has its own story to tell. We cannot convey this diversity in such a confined space. But it is possible to give it a context by examining the global changes which have influenced individual countries and regions. In this way, we will be able not only to place a certain coherence on diverse experiences but also to elaborate on a number of important ideas which have been used over the years by practitioners, politicians and academics to understand the course of events in developing countries and their relation to the global economy. However, be warned: 'simple guidelines are not forthcoming from the bald statistics' (Toye, 1987, p.16); a global view helps our understanding but is not a substitute for detailed knowledge of specific situations.

Professor Hans Singer has this to say about the history of developing countries since 1945:

"Perhaps the story of development is more than just 'one damn thing after another': it is a story of unfolding, of one thing leading to another in a process which can be given some meaning. But the trouble seems to be one of time lags... the development thinkers seem to base their action and thought on experiences of the last-but-one decade or a last-but-one phase, only to be overwhelmed by the inappropriateness of such action and thought in the face of new events and new problems. Is it perhaps a case of a problem for every solution, rather than a solution for every problem?

This seems to come close to the truth. It can be presented pessimistically as always reacting too late and to an obsolete situation; or more optimistically as a learning process."

(Singer, 1989, p.3)

In this chapter we are going to look at some of the proposed solutions to problems and to look at the new issues which emerged, through time, as an outcome of these 'solutions'. As you read, keep the following competing interpretations of events of the last 50 years in the back of your mind. Has this has been a period of:

• steady deterioration for developing countries relative to the developed?

• ups and downs which have followed changes in the international economic climate?

• changing fashions and ideologies which have influenced both the way we interpret the world and the way in which policy makers intervene?

• steady improvement, but temporarily stalled?

• learning but in some cases flawed by previous mistakes?

The chapter is divided into four discrete chunks of time. Although I am not particularly concerned with the 'history of dates', the events of the last 45 years have some significant phases with their own characteristics and from which, as Singer pointed out above, subsequent lessons have been learned and solutions gleaned. So, the structure of the chapter is as follows:

1	Post-war restructuring and Bretton Woods	1945 to early 1950s
2	The 'Golden Years'	1950s and 1960s
3	Debt-led growth	1970s
4	The 'Lost Decade'	1980s

Q Within these four time periods, what are the principal factors which helped or hindered the development process?

Q What are the divergences within the Third World?

In exploring these questions, we can begin to explain why we are witnessing a growing gap between developing countries. The North–South divide has been common parlance. We should now add to this a South–South divide.

11.1 Post-war restructuring and Bretton Woods

By the end of the Second World War, the world economy was in disarray, caused first by the economic crisis of the 1930s and then by the war itself. During the 1930s, rising unemployment and heavy protectionism (see Chapter 6 and Box 11.2 below) resulted in a decline in world trade of 65% in value terms between 1929 and 1933. It was accompanied by a drastic fall in primary commodity prices, the major source of income for developing countries at that time.

This section will briefly examine the agreements initiated in July 1944 at Bretton Woods, New Hampshire, USA, and put in place in the following years. The Bretton Woods agreements are significant: (a) because they stayed intact as an international system until 1973; and (b) because even after this date right to the present, the major institutions it begat (the IMF, World Bank, the UN and GATT) have considerable influence in developing countries.

The Marshall Plan for the restructuring of Europe and Japan in the aftermath of the Second World War was the model for setting up a more ambitious framework for economic stability. A new world system, supposedly based on large-scale international income transfers, was set up at Bretton Woods. The answer to capital scarcity and declining terms of trade was, in the view of the developed country contributors to Bretton Woods, quite simple: move from national-centred economic behaviour to internationally co-ordinated finance and trade. Bretton Woods was a conference of 44 nations. It was dominated by the USA and Britain, while most of the developing countries present were from Latin America (Figure 11.1).

The four Bretton Woods institutions will already be familiar names. The International Bank for Reconstruction and Development (IBRD) – later to become known as the World Bank – was to provide long-term finance for investment, while the International Monetary Fund (IMF) was to be a source of short-term finance to compensate for balance of payments deficits and exchange rate fluctuations. The United Nations (UN) was to be the forum through which international decisions were taken and the means by which international political and military stability was to be maintained. Finally, there was to be an organization which regulated international trade and stabilized world commodity prices. This role was taken on by the General Agreement on Trade and Tariffs (GATT).

Thus, through these institutions, the financial economic and political workings of the world were to be regulated and monitored. Perhaps not surprisingly, Bretton Woods was disproportionately favourable to the continued dominant position in the world economy of those developed countries which later formed the OECD (Box 11.1).

The emergence of other political and economic groupings (such as the non-aligned movement, the Group of 77 and the campaign for a new international economic order) is discussed in detail in Chapter 13. Suffice to say here that they were LDC-centred initiatives and were a direct response to the inadequacies of the Bretton Woods system from the point of view of developing countries.

Despite statements to the contrary at the time, the restructuring of the world economy was predominantly a First World affair. More truly international initiatives were thwarted by a combination of factors.

First, there was only minority representation of developing countries at Bretton Woods; only 18 of the 44 present. Second, there was competition between the USA and UK for economic leadership in the capitalist world. Third, there was a struggle between conservative and liberal economists and politicians. Finally, there was divergence of interest between Third World and developed countries over trade issues.

The representatives from LDCs saw commodity prices as the key trade issue. But no agreement

Figure 11.1   An attempt to co-ordinate the global economy: Bretton Woods, 1944.

Box 11.1 OECD

The OECD, formed in 1961, is the Organization for Economic Co-operation and Development. Its members are drawn from 25 developed countries, mainly from Western Europe and North America, but including Japan, Australia and New Zealand. Today the Group of Seven (G7; USA, Germany, UK, Canada, France, Italy and Japan) is the most powerful political and economic grouping within the OECD.

could be reached with the USA. The latter, no doubt, was reluctant to forego its sources of cheap raw material easily. Most significantly, GATT did little to improve commodity price stabilization, one thing that might have helped developing countries to escape from their vulnerable position in the international economy.

Conservative US foreign policy objectives won the day (well, they *had* won the war!). The developing countries might have achieved more favourable terms in these international agreements if it had not been for simultaneous changes in the US political climate. Immediately after the war, the 'Freedom from Want' ideals of the Roosevelt/Truman era and the liberal New Deal policies of the USA provided an auspicious context within the USA. But these were rapidly replaced by the harsher politics of the Cold War and the McCarthy era. The combined outcome was a system dominated by conservative US foreign policy objectives.

This domination also became evident in the United Nations in subsequent years. It has become a talking shop rather than an international watch dog. The relative cohesion of the UN over the Gulf conflict in 1990–91 is perhaps the exception which proves the rule. The setting up of the UN was an opportunity for LDCs to have a real say in the running of the international affairs but any involvement of the UN in economic matters was opposed by developed countries. In its place, the OECD, dominated by the USA, became the arbiter of trade and dominance in the world. In the words of Hans Singer (1989, p.8), 'So the Bretton Woods system which was meant to walk on four legs (UN, GATT, IMF and the World Bank) was hobbling along on the last two only.'

The United Nations, nevertheless, plays an important role in the international political arena. The composition of the Security Council indicates who wields political power in the world. The most recent round of GATT negotiations, started in Uruguay, similarly indicated who calls the shots in world trade. During the 1980s, the World Bank and IMF redoubled their economic influence over the internal workings of

many developing countries through programmes of structural adjustment (see Box 11.4 below).

Thus the institutions created by post-war restructuring are of continuing significance. Despite their limitations, they have provided an unprecedented continuity in the workings of the global economy.

Bretton Woods sets the context for post-war development in the Third World. Let us now look to the events of subsequent decades and examine how developing countries fared under the new game rules.

11.2 The 'Golden Years' – 1950s and 1960s

The result of post-war restructuring was a boom in the developed countries, based on full employment and low inflation, which was to last for nearly 20 years. The two decades are often referred to as the 'Golden Years' for the USA and Western Europe. How favourable was this new international climate to developing countries?

If nothing else, there was considerable optimism. The seeming success of the system for OECD countries led a number of economists to advocate a sure path to success for developing countries:

> "The process of development consisted…of moving from *traditional* society, which was taken as the polar opposite of the modern type, through a series of stages of development derived essentially from the history of Europe, North America and Japan – to *modernity*, that is, approximately the United States of the 1950s."
>
> (Toye, 1987, p.11; emphasis added)

The means by which 'modernity' was to be reached was *growth*. If an economy could achieve a certain critical rate of growth, the rest would all follow. The means to achieving growth varied. Some of the more influential theories included: savings and investment as the source of growth; development with surplus labour or rural labour as 'cannon fodder' for industrialization; stages of growth towards economic 'take off'; and the 'trickle down' effect where the spoils of growth would gradually filter through to the population at large. In all these, the emphasis was on capital accumulation, the primacy of investment and GNP growth rates as the key indicator of development. With the benefit of hindsight, it might be instructive to look at some aggregate growth rates over time.

Table 11.1 uses World Bank divisions of countries by income levels and by regional groups. The period 1965–73, or the latter part of the Golden Years, is the earliest for which consistent comparative international data are available. Growth rates from 1965–73 show two striking characteristics: First, both low and middle income economies averaged higher growth rates than OECD countries. Second, there are strong regional differences in growth rates, with East Asia in particular averaging much faster growth than either South Asia or sub-Saharan Africa.

Optimism for growth gradually became dampened towards the end of the 1960s. Large-scale official flows of resources to enhance domestic savings and alleviate balance of payments difficulties (from the IMF and World Bank respectively) did not materialize. In their place there was a substantial increase in the flow of private capital in the form of direct foreign investment. Approximately 70% of capital flows into developing countries in the 1960s were from the investments of transnational corporations. This raised the question of how much of the observed economic growth was of an isolated type dependent on the location decisions of these corporations and how much it built up national capacity. Finally, the terms of trade declined again from 1951 to 1965 by some 25% (Singer, 1989, p.13; refer back to Figure 5.2). This was at a time when 70–90% of exports from developing countries were primary commodities and 50–60% of imports were manufactures (Gwynne, 1990, p.41). As a result, balance of payments difficulties were unavoidable.

Or were they? The odds were stacked against the export of primary commodities (minerals, raw materials and agricultural goods such as sugar, cocoa, coffee), so one solution was to diversify into manufacturing. The 1950s and 1960s was a period of rapid industrialization for many developing

Table 11.1 GNP per capita, 1980, and growth rates, 1965 to 1986

Country group	1980 GNP per capita (US$)	Average annual growth of GNP per capita (%)		
		1965–73	1973–80	1980–86
Low-income economies	320	3.6	2.4	4.0
Middle-income economies	1760	4.6	2.4	0.1
Sub-Saharan Africa	570	3.0	0.1	–2.8
East Asia (including China)	420	5.4	4.4	6.6
South Asia	240	1.0	2.0	3.2
Latin America and Caribbean	2000	4.1	2.4	–1.6
OECD member countries	10 750	3.5	2.2	1.9

Source: World Bank (1990) *World Development Report 1990*, Oxford University Press, Oxford, p.160.

Box 11.2 Import substitution industrialization

Import substitution industrialization (ISI) is an inward-looking strategy which consists of setting up domestic industry to supply markets previously served by imports. This domestic industry may be locally owned (state or private) or foreign owned. The latter, direct foreign investment (DFI), will usually involve a series of incentives in order to attract foreign firms to set up production facilities within national boundaries. ISI is often contrasted with export oriented industrialization (EOI).

ISI policies will usually involve the use of relatively high import tariffs, quota restrictions on imports and controlled access to foreign exchange. Such protectionist measures to encourage domestic production are usually combined with disincentives to exporters. The typical pattern of ISI is to start with the production of consumer goods and move to intermediate goods (e.g. parts and components for consumer goods) and then to capital goods (e.g. machines to make parts and components). A major flaw in the strategy is that savings on imports are difficult to attain since substitution in one department (e.g. consumer durables) implies imports for another (e.g. intermediate and capital goods to produce the consumer goods).

ISI is often linked to the argument that new or 'infant' industries need time to develop, but whether such infants grow up or not was the subject of debate in the 1980s. Countries which have followed ISI strategies in a big way include China, India, Brazil, Argentina and Mexico. Nevertheless, most developing countries (like developed countries before them) have employed some degree of import substitution.

Figure 11.2 Import substitution industrialization in Brazil.

countries. It was the era of import substitution industrialization (Box 11.2; Figure 11.2).

Industrialization could not happen in a vacuum. The experience of some Latin American countries from the 1930s to the 1950s showed that, with little access to finance or to protected developed country markets, there was no choice but to look inwards. This generated what Singer called 'export pessimism' and forced countries to look to their internal markets as a means of accumulation.

Industrialization occurred at a different pace in different regions. The larger Latin American countries had begun to industrialize in the 1930s. By the 1960s they already had a substantial industrial base. Along with a considerable measure of direct foreign investment, Latin American industry looked towards internal markets. In contrast, the East Asian newly industrializing countries (NICs; South Korea, Taiwan, Hong Kong and Singapore) were building up industries based on the export of mass-produced consumer goods. India, strongly influenced by the Soviet development model (see Chapter 12), was developing industry with little direct foreign investment and substantial protection. The majority of African countries, with much smaller industrial bases, could not industrialize on a large scale, and could only begin to do so in the following decade when international conditions were very different. These patterns of industrialization reinforced the already existing diversity of developing countries.

As with most aggregate international data, calculations of sectoral distribution of GDP vary (Figure 11.3). Thus Figure 11.3 should be treated as only an approximate measure of change. This said, between 1960 and 1980 there was an unmistakable shift from agriculture to industry. The extent to which manufacturing formed a significant share of GNP varied significantly between countries. In 1965 in India and China, manufacturing accounted for 24% of GNP, while in all low-income countries it accounted for 9% (World Bank, 1990, p.182).

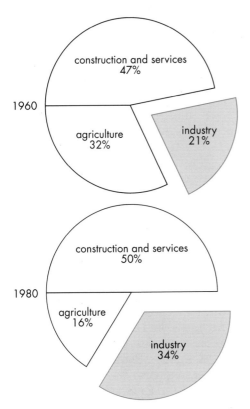

Figure 11.3 Pie charts giving sectoral breakdown of gross domestic product for LDCs in 1960 and 1980 (excluding China).

While growth did occur in at least some developing countries, it became apparent towards the end of the period that by no means all countries were 'taking off' nor was the supposed 'trickle down' taking place. Even when growth did occur, it was not a sufficient condition to establish sustained social and economic development (as defined in Chapter 6). In many cases, income distribution was not narrowing as predicted but widening. Emphasis on growth alone was limited. This was due to any number or reasons, most important of which were limits to ISI, the enclave nature of foreign investment with widespread repatriation of profits and little technological or economic linkages with the rest of the economy, and the neglect of agriculture resulting from an 'urban bias' and massive influxes of food aid (which, some argue, made governments even less inclined to invest in agriculture).

The disillusionment with modernization theories of development took two forms. Among development economists, there was a recognition that developing countries were not and probably never would be like the OECD countries. This was a radical departure from orthodoxy. Seers (1963) argued that OECD economies were a 'special case' rather than the norm and that a separate discipline had to be employed to understand the economics of developing countries. This was one of the earlier recognitions that developing countries showed great diversity and that there were no fixed models for economic development.

At the same time, an equally radical alternative explanation of the continued subordinate position of developing countries was emerging from developing countries themselves. In part, this was a criticism of the legacy of the colonial state in Africa and Asia (see Chapter 10). A more lasting criticism of orthodoxy came from Latin America based on the notion of *dependency* of developing countries (Cardoso & Faletto, 1979). Put simply, the dependency 'school' viewed 'underdevelopment not as a pristine condition of low productivity and poverty but an historical condition of blocked, distorted and dependent development' (Toye, 1987, p.12).

Dependency had two dimensions: political and economic. On the political side, the solution to the perceived negative implications of 'peripheral' status in the world were autonomy and delinking from the 'core' countries. This was associated with nationalist movements and the political cohesiveness that anti-colonial struggles produced. It was also linked to the examples of Cuba, China, Vietnam, Mozambique and other countries which had adopted a broadly socialist development path (see Chapter 12). Even in those countries firmly within the capitalist world, revolutionary movements of rural and industrial workers were widespread in the late 1960s.

The implications of the dependency school were also nationalist. Not only were the prevailing economic relations between developing and developed countries seen as unequal, but it was also perceived that the gap between the two was widening precisely because of this imbalance of economic power.

To sum up, it should be stressed that the so-called Golden Years gave grounds for both optimism and pessimism. Thus Toye suggests that they were:

> "neither the uncomplicated succession of economic take-offs which modernization theory predicted, nor the continuously growing gap in income and welfare between the rich countries and the poor countries prophesied by underdevelopment [dependency] theorists. Instead, there has been a combination of some take-offs, mainly in East Asia, and some severe cases of economic retrogression, mainly in Africa. Thus the polarization that has taken place has done so *within* the Third World, but not between the Third World taken as a group and the developed economies."
>
> (Toye, 1987, p.15)

11.3 Debt-led growth in the 1970s

The 1970s saw a shift away from growth-at-all-costs as the way forward for developing countries towards an emphasis on employment and redistribution with growth. It also saw a substantial increase in the indebtedness of developing countries which for some gave the illusion of development.

Redistribution with growth

In the early 1970s, the International Labour Office (ILO) of the UN published a series of research reports on employment in different countries. This research was prompted by a substantial rural-to-urban migration resulting from the combination of neglect of agriculture in favour of industrial development and the failure of emerging industries to employ much of this labour. Going for growth alone was proving very costly in human terms. Unlike the modernization model where it was assumed industry would absorb surplus agricultural labour, another idea was put forward which contended that rural-to-urban migration would far outstrip the availability of urban employment (Box 11.3; Figure 11.4).

Figure 11.4   Dhaka: the problem of population concentration in urban areas.

Box 11.3 The urban explosion

This is the century of the great urban explosion. In the 35 years after 1950, the number of people living in cities almost tripled, increasing by 1.25 billion. In the developed regions, it nearly doubled from 450 million to 840 million, and in the developing world it quadrupled, from 285 million to 1.15 billion.

In the past 60 years the developing world's urban population increased tenfold, from around 100 million in 1920 to close to 1 billion in 1980. Meanwhile, its rural population more than doubled.

• In 1940 only one person in eight lived in an urban centre, and about one in 100 lived in a city with a million or more inhabitants.

• In 1960 more than one person in five lived in an urban centre, and one in 16 in a city with a million or more.

• In 1980 nearly one person in three was an urban dweller, and one in 10 lived in a city with a million or more.

The population of many of sub-Saharan Africa's larger cities increased more than sevenfold between 1950 and 1980 – Nairobi, Dar es Salaam, Nouakchott, Lusaka, Lagos and Kinshasa among them. During these same 30 years populations in several other Third World cities – Seoul, Baghdad, Dhaka, Amman, Bombay, Jakarta, Mexico City, Manila, São Paulo, Bogotá and Managua – tripled or quadrupled. In-migration has usually contributed more to their growth than natural increase has done. This growth has been far beyond anything imagined only a few decades ago and at a pace that is without historic precedent.

(UNDP, 1990)

The ten largest cities, 1960.

The ten largest cities, 2000 (projected).

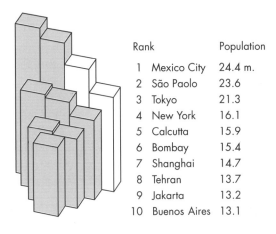

In a developing country

In an industrial country

Rank		Population
1	New York	14.2 m.
2	London	10.7
3	Tokyo	10.7
4	Shanghai	10.7
5	Rhein-Rhur	8.7
6	Beijing	7.3
7	Paris	7.2
8	Buenos Aires	6.9
9	Los Angeles	6.6
10	Moscow	6.3

Rank		Population
1	Mexico City	24.4 m.
2	São Paolo	23.6
3	Tokyo	21.3
4	New York	16.1
5	Calcutta	15.9
6	Bombay	15.4
7	Shanghai	14.7
8	Tehran	13.7
9	Jakarta	13.2
10	Buenos Aires	13.1

As a result, attention shifted to employment-intensive technologies which were seen as more 'appropriate' to situations of extreme labour surplus. Growth with employment was thought to be a means of attaining a more equitable income distribution, reducing poverty, minimizing the potential for political unrest, and so on. As a result of the employment missions, the ILO slogan at the time became 'from redistribution *from* growth to redistribution *with* growth'. There was a simultaneous 'discovery' that women are crucial in economic development. The 'household' (including differences within it) was recognized as a fruitful unit of analysis. Despite this, it is clear that even in the 1990s recognition of the position of women still has a long way to go (see Chapters 3 and 15).

It took some time for academics and international organizations to realize that 'unemployment' was a misnomer since most of the world was hard at work (Chapter 3)! The 'informal sector' (those who were not in formal employment) became the focus of attention. Recognition of different patterns of labour utilization was an important step forward from the Keynesian notion of full (formal) employment. The ILO, rather than viewing the mass of people who were not in formal employment as a problem (of overcrowded cities of vagrants working on the margins of illegality), saw them as a resource, a view which resurfaced in the 1980s (Soto, 1989).

Subsequently, some argued that the informal sector, defined to include such a wide array of income-generating activities was perhaps too broad a concept to be useful. Nevertheless, it seemed to be a step in the right direction that, at least, there was recognition that these economic activities were taking place.

Consideration of the 'basic needs' of the world's urban and rural poor puts a very different light on the meaning of development than that previously linked to macro-economic growth. But the two cannot be separated too easily.

Debt-led growth

The suspension of the free convertibility of the US dollar to gold at fixed exchange rate occurred in 1971. This change was to have important financial ramifications for developing countries, particularly since it was linked to a growing structural crisis in the OECD (including growing protectionism breaching GATT agreements, increasing unemployment, etc.). The combination of the two threw the system of world trade into confusion.

The assertion of oil power by the oil-producing cartel, OPEC, in 1973 in the 'first oil crisis' added fuel to the fire and radically changed the terms of trade (for the worse) of those dependent on oil imports.

This can be seen from Figure 11.5. After the sharp oil price increase in 1973, oil prices stayed high for the following decade. This coincided with a long-term decline in other commodity prices, also shown in Figure 11.5. The combined impact was particularly serious for non-oil producing, primary commodity exporters, i.e. many developing countries.

As OPEC savings surpluses grew from the increases in oil prices, large deposits were placed in the commercial banks. On account of the economic recession in OECD countries, partly brought about by the oil crisis, the demand for credit declined (something which in turn raised liquidity in the international banking system). Commercial banks began to turn their attention to other regions of the world and saw particularly good markets in the Third World and, to a lesser extent, Eastern Europe. As a result, commercial and official lending to Third World countries grew substantially (Figure 11.6).

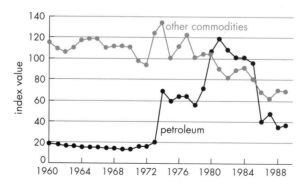

Figure 11.5 Weighted index of commodity prices (1979–81 = 100).

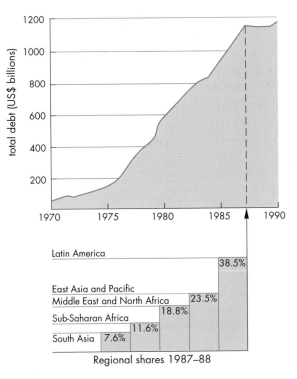

Figure 11.6 *Growth and regional distribution of developing countries' debt.*

(rapidly losing export markets and finding a mounting import bill) to resist. Then the 'debt trap' was sprung. By the late 1970s, the OECD began to adjust to the reality of recession. One outcome was continually rising interest rates which had serious consequences for large-scale borrowers and their debt repayments (Figure 11.7). As the 1980s opened, prospects for all but a few developing countries looked bleak.

Figure 11.7 *The debt trap.*

OPEC made oil surpluses 'available' to LDCs through what was called the eurodollar market – oil dollars deposited in private European banks. After 1974 there was a rash of borrowing, encouraged by commercial banks which did not know what to do with so much cash. For a while it appeared that this was the flow of Marshall-type finance that had been deemed necessary at Bretton Woods. There was much talk of a New International Economic Order (NIEO) based on recycled oil money, but it remained only talk. The most comprehensive statement of this is to be found in *The Brandt Report* published in 1980 which generated controversy across the political spectrum (Brandt, 1980). Some considered this as an opportunity lost. Others saw it as a pipe dream or a thinly disguised strategy for developed countries to pull out of their own growing economic crises in the name of interdependence.

Nevertheless, at the time, recycled OPEC dollars were 'cheap money' (at low interest rates) which was hard for many developing countries

11.4   The 1980s – development in reverse?

The 1980s has been the decade of neo-liberation. Observers also talk about it as the 'Lost Decade' for developing countries. Many believed that up to the 1970s there was a measure of human progress – however slow – which gave indications that it would continue. So what went wrong? And for whom was the decade lost? Some answers are proposed in Chapter 13, where global institutional changes are examined. In this section, we will discuss the economic and social reversals of the 1980s.

The debt crisis, as elaborated above, is crucial here. In attempts to reduce growing inflation, OECD countries slowed down their economies, thereby depressing prices and demand for commodities, and allowed interest rates to rise. For those developing countries which either relied on commodity exports or on borrowing as a source of foreign exchange, this turn-round by OECD countries had serious implications. For those LDCs which relied on commodity exports and borrowing, the consequences were dire. Thus, in the words of the South Commission: 'a large part of the cost of controlling inflation and introducing structural change in the North was borne by the South. Developing countries had to pay out more and more to service their debt while receiving less and less for their exports. As these contrasting movements aggravated their financial difficulties, commercial banks decided to stop lending them new money, and the result was the international debt crisis of the 1980s' (South Commission, 1990, p.56).

By 1983, OECD countries were making something of a recovery (by some indicators) although this was not on the scale of the 1960s. More significantly, and unlike the 1950s and 60s, the recovery did not generate improvements in the international economic environment for developing countries. This is shown by some of the dramatic reversals for many developing countries during the 1980s.

- *Debt repayments*. Continued high interest rates, combined with a decline in commercial bank lending, resulted in the paradox that developing countries were paying out more finance in service payments than they received as borrowing. Figure 11.8 shows that there was a substantial net outflow (service payments less borrowing) from developing countries after 1983.

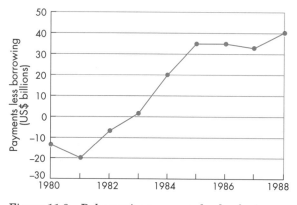

Figure 11.8   Debt service payments by developing countries less external borrowing, 1980–88.

- *Direct foreign investment* declined by some two-thirds in the early 1980s, and there continued to be a net outflow of profits from LDCs (i.e. after taking into account new investment).

- Non-oil commodity *prices* continued to decline rapidly through the 1980s. Figure 11.5 above shows that the the real price of the 18 main non-oil commodities fell by 25% from 1980 to 1988.

- *Growing protectionism* in the OECD further mitigated against other developing country exports, such as auto parts, steel, electronics, textiles, petrochemicals and agricultural products. This was perhaps the greatest irony for developing countries in the 1980s: while the neo-liberal ideology of free markets took firm root, its proponents – OECD countries – were making their own economies more protected, not less. British readers in particular may recall repeatedly hearing the argument that cheap exports from the Third World were threatening UK industry! The 'new' protectionism (so-called to distinguish it from previous versions such as in the 1930s or 1950s/1960s) was based on euphemisms such as 'voluntary export restraints' and 'orderly market arrangements', which meant little more than protectionism in favour of the OECD through a back door while the formal free trade decisions of GATT could remain intact.

Barring some significant examples (principally the East Asian NICs and, to some extent, China) the development reversals of the 1980s were either severe or very severe. In the former category falls much of Latin America while the latter includes much of sub-Saharan Africa.

For many developing countries, there was a combination of declining international demand, increasing protectionism in the OECD, deteriorating terms of trade, negative capital flows, continuing high interest rates, and unfavourable lending conditions. The signs of progress up to the 1970s ground to a halt and in many cases went into reverse. Per capita national incomes in Latin America and Africa declined,

investment declined (resulting in the deterioration of infrastructure in transport, communications, education and health care), and
unemployment and underemployment grew.

Least developed countries (LLDCs) grew in
number from 31 to 42 countries in the 1980s.
LLDCs are defined by the UN Development
Programme as:

> "A group of developing countries estab-
> lished by the United Nations General
> Assembly. Most of these countries suffer
> from one or more of the following con-
> straints: a GNP per capita of around $300
> or less, land-locked, remote insularity,
> desertification, and exposure to natural
> disasters."
>
> (UNDP, 1990)

They were the most adversely affected since
they are the most reliant on exports of primary
commodities and on imports of basic inputs
and capital goods to keep their economies
going. Official development assistance to these
countries has, nevertheless, been pitifully low.
Despite a call by the UN in 1981 for assistance
to reach 0.15% of donor GNP, the actual level in
1988 was only 0.09%.

Economic decline has had social consequences.
During the 1980s, conditions worsened on most,
if not all, of the social indicators listed in Box
11.6 at the end of this chapter. Disaffection has
been compounded by recourse to oppression and
military force which, in turn has added to loss of
support and legitimacy for many governments.

Interventions by international financial agencies, particularly the World Bank and the IMF,
appear to have hindered rather than helped the
situation. Stabilization and structural adjustment policies (Box 11.4) have been imposed as
conditions of financial assistace. Adjustments
have to be recognized as necessary by many of
the affected Third World governments in order
to claw their way out of crisis. Nevertheless, the
manner in which these adjustments have been
imposed has been criticized. The South Commission has been one of the more moderate critics in
recent years. This is their view of the impacts of
adjustment policies:

> 'in the adjustment process of the 1980s,
> these needed reforms were frustrated by
> an unbalanced international approach to-
> wards structural adjustment and by the
> conditionality prescribed by the interna-
> tional financial institutions. The macro-
> economic policies – in particular fiscal and
> exchange rate policies – virtually forced
> upon developing countries as part of pro-
> grammes for stabilization and structural

Box 11.4 Adjustment policies

'Adjustment' policies describe all policies designated to reduce the basic imbalances in the
economy, both on external [balance of payments]
accounts and in domestic resource use. It is
helpful to distinguish *stabilization* policies from
other types of *adjustment* policies. Stabilization
policies consist in reducing imbalances in the
external accounts and the domestic budget by
cutting down on expenditure (by the government and by firms and households), and reducing credit creation and the budget deficit.
Stabilization policies are therefore deflation-
ary, and tend to have rapid effects on the balance of trade through a reduction in imports.

Other adjustment policies are designed to
change the structure of the economy (i.e. to
achieve *structural adjustment*), so as to improve the balance of trade and the efficiency of
the economy over the medium term. The intention of these adjustment policies is to expand
the supply of tradeables, increasing both exports and import-substitutes. IMF policies
are…mainly, but not exclusively, stabilization
policies, while the World Bank Structural
Adjustment Loans (SAL) are designed to secure adjustment of the medium term.

(Cornia, 1987, p.48)

adjustment were geared to achieving a quick, short-term improvement in the balance of payments. Safeguarding the interests of international commercial banks even at the cost of a severe economic contraction thus became the primary concern of international strategy on debt management.

Further, the programmes for stabilization and adjustment pressed upon developing countries did not provide for sufficient external financial support to permit adjustment to occur and endure without choking their growth. The programmes were based on unduly optimistic assumptions about the speed at which structural maladies could be corrected. In addition, they were generally shaped by a doctrinaire belief in the efficacy of market forces and monetarist policies. This combination of priorities and policies aggravated the developing counties' economic woes and social distress in a number of ways.

In particular, the complete disregard of equity in prescriptions for structural adjustment consisting in cuts in public spending and changes in relative prices [e.g. for basic foodstuffs] had devastating effects on vital public services like health and education, with especially harmful consequences for the most vulnerable social groups.'

(South Commission, 1990, p.67)

By all accounts, such social belt-tightening (where most were already down to the last notch) had few visible positive economic outcomes. On the contrary, 'the application of such policies accentuated the maldistribution of income within developing countries, while in many cases their beneficial impact on public finances was negligible – and is certainly outweighed by their long term economically detrimental effects' (South Commission, 1990, p.68) (Figure 11.9).

If the combined impacts of debt and inflation have been at the heart of Latin America's troubles in the 1980s, in Africa a broader set of adverse circumstances has resulted in a much deeper crisis.

- Fourteen of the *poorest 20 countries* in the world are in sub-Saharan Africa, many of which became poorer in the 1980s, with per capita GNPs below US$300 in 1987. Meanwhile debt increased dramatically. In absolute terms, the region's total external debt rose from US$6 billion in 1970 to US$134 billion in 1988, an amount equal to sub-Saharan Africa's total GNP or to 3.5 times its total export earnings.

- *Military spending* increased between 1972 and 1987 almost in direct relation to cuts in *social welfare* spending. War was commonplace. From 1945 to 1989 there were more than 30 wars in sub-Saharan Africa. (Box 11.5 gives information on the arms trade, which of course relates to other LDCs as well as those in Africa.)

- Many countries continue to rely on one *agricultural commodity* for more than 40% of export earnings. The real prices for these commodities dropped by more than 40% through the 1980s.

- *Desertification* of 650 million square kilometres in the last 30 years is said to have affected the livelihoods of 60 million people, forcing at least 10 million to leave home as a result.

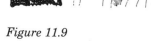

Figure 11.9

234

Box 11.5 The arms trade and developing countries

The arms trade between North and South became an important political issue after the Second World War, as the Cold War unfolded and infant nation states in the Third World became involved in superpower politics. This involvement took many forms but economic and military aid was particularly important. Third World countries accepted economic and military aid principally because they had little choice. Apart from the desperate economic conditions prevailing throughout the Third World, which precluded for most the independent creation of technologically advanced defence forces, the majority of fledgling states throughout Africa and Asia inherited or created by their new, independent existence national security problems and concerns. For both regional and international reasons, the demand for military aid was high during the 1950s and 1960s.

For Third World countries, an unprecedented access to all types of defence equipment following the commercialization of the arms trade at the end of the 1960s led to sharp rises in both the resources devoted to procurement and the quality of the equipment received. Thus, during the 1970s world-wide arms sales grew from $9.4 billion in 1969 to $19.8 billion in 1978 (constant dollars) of which approximately 75% was sold to the Third World.

During the 1980s, the rate of growth of arms imports by developing countries slowed down.

However, as we can see from the figure in this box, arms remained a significant item of international trade.

(Chris Smith, Institute of Development Studies, Brighton, personal communication)

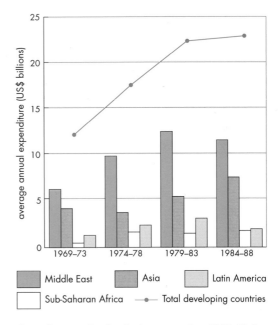

Arms imports by developing countries, 1969–88 (at constant 1985 prices).

A depressing picture indeed! The experience of the 1980s in many regions of the world points to the accuracy of its description as a 'lost decade'. Nevertheless, there are glimmers of hope.

First, social and economic crisis is not universal. Some economies and regions, on a greater or more moderate scale, weathered the decade and even thrived. In particular, the South-east Asian countries prospered in economic (if not in political) terms (see Table 11.1). South Korea, Taiwan, Hong Kong and Singapore now have the USA, Europe and Japan seriously worried over the extent of their international competitive-

ness. Even the so-called second-tier NICs (such as Thailand, Malaysia, the Philippines and Indonesia) registered remarkable economic growth through the 1980s. Smaller scale economic successes can also be seen elsewhere such as in Sri Lanka and China. In short, macro-economic indicators hide a great and growing diversity among developing countries.

Second, and thinking back to the quotation from Hans Singer at the beginning of this chapter, the negative experiences of the 1970s and 1980s appear again to have engendered a reassessment of development thinking. In whichever area of development one cares to mention (rural development, aid, industrialization, employment, finance and so on) it has become difficult to stick to the dogmas of either decade. As Chambers points out, in the 1970s 'the solution to rural poverty was not less government but more' and in the 1980s 'the solution to the problems of development was not more government but less' (Chambers, 1989).

Chambers goes on to say that:

> 'Both ideologies, and both sets of prescriptions, embody a planner's core, centre-outwards, top-down view of rural development. They start with economies not people; with the macro not the micro; with the view from the office not the view from the field; and in consequence their prescriptions tend to be uniform, standard and for universal application.'
>
> (Chambers, 1989, p.6)

Thus, recent contributions to the development debate (for example, Colclough & Manor, 1991 and Drèze & Sen, 1989) have taken a more pluralistic approach to the problems of the poor in developing countries. They cut across the more rigid ideological boundaries of the 1970s and 1980s, which in many cases came up with most inappropriate prescriptions for specific situations.

Even the World Bank has been forced to rethink where it is going. Its steadfast attachment to rigid policies such as market-led structural adjustment programmes throughout the 1980s has weakened. Its emphasis on poverty in the *1990 World Development Report* was, for the Bank, an unexpected shift of position. There is still substantial support in the Bank for policies of economic liberalization and moves afoot to increase lending to the private sector. Nevertheless, there is recognition that liberalization alone is an inadequate response to the social and economic complexity of LDCs. An emphasis on relieving poverty and improving education, nutrition and health care is a welcome step from such a powerful international institution.

It appears that the five points raised at the beginning of this chapter on how we might view development in the last 45 years have led us to no single conclusion. We have witnessed a deterioration for some countries while for others there has been socio-economic improvement. Many of these experiences have reflected changes at an international level but we have also witnessed localized crises at times of international boom and vice versa. Ideologies and ways of viewing the world have certainly changed over the years, but it is not clear in all cases whether policy interventions have been pro-active or reactive. Finally, we look set for another round of learning in the 1990s as a result of the crisis-ridden decade of the 1980s.

If one thing has become clear, it is that diversity between developed and developing countries and between developing countries themselves has become more pronounced in the last 45 years. The growing South–South divide has important implications for the way we understand the catch-all term 'development'.

A useful way of assessing the last 45 years is with the following 'balance sheet' of human development. This chapter has emphasized the economic aspects of development. Economists sometimes forget that behind economic change there are people and livelihoods. Box 11.6 gives a summary of human progress and deprivation. I will leave readers to draw their own conclusions.

Box 11.6   Balance sheet of human development

Human progress	*Human deprivation*

Life expectancy
- Average life expectancy in the South increased by a third during 1960–87 and is now 80% of the North's average.

Education
- The South now has more than five times as many students in primary education as the North: 480 million compared with 105 million.
- The South has 1.4 billion literate people, compared with nearly one billion in the North.
- Literacy rates in the South increased from 43% in 1970 to 60% in 1985.

Income
- Average per capita income in developing countries increased by nearly 3% a year between 1965 and 1980.

Health
- More than 60% of the population of the developing countries have access to health services today.
- More than 2 billion people now have access to safe, potable water.

Children's health
- Child (under five) mortality rates were halved between 1960 and 1988.
- The coverage of child immunization increased sharply during the 1980s from 30% to 70%, saving an estimated 1.5 million lives annually.

Food and nutrition
- The per capita average calorie supply increased by 20% between 1965 and 1985.
- Average calories supplies improved from 90% of total requirements in 1965 to 107% in 1985.

Sanitation
- 1.3 billion people have access to adequate sanitary facilities.

Women
- School enrolment rates for girls have increased more than twice as fast as those for boys.

Human deprivation

Life expectancy
- Average life expectancy in the South is still 12 years shorter than in the North.

Education
- There are still about 100 million children of primary school stage in the South not attending school.
- Nearly 900 million adults in the South are illiterate.
- Literacy rates are still only 41% in South Asia and 48% in sub-Saharan Africa.

Income
- More than a billion people still live in absolute poverty.
- Per capita income in the 1980s declined by 2.4% a year in sub-Saharan Africa and 0.7% a year in Latin America.

Health
- 1.5 billion people are still deprived of primary health care.
- 1.75 billion people still have no access to safe source of water.

Children's health
- 1.4 million children still die each year before reaching their fifth birthday.
- Nearly 3 million children die each year from immunizable diseases.

Food and nutrition
- A sixth of the people in the South still go hungry every day.
- 150 million children under five (one in every three) suffer from serious malnutrition.

Sanitation
- Nearly 3 billion people still live without adequate sanitation.

Women
- The female literacy rate in the developing countries is still only two-thirds that of males.
- The South's maternal mortality rate is 122 times that of the North's.

(*Source:* UNDP, 1990)

12

SOCIALIST MODELS OF DEVELOPMENT

ANDREW KILMISTER

Socialism has been one of the most influential movements of the post Second World War world. One of the main processes that has structured this world has been the attempt by a significant number of countries to follow a socialist path to, or model of, development. These attempts and the consequences are the subject of this chapter.

Q What are the basic tenets of socialist development? What are the historical processes that have shaped socialist development in the USSR? How has this model been adapted in China and influenced development in other Third World countries?

The issues covered in this chapter are treated in much more detail in the following books: Post, K. & Wright, P. (1989) *Socialism and Underdevelopment*, Routledge, London, and Ellman, M. (1989) *Socialist Planning* (2nd edition), Cambridge University Press, Cambridge.

12.1 Socialist development

Socialism as a system of ideas and as a political movement has taken a variety of forms, and the nature of a 'truly socialist' society is hotly debated. Despite this diversity the search for a socialist road to development has resulted in a number of central features that unite the different countries following this path. This analysis enables us to think of a distinctive model that can be described as a socialist model of development. It is based on five economic characteristics and one political characteristic.

Economic

1 State ownership of large- and medium-scale industry, with relationships between individual enterprises governed by plans laid down from the centre rather than operating through the market. In particular, investment projects and the level of investment carried out in the economy are centrally planned.

2 Extensive state control over foreign trade and investment.

3 State intervention in the labour market and in the hiring and firing of labour by individual enterprises, often leading to virtual full employment in the economy.

4 State control over retail and wholesale prices, often associated with shortages of consumer goods at the ruling subsidized prices.

5 State intervention in agriculture and in the relationship between town and country, often taking the form of collectivization of farms and delivery of collective farm produce to state marketing agencies at fixed prices.

Political

6 In general, socialist societies are ruled by a single party, claiming to be a Marxist–Leninist party ruling in the name of the working class and peasantry and occupying a 'leading role' in the society. Organized political opposition is not tolerated.

You may want to think about what, if anything, has been left out of this list. Are there any important characteristics of socialist societies which have been ignored?

Clearly, the model of socialism outlined above diverges in several ways from the ideals around which the socialist tradition was formed and has developed. Part of the purpose of this chapter is to explain how this divergence took place and to outline the various factors which led to the ideals of socialism being realized in this particular way rather than some other. This is not meant to imply that this model represents the only direction that socialist development might take in the future. Some writers have argued that the divergences between what has occurred in societies such as those outlined above and 'true socialism' is so sharp that these countries are not really socialist at all, and should instead be described as 'centrally planned' or 'state socialist'. In this chapter I shall not attempt to judge this debate. I shall concentrate on analysing these societies as they are without prejudging whether they live up to any particular ideas or standards. However, when reading the account it will be useful to keep in mind what elements of socialism as a political goal have not been attained by these societies, and how failure to attain these elements has affected their pattern of development.

The model of socialism outlined above has, since the mid-1970s, been in a state of crisis. In the late 1980s and early 1990s this crisis became acute. Two socialist countries, East Germany and South Yemen were absorbed, through a process of national unification, into capitalist neighbours. It appears likely that a third, North Korea, may go the same way in the not too distant future. Several socialist countries in Eastern Europe have opted to try to institute a capitalist market economy and have explicitly abandoned state socialism. The two largest socialist societies, the USSR and China, face profound political problems. Asian socialist countries (China, Vietnam, Cambodia) have instituted significant economic reforms, dismantling much of the centralized planning apparatus and strengthening market relations, particularly in the countryside.

However, this in no way diminishes the importance of studying these societies and the socialist model that they embodied. It remains important to ask a number of questions about that model. Why did it develop in the way that it did? What lessons does it hold for other developing countries? What might the future bring for these societies? We have to ask these questions bearing in mind that the socialist model is now in a state of flux, and that we are not analysing a static framework but something which is changing and developing.

12.2 The dilemmas and contradictions of Soviet socialist development

The concept of the socialist model of development would have seemed extremely strange to Marx, Engels and their immediate followers. The majority of socialists in the nineteenth century thought that socialism was something that would happen in countries with a significant degree of industrialization and a sizeable working class, and that the socialist revolution would be international. The idea of relatively poor countries undergoing a socialist revolution and then attempting to institute development policies within the framework of the nation state was not one which was widely considered prior to the Bolshevik revolution in 1917. The most far-reaching attempt to consider these issues was the analysis developed by Trotsky in response to the 1905 Russian revolution. Trotsky argued that socialist revolution was not only possible in a country such as Russia, but was also necessary. In developing countries, dominated by imperialism, the working class were the only group in the society that could carry out even basic democratic reforms and lay the basis for industrialization. For this to happen the revolution in such countries had to be 'permanent' in two senses.

First, each country's democratic revolution (the equivalent to the Civil War of the 1640s in England or the 1789 French revolution) would have to be transformed into a socialist revolution. Secondly, that nationally based socialist revolution would have to become international. Trotsky's analysis was combined by the Bolsheviks with Lenin's account of a world economy dominated by imperialism, an imperialism which distorted and suppressed national economic development in the non-imperialist and colonial countries. Together the two strands of thought provided the inspiration for the attempt to institute a socialist transformation in the USSR.

The 1917 revolution appeared to offer striking confirmation of Trotsky's analysis. The responsibility for industrializing and developing the USSR was now placed with the Soviet working class and the Bolshevik party. However, the revolution never became permanent in the second sense. It remained a national revolution and the Soviet state was forced to forge a new model of development in extremely beleaguered circumstances of civil war, foreign intervention, and comparatively low levels of economic development. The questions posed by this situation have marked all subsequent attempts at socialist development strategies. There were four interlinked questions which the Bolsheviks faced as they dealt with the heritage of the revolution.

1 What relationship would a socialist society such as the USSR have with a generally hostile and more developed outside world?

2 Where would the resources for industrialization and development come from?

3 What would be the relationship between the working class (in theory the main beneficiaries and leaders of a socialist project) and the most numerous group in the population, the peasantry?

4 What would be the role of the state and the government in building a socialist society?

The dilemma faced by the Bolsheviks was something like this. It was taken as a precondition for socialism that industrialization should take place. Yet who should finance this industrialization? In the absence of any significant group of capitalists or landowners in the USSR (these had all been expropriated) only three possibilities suggested themselves. Either resources could come from abroad, via foreign investment, or from the working class or from the peasantry. Yet foreign investment was unlikely to be forthcoming for political reasons. The working class was too small to provide the necessary resources, and anyway the contradiction of a workers' party financing industrialization on the backs of workers was too acute. This seemed to imply that finance for industrialization would have to come from the peasantry. However, this carried with it the risk that the Bolsheviks would alienate the peasantry who were the largest group in the country and on whom they depended for food and other agricultural products. The whole question thus arose of the extent to which a party and government based on the working class, which still constituted a minority of society, could intervene in the development process and shape that process, without undemocratic consequences. On the other hand, since the working class in the USSR only had political representation through party and state organs, party and state involvement in industrialization seemed essential if industrialization was to serve the interests of workers.

The Soviet industrialization debate

Initially, the policies of the Bolsheviks after 1917 were formulated very much in response to pressures from outside the party. Large-scale industry was nationalized in response to factory occupations by workers, while land was distributed to the peasants in response to land occupations (Figure 12.1). Then in the civil war of 1918–21, during the period of 'war communism', the government established a system of total control over the economy of that part of the country which they still governed, as part of the mobilization of resources for the war effort. One current of opinion within the party, represented

Figure 12.1   Propaganda for the transformation of Soviet agriculture.

by Nikolai Bukharin and Evgeny Preobrazhensky, saw war communism as representing an opportunity for an immediate transition to a planned, moneyless socialist economy. However, by 1921 it became clear that the system was not workable in the long term. This was primarily because of severe peasant unrest at the requisitioning of grain. In 1921 Lenin instituted what was known as the New Economic Policy (NEP), the main features of which were state control of large and medium scale industry, finance and foreign trade, together with the restoration of market relations between town and country. After paying a tax in kind, peasants could market the rest of their output relatively freely. Politically the Bolsheviks emphasized the alliance or *smychka* between the working class and the peasantry.

The immediate effect of NEP was positive and allowed for the rebuilding of the shattered economy following the civil war. However, by the mid-1920s the question of how further industrialization was to be financed began to become critical. As the Soviet Union moved from reconstructing the old pre-war economy to new development, considerable resources were required. Various competing viewpoints emerged in a fascinating and wide-ranging debate (Erlich, 1960; Day, 1977).

Bukharin argued that industrialization was not possible without the consent of the peasantry. The peasantry could not be required to provide the resources for industrialization, and the pace of industrialization would have to be slowed to such a level that the growth of the economy generated enough resources for a steady industrial growth without placing a burden on the countryside. The state was extremely circumscribed in the role it could play and should not interfere with the equilibrium between town and country. Following the theory of 'socialism in one country', Bukharin believed that the USSR could maintain relatively self-sufficient development.

Preobrazhensky favoured taxing the peasantry in order to provide resources for a faster growth of industrial production and claimed that this would in turn benefit the countryside by increasing the flow of goods from town to country and thus boosting agriculture. The state and party were entitled and obliged to shape the paths of industrialization. Following Trotsky and the Left Opposition within the party, he continued to see the outside world, despite its capitalist nature, as a possible source of resources for the development of the USSR and hoped for socialist revolution elsewhere before too long.

Other figures such as Groman and Bazarov attempted to steer a middle course between these two positions. In addition, Chayanov developed a distinct approach to the analysis of the peasantry, inspiring a school of writers who argued that the peasant economy had to be understood in its own terms and that using this understanding to further agricultural development was a precondition for progress in the USSR.

By 1928 the development of the USSR had run into severe problems. It was not proving possible to provide sufficient incentives to the peasantry to produce enough grain to feed the towns. Consequently, industrial development was faltering. The problem appeared to have no easy solution. If the line of the Left Opposition were to be followed and the peasantry were to be taxed, they would be even less likely to support the industrialization effort. Yet without more resources industrialization could not proceed fast enough to modernize agriculture.

Stalinist collectivization and industrialization

The policies of Stalin in the early 1930s represented a response, an exceptionally brutal response, to the problems outlined at the end of the previous section. The difficulty of obtaining grain from the peasantry was met by forcible collectivization and grain extraction. It might appear that Stalinist development involved making the peasantry provide the resources for growth. For many years this was what people believed had happened; however, recent research reveals a slightly more complex picture. The sheer weight of destruction of resources, particularly livestock, in the countryside which resulted from collectivization, and from peasant resistance to collectivization, was so great that the agricultural sector was not able to provide a surplus which would finance industrial growth. In fact, in the early 1930s the reconstruction of agriculture was more of a drain on the towns than before. However, what collectivization did provide was a massive flow of labour from the countryside to the towns, as rural society was shattered. This flow of labour helped to limit industrial wages even as the economy grew tremendously quickly. As the Soviet government quickened the pace of growth, often financing this by printing money, inflation rocketed and real wages fell dramatically. In this way Stalinist industrialization was financed not just by the peasantry, but, ironically, by the working class in whose name Stalin claimed to rule. (Ellman, 1975; for a dissenting view see Morrison,

1982; a classic analysis is Rakovsky, 1981, first written in 1930).

Collectivization in the countryside was matched by the growth of industrial planning, particularly the famous Five Year Plans, in the urban areas. The two together acted as a powerful means for cementing the dominance of the party and state over the rest of society. The planning system also acted as a means of mobilizing resources and directing them into priority areas for industrial growth. Under the banner of the doctrine of Socialism in One Country, foreign trade dwindled sharply. In this way the socialist model of development outlined at the opening of this chapter began to crystallize. While urban unemployment had been a major problem under NEP, the sheer pace of growth in the USSR in the 1930s led to virtual full employment by the end of the decade, and this was maintained during and after the Second World War. As the planning system became institutionalized, the role of money lessened to become purely a means of accounting for transactions, and goods were allocated to firms through the planning mechanism rather than the market. The planners gradually extended their control to the prices of final consumer goods, as well as industrial goods, and the distinctive model of socialist development became complete. This model answered the questions faced by the Bolsheviks in the 1920s starkly and clearly: the relationship of the USSR to the rest of the world was essentially hostile, and independence was a priority. The state and party were to shape the industrialization process in the name of the working class, and while nominally workers and poor peasants were in alliance against rich peasants (or *kulaks*), in reality the working class and peasantry together were to provide the resources for industrial development (Figure 12.2).

Figure 12.2   *Two images of industrialization in the USSR: (left) an idealist image of Stalin at a new dam; (above) a grim reality, 1931.*

While it was tremendously harsh and wasteful, this model did allow for a dramatic industrialization effort in the USSR. The economy created was able to withstand Nazi Germany and, following the Second World War, to extend the framework created by Stalin to much of Eastern Europe. In this way the model of socialist development which had emerged from Stalinist industrialization became attractive, not only to communist parties outside Europe, but also to nationalist movements who saw it as the only feasible path to industrial development and economicindependence in a world dominated by imperialism.

12.3 The Chinese path to development under Mao

The first, and most important, test of the socialist model of development came with the Chinese revolution of 1949. Initially, the Chinese communist party followed a path very similar to the Soviet model, though with much less brutality and conflict. The first Chinese Five Year Plan was instituted in 1952, and collectivization was carried out in 1955–56. It is important to recognize the sheer scale of what the Chinese achieved in these years, including the largest land reform in history (a classic analysis is William Hinton's *Fanshen*, published in 1972). However, it quickly became clear that the Chinese case was significantly different from that of the USSR, and that the Soviet model of development could not simply be transposed to China. There were four main points of difference, described in Box 12.1.

The sum total of these differences was to make it extremely difficult for the Chinese to carry through a similar process of industrialization to that which had occurred in the USSR. The Chinese party arose out of, was dependent for its support on, and was based in rural society. Further, rural society and the countryside carried even more weight in terms of population

---

Box 12.1 The characteristics of the Chinese revolution

1   Despite some areas of development, the Chinese economy in 1949 was even weaker than the economy inherited by the Bolsheviks in 1917, with less industrial development. (For a more detailed account see Riskin, 1988, ch.1.)

2   The Chinese revolution was essentially a peasant revolution in a way the Russian revolution was not. The Chinese communist party had begun, in 1921, as a predominantly urban party, on the Russian model. However, the vast majority of the urban cadres of the party had been massacred in 1927 by the Chinese nationalist movement, the Kuomintang, under Chiang Kaishek. After 1927 the leadership of the party had been based in the rural areas, under Mao Tsetung. By 1949 both leadership and membership of the party were mainly peasant in origin.

3   The Chinese revolution was very much a national revolution, arising in large part out of the struggle against Japanese invasion and colonialism, and against imperialist domination from other large powers. During the nineteenth and early twentieth centuries Britain and France among others had forced trading treaties on the Chinese, including provision for the import of opium to China. China was still to a large extent not yet unified and under the sway of competing warlords when the Japanese invaded Manchuria in 1933, moving southwards in 1937.

4   The Chinese revolution arose out of a long period of war and civil war. When the Chinese communist party came to power it was in many ways a military organization with few formal democratic structures. On the other hand, because of long dependence on the rural population during wartime, it had achieved a considerable amount of sensitivity to changes in the mood of its supporters, while having few formal channels through which such changes could be expressed (Selden, 1971; Snow, 1972, is an eyewitness account).

and contribution to the economy in China than in the USSR in the 1920s. Industrialization based upon the break-up of rural society was not a feasible project in China. The working class in China was clearly too weak to provide the resources needed to industrialize the country. In the 1950s considerable assistance was forthcoming from abroad, from the USSR. However, by the late 1950s it was becoming clear that the USSR would not and could not provide enough resources to guarantee industrialization in China, and also that such resources as it did provide might entail an unacceptable loss of national independence in decision-making for the Chinese. The Chinese party was thus faced with trying to initiate development in China within severe constraints, both political and material, and within a framework of little if any formal democracy, but a high degree of informal responsiveness to the view of their supporters. In this context, development in China under Mao tended to take the form of a series of campaigns, the most important of which was the 'Great Leap Forward' of 1958–59, in which mass mobilization of the population acted as a substitute for formal political involvement, and was used as a way of trying to overcome the severe material constraints faced by the government (Maitan, 1976).

Collectivization in China and the problems encountered have recently been the subjects of lively debate. There are disputed views as to the extent that collectivization seriously weakened communist support in the countryside. (Nolan, 1976; Selden, 1984; Nolan, 1989). However, there is agreement that collectivization in China did not create adequate resources for industrial development. The Great Leap Forward was an attempt by Mao to solve this problem. It had two main features.

First, the problem of relations between town and country was to be solved by dissolving the barrier between the two. Responsibility for production was to be decentralized radically to more or less self-sufficient territorial units ('people's communes') which would manage both industrial and agricultural production. There would be some agricultural production in urban areas and a great deal of small-scale industrial production in rural areas (for example, the famous 'backyard steel production'). In this way, through breaking down the division between town and country ('walking on two legs') the problem of financing industrialization would be overcome.

Secondly, the Great Leap Forward involved a massive mobilization of the population through propaganda and example, calling on the Chinese people to speed up development almost as an act of will (Figure 12.3).

The Great Leap Forward represented a distinctively different approach to socialist development from the Stalinist path, as it responded to China's particular problems. However, despite embodying many important ideas which have reappeared in China and elsewhere later, it was unsuccessful. The quality of the industrial goods produced in the rural areas was not good enough, and much of the backyard steel production, for example, was unusable. Industrial output figures were hugely exaggerated. Agricultural disruption led to a calamitous famine in 1959–61. The rest of the party leadership abandoned the Great Leap and forced Mao into a 'backseat' role in economic policy, making him little more than a figurehead.

Following the abandonment of the Great Leap Forward, the Chinese reverted to a model of development which resembled much more closely the Soviet model. However, much more attention was paid to providing resources for the countryside and ensuring balanced growth between town and country. Despite the violent political upheavals of the late 1960s and the cultural revolution which followed, as Mao attempted to regain the power from which he had been ousted, economic policies remained remarkably constant until the mid-1970s. The Maoist faction which attained dominance in the cultural revolution did not succeed in implementing any significantly different model of development. It has been argued that the rebellion in 1971 by Lin Piao, the Minister of Defence and Mao's chosen successor, was aimed at giving a higher priority to industrial development, particularly electronics; however, this rebellion was

Figure 12.3   Painting the peasants' vision of a better life in China.

crushed and development continued along the same path (Maitan, 1976, ch.14; Eckstein, 1977, pp.59–63).

12.4   The crisis of development in the USSR and China

In retrospect, the mid-1970s appear to be a turning point in both the USSR and China for different reasons. From 1975 onwards there was a decisive slowdown in the rate of growth in the USSR. The reasons for this are undoubtedly complex, but one central factor was widely seen as operating by observers both in the USSR and elsewhere. The Stalinist model of industrialization had proved effective as a means of mobilizing resources for the initial stages of industrialization. However, while it worked, although crudely, for 'extensive' industrialization, which was dependent on drawing new resources of both capital and labour (women, rural workers) into industry, it was much less well suited to 'intensive' industrialization which depended on technological development, increasing levels of skill and using existing resources more efficiently. By the mid-1970s there were few unused resources in the USSR for further extensive development, and intensive development had now become crucial. Further, as the population became urbanized and educated, the crude mechanisms of labour control and political repression became ineffective and labour productivity ceased to grow. This was accentuated by social problems, such as widespread alcohol abuse, and by the drain on resources caused by the stagnant agricultural sector. Finally, technological development appeared to require a greater openness to the West in terms both of foreign investment and imports of technology, with possible accompanying political changes. These issues provided the background for the changes which occurred in the USSR after the election of Mikhail Gorbachev as General Secretary of the Communist Party in 1985 and the initiation of the policy of *perestroika* or reconstruction.

Whereas in the USSR the problems of the socialist model of development related to the fact that a measure of development had taken place and in turn had thrown up new problems, in China the problem was that the model had failed to ensure industrialization and development. By the time of Mao's death in 1976 industrial growth rates were falling significantly, and agricultural production was barely keeping pace with population growth. In addition, the situation of the poorest people in the countryside was such that Chinese official estimates in 1980 indicated that about 100 million peasants depended on state relief (McFarlane, 1984, p.25).

The response of the Chinese Government to these problems was to institute a programme of thorough economic reform. Initially the reform was centred in the countryside, and involved breaking up the communes, which were later abolished, and basing agriculture on small-scale family production. This 'production responsibility system' took several forms, but increasingly families came to act as *de facto* private peasant producers, leasing land and some equipment from the state. Small-scale rural industry was again encouraged, but now on the basis of production for the market. From 1984 onwards the pace of reform quickened in the urban industrial sector, with enterprises becoming linked much more through market relationships, and with

responsibility for pricing and production decisions being decentralized to individual firms. China began to encourage foreign investment, particularly in 'Special Economic Zones' where incoming investors were to be attracted by cheap labour, freedom from taxes, and investment incentives.

The Chinese communists regarded these reforms as consistent with a socialist road to development. However, they had drastically changed their conception of the timescale required for such a road to be followed. They now saw a long period of market-based industrialization as a necessary first step on the way towards a socialism which now lay decades into the future.

Early events appeared to favour these reforms. Rural production, both in agriculture and industry, grew significantly after 1978. While inequalities emerged in the countryside, the narrowing of differentials between regions and between town and country which resulted from increased living standards in the country may well have reduced inequality in China as a whole. Industrial production also quickened. However, by the mid-1980s the Chinese economy was under severe strain. Inflation had become a major problem. Investment in the Special Economic Zones was not feeding through enough to the rest of the economy, and nationalist sentiment was becoming more prevalent. There had been problematic shifts in rural production: away from grain production (where prices were fixed by the state to protect urban workers) to other crops, and away from food production as a whole to small-scale rural industry. Youth unemployment was a problem in the cities, as were demands for higher wages and better conditions from students and urban workers. Speculation grew as people bought goods at fixed state prices in one market and then sold them on the free market at higher prices.

These economic problems, and the associated lack of political democracy, have led to a profound crisis for socialism in China. Equally, the USSR is now undergoing massive upheaval. However, it is important to grasp the differences between these two cases. In the USSR the socialist model of development has in a sense achieved its object and outlived its usefulness, whereas in China this model has had to be revised fundamentally as it failed to ensure industrialization and development. The two crises together, however, carry significant implications for those who wish to apply the socialist model of development elsewhere in the world. In the next section we explore these implications.

12.5   Socialist development in a wider context

China was by no means the only country in the post-war world where socialist development proved a powerful attraction. There are three main groups of countries where the socialist model was either adopted or was a major influence on political and economic developments. I shall look at them in turn.

First, and least numerous, were those countries which had escaped formal colonization and which remained largely separated from the world economy, ruled by archaic monarchies or traditional regimes. In these societies there was no indigenous capitalist class which appeared able to initiate development and no sign of development occurring as a result of foreign penetration. The only agency that appeared capable of bringing about any development was the state. Consequently, socialist ideas, with their stress on state-sponsored industrialization, appeared attractive to groupings in the intelligentsia and the army. Examples here are Afghanistan (Halliday, 1978, 1980) and Ethiopia (Halliday & Molyneux, 1981). However, the record of such revolutions has not been good. The parties or groupings which carried them out have had insufficient social backing, particularly in the rural areas, and this has been compounded by extensive social and cultural resistance to the idea of a unified interventionist state in these societies. Further, ill-judged intervention by the USSR and foreign aid for rebel forces in Afghanistan, and secessionist movements in Eritrea

and Tigray within the Ethiopian state, weakened these governments to the point where survival rather than social transformation was a priority. The only successful example of this kind of revolution, now also undergoing great changes, remains the earliest, Mongolia.

The second grouping comprises those countries where socialist ideas became dominant in national liberation struggles against colonialism. This was particularly the case where decolonization was delayed and only achieved after fairly prolonged guerilla struggle (South Yemen) or protracted war (Angola, Mozambique and the other Portuguese colonies in Africa). It was also the case where the decolonization process was blocked by foreign intervention and national division and partition (Korea, Vietnam). Related to this grouping, but slightly separate, are certain countries in the Caribbean and in Central America. Socialist ideas were able to gain predominance in popular national movements against weak national capitalist classes, in Cuba and, in a different way, in Nicaragua. While these were not formally national liberation struggles against overtly colonial regimes, they had many of the characteristics of such struggles because of the overwhelming dominance of the USA in the political and economic life of these countries. This meant that the ruling groups in these countries had a relatively small degree of real independence from the dominant imperial power in the region, the USA, and relatively weak roots in the society. The struggle against these ruling groups took on many of the characteristics of an anti-colonial struggle.

The record of socialist development in these cases has been more successful, but has still encountered major problems. External hostility has taken the form both of military intervention and subversion and of denial of aid and investment. This has led in some cases to considerable dependence on the USSR for material support, which now appears rather fragile. Relationships with the peasantry have not always been easy, and there have been major reassessments of relationships with the peasantry and of collectivization in Vietnam, Mozambique and

Nicaragua. Most fundamentally, the extent of significant industrial development and growth of material living standards in these countries has been limited. The most impressive industrial development in quantitative terms is probably that of North Korea, heavily contingent on a remarkable endowment of raw materials and minerals and marred by a grotesque personality cult centred around the leadership of Kim Il Sung (Halliday, 1981). Even here, neighbouring South Korea has clearly outstripped the north in industrial production. As mentioned above, South Yemen has dissolved into North Yemen, Vietnam has embarked on a major economic reform along Chinese lines, and Angola and Mozambique have moved dramatically towards a market-based economy. In Nicaragua, the Sandinista government, weakened by external military intervention, civil war and economic collapse necessitating austerity measures and cuts in real wages, lost the election in February 1990.

In the third group of countries, external factors have also played a major role. Although socialist or communist parties did not come to power after decolonization, the nationalist movements which did entered into loose alliances with socialists and appeared to adopt socialist policies. Such movements were often based around state employees: soldiers, teachers, doctors, or associated professional people such as lawyers, who shared the outlook that the state should be the prime motor of development. Emerging out of anti-colonial struggles, state control of the economy appeared to be a guarantee of national independence and unity in a hostile world. Consequently, these movements and governments adopted many of the formal trappings of the socialist model of development (five-year plans, nationalization, control over finance and foreign trade) and in return received broad support from national communist parties, and from the USSR. Examples include Egypt under Nasser after the later 1950s, Iraq after the revolution of 1958, Indonesia in the 1950s under Sukarno, India under Nehru after independence, various different forms of 'African socialism', and many others. The result of this has not been

encouraging for proponents of socialist development. In these cases, almost without exception, the rhetoric of socialism has now been dropped and the countries in question have opted for capitalist development of one kind or another, involving free access for multinational capital and the institution of market relationships within the economy. In several cases denationalization and privatization are also on the agenda, or have already occurred. At best, communists and socialists have been marginalized, as in India, to relatively ineffective bases in regional government; at worst, as in Indonesia, they have been massacred and the old nationalist government has been replaced by authoritarian government from the Right.

The reasons for this sorry story are many and complex. However, one central factor must be taken into account. With the increasing internationalization of the world economy it can now be questioned whether the kind of state-centred, nationally independent development which these countries attempted to initiate in the immediate post-war years is any longer possible. In particular, state-centred development runs up against two major disadvantages. First, the resources that individual nation states can put into industrialization now pale beside those of multinational companies. Secondly, individual countries' economic policies are now limited by an ever more developed international financial system which can direct and redirect monetary flows between countries with exceptional speed. In these circumstances, state-sponsored development appears increasingly unattractive, and even utopian, compared with opening up the country to a world economy which can dictate the allocation of incalculably more resources than relatively poor nation states.

For all these reasons, then, the socialist model of development, not just in the USSR and China but also elsewhere in the world, is in danger of appearing an unattractive and unviable model with little, if any, immediate future. But sources of renewal for socialist ideas and projects do exist, and these are examined in the next section.

12.6 Renewal of the socialist model

Two distinct, but not exclusive, concepts of socialist renewal have been put forward as responses to the problems outlined earlier in this chapter.

First, there is the viewpoint recently outlined by writers such as Giovanni Arrighi (1990) and Fred Halliday (1990a), which sees the main relevance of socialism as having shifted from the provision of a path for poor countries towards independence and development back towards the classical conception of socialism as the means towards the criticism and overthrow of developed capitalist systems. According to this view, socialist movements will once more have to become international in character, and will focus on transcending industrial capitalism. The main support for such movements will thus come from the working class and other oppressed groups in the advanced capitalist countries. This view, however, appears to leave little role for socialism outside those countries. The danger arises that socialism will become seen as a Eurocentric ideology, and that, while it may criticize imperialist domination of other countries, it will have little purchase on challenging that domination in any concrete way.

It is at this point that a second concept of socialist renewal becomes important. This viewpoint stresses the kind of industrialization and development which has taken place outside the main capitalist countries in the last 30 years. While highly uneven, exploitative and precarious in many ways, this development has led to the growth of sizeable working classes and national capitalist classes in many countries. These countries have undergone a considerable amount of industrialization and are no longer strictly comparable to the relatively poor and predominantly agricultural societies in which the socialist model of development has previously mainly been tried. Examples include South Korea, Turkey, Mexico, Brazil, Iran, South Africa, the Philippines; some would add Eastern European countries to the list, such as Poland. In such countries, socialism remains a potential force. Such countries have resources which could be used to finance a socialist political project, and a working class which is a potential beneficiary of and support for such a project, yet in most cases they remain excluded from equal participation in the world economy with the developed capitalist countries. In such circumstances socialism once more can appear as a potential guardian of national independence and development, and also of equality and humanity.

Strong movements of the Left have emerged in all the countries mentioned above in the last 15 years; however, the form they have taken has differed sharply either from traditional communist parties or national liberation movements. The most salient difference is in the emergence of mass political trade unionism as the backbone of socialist politics, as has been seen in Brazil from 1978–79 onwards (Beecham & Eidenham, 1987), in South Africa from 1973 (Friedman, 1987), in South Korea from the mid-1980s (Asia Labour Monitor, 1988), in Turkey prior to the military coup of 1980 and in the Philippines both before and after the fall of the Marcos regime (ILR, 1988). In the Eastern European context a similar phenomenon was the strike wave of 1980 and the emergence of Solidarity. However, trade union movements in Brazil, South Africa and the Philippines, for example, have realized that trade unionism on its own cannot achieve their goals. Consequently, they have linked up with wider political movements and this has resulted in a constant tension between explicitly socialist goals and other political projects, often based on some form of nationalism. The most favourable example for socialists is probably Brazil, where the trade unions helped to form a workers' party, the PT, which is now a potent political force in the country. In South Africa the majority of trade unions have aligned themselves with one or other of the national liberation movements such as the ANC. Other cases have been less auspicious so far. In South Korea the political opposition remained divided, allowing the old military regime to engineer a smooth succession despite the 1987 strike wave. In Mexico, the 'populist' movement led by

Cuauhtemoc Cardenas had a relatively weak base amongst workers, due to the incorporation of the trade union movement over many years by the ruling party, the PRI. Consequently, it has so far proved unable to resist the implementation of free-market based austerity policies. In other countries, trade unions have existed in an ambiguous and problematic relationship with more traditional left-wing forces: in Turkey the communist party; in the Philippines a communist-led guerilla movement, the New Peoples Army. In neither case has the Left been able to achieve dominance in the society as a whole. Most tragically, in Iran the workers' movement and workers' councils were forcibly disbanded by the Khomeini regime following the revolution of 1979, and in Poland Solidarity, weakened by the 1981 declaration of martial law, has turned towards a free market programme and lost much of its support in the factories (Callinicos, 1987).

The potential for a renewal of the socialist project in these 'middle income' countries remains, then, as yet unfulfilled. However, the movements that have emerged do offer the prospect of such a renewal. The socialist idea retains much of its power in these countries. The problem of providing the resources for development is less acute as a result of the industrialization which has taken place. The problem of relationships with the peasantry is also less acute, though still important, as shown by the role of private farmers in Poland. However, Brazil has shown the possibility of new kinds of alliance between workers and rural inhabitants – for example, over environmental issues such as deforestation (Treece, 1988). In some countries, particularly South Africa, the independent peasantry is now much weakened and largely absorbed into capitalist agriculture, where it faces many of the same issues as urban workers (Krikler, 1987). Such countries, then, are sufficiently different from the countries where the socialist model of development has previously been tried to hold out the possibility of a rejuvenation of that model in the future. Paradoxically, at a time when many have doubted the continued importance of the industrial working class as a motor for change in Western society, it appears more numerous and assertive in many developing economies than at any previous point. Furthermore, the political culture of these countries, with considerable experience of autonomous organization now being accrued, is in many ways much more favourable to vibrant socialist movements which can retain independence of the state than that ruling in Russia in 1917 or China in 1949.

One issue remains a major difficulty for these movements. That is the question, stressed earlier, of the possibility of national independent development in an increasingly independent world. As yet none of the movements described above has succeeded in putting forward a distinctive model of the relationship between a socialist economy and the world system. It is the absence of such a distinctive perspective which leads to the necessity for political collaboration with other Left-inclined movements of the kind outlined above. However, such movements of a nationalist or 'populist' persuasion again run the risk of failing to articulate a convincing model of national economic development, just as their predecessors of the 1950s and 1960s in Egypt, Indonesia, Iraq and elsewhere failed, and also risk being drawn back into the orbit of multinational capital and international finance. In short, without a clear conception of the possibilities for socialist development in an integrated world economy, it becomes harder for socialist movements to assert a class interest against cross-class national movements, and harder for national movements to assert a national interest against multinational forces.

It would seem that socialist renewal is no longer centred around the idea of a socialist model of development, but rather around the idea of socialist interventions, up to and including the taking of state power, in an interdependent world economy. However, the success of such a renewal appears to depend on the ability to articulate a convincing conception of the limits and possibilities for nationally based movements in such a world economy. This conception, if it arises, is likely to depend on the emerging community of interests between workers in an interdependent world.

Summary

The socialist model of development emerged out of a particular historical experience, that of the USSR following the Bolshevik revolution. However, the problems faced by the Bolsheviks were of a general nature, and the industrialization achieved in the USSR thus appeared to offer lessons for other countries attempting to develop and industrialize. In part, this was true. However, the model proved to be of more limited applicability than had been hoped, particularly in the Chinese case but also when applied elsewhere. In addition, as the world economy has become more interdependent, the very concept of a nationally based socialist road to development has been called into question. That interdependence though, and the global industrialization which accompanies it, raises the possibility of the renewal of the socialist model, both through reassertion of socialism in the developed world, and through new movements in the industrializing world, which no longer operate under such severe material constraints, and which have a powerful base of support in a numerous working class.

UNDERSTANDING DEVELOPMENT IN THE 1990s

13

THE THIRD WORLD IN THE NEW GLOBAL ORDER

ANTHONY McGREW

Figure 13.1   Presidents Bush and Gorbachev at the signing of the CSCE Joint Declaration, 1990.

On 19 November 1990 the member states of the North Atlantic alliance and the Warsaw Pact met in Paris to agree a joint declaration officially confirming the end of the Cold War. This historic declaration not only affirmed 'the end of the era of division and conflict which has lasted for more than four decades' but also that East and West 'are no longer adversaries' (CSCE Joint Declaration, 1990). The language of this agreement signified the end of the post-war era and the termination of the global struggle between two opposing socio-economic systems: capitalism and communism (Figure 13.1).

A new era in world politics has begun, one that is no longer haunted by the twin legacies of 1917 and the 1945 Yalta accords, events which conspired to create a divided world. Instead Western and Soviet diplomatic energies are today expended upon the construction of a new, post Cold War global order. Underlying this effort is a realization that the complex and interdependent character of the modern world system demands both new 'thinking' and, increasingly, new mechanisms of global management and cooperation. World politics appears to be in transition, rapidly being divested of old enmities,

Acronyms used in this chapter are spelt out in full in the List of Acronyms, Abbreviations and Organizations at the end of the book.

alignments and modes of thinking but confronting an uncertain and unpredictable future. This environment of change and uncertainty has imposed a particular burden on Third World states. While new political opportunities exist for advancing the collective interests of the Third World in the emerging post Cold War order, equally, new threats and challenges are posed to Third World development strategies.

The third part of this book aims to introduce a range of tools for analysing development in the contemporary world. In subsequent chapters it will become clear how much importance should be attached to asking questions about how development relates to the state and to technology, to analysing development in terms of gender relations, and to 'taking culture seriously'. In this chapter we explore how opportunities and strategies for development in the 1990s are likely to be conditioned by changes in the global order.

Q Is the idea of the 'Third World' as a political entity still valid, or does the new global order mean 'the end of the Third World'?

Q What are the implications of contemporary changes in the global order for development in the 1990s?

This chapter reviews the dynamic of the North–South relationship in the post-war period, in the context of the institutional framework described in Chapter 11. It then discusses the global political and economic forces which have conditioned Third World development, as well as the collective attempts of Third World states to refashion the global order in accordance with their own needs and interests. Finally, it sets out the emerging characteristics of the contemporary global system and assesses the distinctive character of the development opportunities and challenges of the 1990s.

13.1 The post-war global order

The very category of the 'Third World' was defined by the existence of the Cold War conflict between East and West, the First and Second Worlds respectively. Invented primarily as political and journalistic shorthand to describe an emerging coalition of newly independent states, the term is today embedded in the discourse and diplomacy of international relations. Yet, as will become apparent, it is doubtful whether the term retains any convincing political or conceptual validity in the contemporary world, not least because of the demise of the Cold War which gave it its political currency. What is incontrovertible, however, is that from the 1950s through to the late 1970s, common economic problems, the desire to stand outside the East–West conflict, and the commitment to restructuring the global economic system provided the political motivation for these newly independent states to operate collectively as a bloc in the international arena. How did this Third World coalition arise? What forces nurtured its existence?

Throughout the post-war period, until the early 1990s, the central axis of world politics was East–West competition, the 'Great Contest' between capitalism and communism (Halliday, 1987). In the aftermath of the Second World War, unable to agree upon the shape of the post-war global order, the two superpowers, the USA and the Soviet Union, established their own respective alliance and bloc systems. In the process, Europe was divided and the Cold War imposed its own dynamic on international politics. Whether global peace or war prevailed depended entirely upon how the two superpowers managed their bilateral relationship. This situation was compounded by the existence and expansion of their respective arsenals of nuclear weapons.

Accordingly, military power was highly concentrated in the hands of both superpowers. No other state or even group of states could challenge the military supremacy of either of the two Great Powers; nor could either power hope to defeat the other. The result was a military stalemate, a bipolar division of the world into two armed camps in which the stability of the whole global order was contingent upon how well the superpowers could manage their rivalry. Not surprisingly, the global agenda reflected a pre-occupation with East–West issues. One consequence of this was that other important matters, such as North–South relations, tended to be marginalized or manipulated by both superpowers in pursuit of their own strategic advantage.

For much of the post-war period, therefore, the Third World only became politically salient in so far as it impinged upon or became implicated in the global rivalry of the two superpowers. Indeed, the superpowers exported their competition to the newly independent states as each sought to prevent the other from extending its influence. From the 1960s onwards the Third World became a primary arena for superpower rivalry, with many unwelcome consequences for those states unfortunate enough to be implicated in the 'Great Contest'. As Halliday notes:

> "...with over 140 conflicts, costing over 20 million lives, the Third World has remained an area of conflict...in which the USA and the USSR have repeatedly been involved."
>
> (Halliday, 1989, p.13)

Achieving statehood in a tightly bipolar world rapidly sensitized many newly independent nations to their subordinate status in the global power hierarchy and encouraged a desire to avoid overentanglement with either superpower. Combined with a shared reaction against colonialism and the confronting of common development problems, a collective consciousness emerged, nurtured and articulated through the formation of the non-aligned movement (NAM). From its inception following the Bandung Conference in 1955, the NAM operated as a kind of international pressure group for the Third World. Composed mainly of newly independent states, the NAM sought to construct a political coalition and encourage solidarity among the emerging nations in their struggle against new forms of domination – in particular, superpower interventionism. In effect, the NAM functioned in its early years as an international coalition promoting the interests of the Third World in a system dominated by the East–West conflict. It sought to restructure the international system to accommodate the development needs of the poorer and newly independent states. Most importantly, however, it created the political conditions which nurtured and legitimized the formation of a Third World bloc within the global system. Out of the NAM emerged the Group of 77 (G77 – now over 120 states) which brought together the majority of Third World states as a voting bloc within the United Nations (and its associated agencies). The G77 functioned, and continues to do so, as a counterbalance to the influence of the Western industrial states within the UN and other global institutions, since on most issues it can mobilize a voting majority. Through the NAM and the G77, the Third World gradually established a collective identity and emerged as a significant political force in the global arena.

From the early 1960s until the early 1980s, this Third World bloc used its majority voting power within the UN to redirect the global political agenda away from East–West issues to the needs of the newly independent states. North–South issues achieved increasing primacy within the UN and its specialized agencies. In 1960 the UN General Assembly passed the 'Declaration on the Granting of Independence to Colonial Countries and Peoples', thereby establishing the legitimate right to national self-determination for all colonial peoples. Besides seeking to promote and defend its rights, this new grouping also sought to transform the relationship between the economically advanced and the underdeveloped nations through the creation of new international or intergovernmental organizations (IGOs), such as the UN Conference on Trade and Development (UNCTAD). As a consequence, a new axis of conflict in world politics emerged: the North–South confrontation.

13.2 The 'Third World' and attempts to restructure the global order

From the 1960s through to the late 1970s the political achievements of this Third World bloc were nothing short of spectacular. Even staunch critics of 'Third-World-ism' – that body of thinking which argues for global primacy to be given to Third World needs – have recognized that the bloc's cohesiveness and sense of collective identity were impressive: 'That this unity could be maintained in the light of the greater political, economic, and cultural heterogeneity amongst the G77 was remarkable' (Michalak, 1983, p.28). Yet for the majority of Third World states this collective political power has purchased only meagre benefits in terms of creating favourable conditions for development. As Samir Amin observed: 'If the 1960s were characterized by the great hope of seeing an irreversible process of development launched throughout what came to be called the Third World...the present age is one of disillusionment' (Amin, 1990a, p.1). What then has been the fate of Third World attempts to refashion the global order and why?

The primary objective of both the NAM and G77 was never simply to counteract superpower dominance within the global system. Both also sought to remove the many barriers to Third World development. This translated into a desire to restructure the institutions, relationships and mechanisms which governed the world capitalist economy in which, for historical reasons, these states were inescapably embedded. The purpose of this restructuring was to redress the inbuilt inequalities of the system in order to provide the foundations for effective national development. Accordingly, the G77 sought changes to the governing arrangements which had been fashioned in the early post-war period (see Chapter 11) to manage trading, financial and technological relationships between states within the world capitalist order. These arrangements for international governance were based on what Chapter 11 called the four Bretton Woods institutions. At least three of these (GATT, the IMF and the World Bank) oversaw regulatory regimes that in practice embodied the interests and values of the dominant Western states, most obviously the USA as the hegemonic capitalist power.

As a result they largely failed to reflect what the G77 perceived as the legitimate development needs of Third World nations. Indeed, the majority of developing states regarded these regimes with deep suspicion as reinforcing the underlying structures of economic dependence which maintained the subordination of the South to the affluent North. In effect, most Third World states considered that all these regimes undermined rather than assisted the prospects for real development. Accordingly, the dominant political strategy of the G77 involved exploiting its voting power within each international agency in order to restructure these global regimes in ways which reflected the development requirements and priorities of the Third World. This strategy, it was believed, was a crucial first step towards correcting the structural inequalities perpetuated by the logic of the prevailing world capitalist order.

Among the initial political 'achievements' of this G77 strategy were the revisions to the GATT regime in 1964. These revisions recognized for the first time the special status and needs of the developing countries in the international trading system. A second achievement was the creation of UNCTAD, which was to be the primary forum within which North and South could negotiate on global economic reform. Building upon its success, the G77 set about negotiating with the North (within UNCTAD) a complete system of trade preferences which would reduce or limit tariff barriers and import restrictions on the Third World's exports. This was a significant development since it challenged directly the 'free trade' ideology which was fundamental to the 'embedded liberalism' sustaining and legitimizing the post-war world capitalist order. Somewhat surprisingly, the negotiations were successfully concluded and in 1970 the Generalized System of Preferences (GSP) was established, although it did not take effect until 1976. But these initial reforms were effectively a prelude to demands for a more comprehensive as well as radical restructuring of North–South relations.

In 1974, exploiting the new-found commodity power of the OPEC oil cartel, the G77 and NAM persuaded the Sixth Special Session of the UN General Assembly to agree a Declaration on the Establishment of a New International Economic Order (NIEO). This, together with a programme for action and Charter of Economic Rights and

Duties of States, established a radical agenda for change. In crude terms, the NIEO aimed to replace the dominance of the market and neo-liberal principles in determining economic relations between North and South with, as Krasner (1985, p.5) suggests, authoritative international regimes based upon the principles of equality and justice. Through commodity agreements, more preferential trade arrangements, increased aid, debt relief, controls over multinational corporations, and the democratization of all the key international financial institutions (such as the IMF and World Bank), the NIEO offered the prospect of ameliorating some of the structural impediments to Third World advancement. Alongside demands for a NIEO there were also demands for a New International Information Order (NIIO) as Third World governments began to recognize the industrial and political significance of the emerging information technology revolution.

But this challenge to the dominance of the Western industrialized states with respect to the management of the global economy failed to deliver any significant 'victories' for the Third World. As Halliday somewhat harshly observes:

"The campaign for a NIEO, launched on the wave of enthusiasm following OPEC's success in 1973, achieved none of its stated goals and was, by the early 1980s, a stalled campaign, in practice as dead as the League of Nations: impatiently rejected by Washington, politely but ineffectively endorsed by Moscow."

(Halliday, 1989, p.21)

Why did this campaign fail? This is an important question given that the 1970s was a period which 'offered a unique window of opportunity for the Third World because it was a period characterized by Third World control of major international forums but continued Northern (i.e. Western) commitment' (Krasner, 1985, p.11).

It would be historically inaccurate to imply that the diplomacy surrounding the NIEO, which extended well into the 1980s, produced nothing substantive. A Common Fund and a Compensatory Finance Fund were established, both designed to assist primary commodity producers through price stabilization and support mechanisms. Moreover, the politics of the NIEO encouraged the formation of regional arrangements, such as the 1975 Lomé Convention between the European Community (EC) and the African, Caribbean and Pacific (ACP) states, which embodied a number of progressive proposals for international economic reform. But despite these 'achievements' the fundamental objective of the NIEO campaign, to restructure rather than simply to reform the international economic system, was never achieved. The failure of this campaign signalled the close of an era in North–South relations, for it symbolized the demise of 'Third-World-ism'.

Four factors contributed to this situation. First, the already fragile political cohesion of the Third World bloc began to evaporate rapidly as the changing political climate of the 1980s intensified existing internal contradictions. Conflicts of interest between oil producing states and oil consuming states, between 'newly industrializing countries' (NICs), and the 'Fourth World' countries or LLDCs (see Chapter 11), between Islamic fundamentalist and non-Islamic states, combined with deepening regional instabilities and political revolutions, shattered the pragmatic consensus which had sustained the unity of the G77 throughout the 1960s and early 1970s. Moreover, growing economic differentiation within the Third World, most evident in the rise of the NICs, undermined the 'dependency ideology' which 'provided the basis not simply for unity among Third World states but for the very idea of the Third World itself' (Krasner, 1985, p.88).

Secondly, the fragmentation of the Third World coalition occurred just as many industrialized states, notably the USA and the UK, embraced monetarist and market-oriented philosophies of international economic management. Associated with this was an aggressive anti-Third-Worldism. This was expressed both in the West's failure to implement agreed reforms (e.g. the Common Fund) and the (temporary) US retreat, in the early 1980s, from multilateralism, which as Mayall notes: 'immediately undermined the basic assumption on which the Third World reform movement was based, namely that numerical voting power could partially offset relative weaknesses in economic and military capabilities' (Mayall, 1990, p.144).

Thirdly, economic recession in the early 1980s, combined with a process of global industrial restructuring, sealed the fate of the Third World challenge to the prevailing order. Whereas the political momentum for reform gathered pace during the 1960s and 1970s at a time when the Third World was doing relatively well economically, it dissipated rapidly as the combined effects of recession, economic restructuring and debt absorbed the political energies of the majority of Third World governments.

Fourthly, the 'end of the Third World' as a global political coalition can only be understood in the context of a series of fundamental changes in the character of the world order, which became vividly apparent in the early 1990s with the termination of the Cold War. Indeed, it was no historical accident that talk of the end of the Cold War, the end of the Third World, and even the 'end of history' surfaced at much the same time, for such discourses reflected a common belief that world politics was in a state of transition, that civilization was on the cusp of a profound transformation. But is this an accurate description of the contemporary historical condition? Should we be composing an epitaph or a new epithet for the Third World?

13.3 Characteristics of the contemporary global system

Writing in 1974, well in advance of the many profound changes which occurred in world politics in the late 1980s, Seyom Brown argued that:

> "The alignments and antagonisms of the recent past are shifting ground, and the structures premised on their stability appear to be crumbling. Even the bedrock of the international system, the sovereignty of nation states, is subject to severe erosion."
>
> (Brown, 1974, p.1)

Events have transpired to make Brown's argument appear decidedly prescient, for the two dominant axes of post-war international politics, the East–West conflict and the North–South conflict, no longer appear to define contemporary global political reality. The virtual abandonment of communism in the East and the relative decline of the USA in the West have brought the Cold War to an end and laid to rest the ideological conflict between capitalism and communism. Similarly, economic crisis and restructuring within the world capitalist system have contributed enormously to the diffusion and displacement of the North–South confrontation. In addition, the increasing significance of global and transnational policy problems, such as the environment, has reinforced the erosion of these 'old' alignments and enmities. However, such changes by no means promise a more peaceful, co-operative or stable world order.

The dominant characteristics of the global system in the 1990s are superpower decline, globalization, complexity and diversity, the possible decline of the nation state, and processes of international governance. Let us look at each in turn.

Superpower decline

Third World states in the 1990s confront a highly complex and differentiated global order. This, as will become more evident, has critical ramifications not only for national development strategies but also for collective attempts to redress global inequalities. In particular, the termination of the Cold War has confirmed the end of a bipolar world, for although the Soviet Union and the USA remain military superpowers, political and economic power in the global system is today more widely dispersed than at any point in the post-war epoch. Even in the military sphere the relative position of both powers is likely to be eroded, as mutually agreed arms reductions and the scaling down of their military establishments significantly reduce their military capabilities. Such change is already apparent in the global military order where the relative dominance of both superpowers has been progressively eroded over the years.

The relative decline in the military supremacy of both superpowers is evidenced in, among other things, the spread of sophisticated military capabilities and indigenous military production to the Third World, as shown in Table 13.1 (see also Box 11.5 in Chapter 11 on the arms trade). To a degree, this was demonstrated in the 1991 Gulf War in which an unprecedented Western military force was apparently required to defeat Iraq.

Moreover, despite continuing academic debate, there is an emerging consensus that the Cold War came to a rather abrupt end partly because neither superpower, but most obviously the Soviet Union, could sustain the domestic burdens of global rivalry. The relative decline of both powers implies the end of bipolarity, for power is no longer predominantly concentrated in Washington and Moscow. Despite some apparent US claims to sole dominance, the contemporary

Table 13.1 Third World military capabilities: number of Third World producers of at least one major weapons system, 1960 and 1980 (with examples of types of major weapons produced)

1959–60	1979–80
Argentina (lightplane)	Argentina (all types)
Brazil (lightplane, trainers)	Brazil (all types)
Burma (corvette)	Chile (tanker)
Chile (light trainer)	China (all types)
China (all types)	Colombia (lightplane)
Colombia R. (LC)	Dominican R. (LC)
Egypt (trainer)	Ecuador (corvette)
India (trainers and engines, transport planes)	Egypt (ATM, PB, helicopters)
Indonesia (trainer)	Fiji (surveyship)
Israel (trainer)	India (all types)
Korea, N. (PB)	Indonesia (trainer, helicopter, LC)
Mexico (CPB)	Israel (all types)
Peru (tanker)	Korea, N. (PB, sub)
Total = 14	Korea, S. (fighter, APC)
	Malaysia (PB)
	Mexico (CPB)
	Nigeria (helicopter)
	Pakistan (helicopter, trainer, ATM)
	Peru (frigate, surveyship, tanker)
	Philippines (CPB, lt. transport, helicopter)
	Singapore (PB)
	Taiwan (fighter, AAM, ShShM, helicopter, PB, frigate)
	Thailand (PB)
	Venezuela (PB)
	Total = 26

Abbreviations

AAM, air-to-air missile;
APC, armoured personnel carrier;
ATM, antitank missile;
CPB, coastal patrol boat;
LC, light craft;
PB, patrol boat;
ShShM, ship-to-ship missile;
sub, submarine.

Source: Data from Newman, S. G. (1984) 'International stratification and Third World military industries', International Organization, 38(1), pp.167–98.

global system may be more accurately described as multipolar or polycentric, because of the existence of new centres of power, such as Germany, Japan, and even the EC. This shifting structure of power, as will become apparent, has significant implications for the development prospects of all Third World states.

Globalization

In parallel with the changing structure of power, processes of **globalization** have intensified in recent years. As the earlier chapters on hunger, diseases and the environment have shown, government policies or private activities in one part of the world can have ramifications for people's fate anywhere across the globe. Far from being an abstract concept, globalization articulates one of the dominant characteristics of modern existence. A moment's reflection on the contents of our own kitchen cupboards would underline that, simply as passive consumers, we are very much part of a global network of production and exchange (Figure 13.2).

> **Globalization:** The forging of a multiplicity of linkages and interconnections between the states and societies which make up the modern world system. The processes by which events, decisions, and activities in one part of the world can come to have significant consequences for individuals and communities in quite distant parts of the globe.

Of course, globalization does not mean that the world is becoming more politically united or culturally homogeneous. Indeed, globalization is highly uneven in its scope and highly differentiated in its consequences. For example, urban life in the capital cities of most Latin American countries is perhaps much more deeply implicated in global processes than, for instance, is rural life in the Orkneys or Shetland Isles. Nor can globalization simply be equated with the globalizing tendencies of modern capitalism. That there is one world economic system is becoming increasingly apparent following the

Figure 13.2

collapse of communism. Nevertheless, globalization has significant military, political, legal and cultural dimensions. For example, globalization has affected southern African music (see Chapter 19), as well as the spread of concepts like liberal democracy (Chapter 14) or technical efficiency (Chapter 16).

It is important to recognize that the dominant processes of globalization have intimate connections with processes of Westernization. Moreover, far from being a completely novel or predominantly contemporary phenomenon, a globalizing imperative has been evident in previous periods of history, and is perhaps most powerfully visible in nineteenth-century imperialism (see Chapters 8–10). Globalization in the 1990s is neither an historically unique process nor necessarily the harbinger of a world society.

Complexity and diversity

Connected to the erosion of bipolarity and the impact of globalization is the awesome complexity of the contemporary global system. A plethora of issues, from arms control, ozone depletion, debt, drugs, terrorism, to currency crises, space exploitation, human rights, hunger and AIDS, crowd the global political agenda. Closely associated with this has been the incredible expansion in the range and number of agents and agencies on the global stage. World politics can no longer be pictured as foreign relations between governments. Rather, it is perhaps best visualized, as Burton (1972) suggests, as a complex 'cobweb' of political interactions, transnational corporations (e.g. IBM, ITT), international governmental organizations (IGOs – e.g. UNWRA, WHO), international non-governmental organizations (INGOs or just NGOs: for example, Greenpeace, Amnesty International, Red Cross), resistance movements, some of which have gained quasi-governmental recognition internationally (e.g. PLO, ANC), regional organizations (e.g. EC, ASEAN), internationally organized religions (e.g. Islam, Roman Catholicism), local communities and local NGOs, and even private individuals (Table 13.2; see also Table 13.3).

Table 13.2 Growth of state and non-state actors since 1945

	States (UN members)	IGOs	INGOs
1946	55	–	–
1956	80	132	973
1960	100	154	1255
1972	132	280	2173
1981	157	337 (702)	4265 (5133)
1989	159	300 (1489)	4621 (9712)

Source: Saur, K. G. (1989) Yearbook of International Organizations, Munich.

Note: For purposes of comparison, the first figures under IGOS and INGOS are for organizations defined by the Yearbook as 'conventional'. The figures in brackets are defined as 'other international bodies' and include: 'organizations emanating from places, persons, other bodies; organizations of special form; internationally oriented national organizations.'

Decline of the nation state?

The idea of the sovereignty of states within their own borders has been a paramount principle of international relations for some time, at least within the framework of international law. However, for the majority of Third World governments, vulnerability to external forces has been a dominant reality shaping development prospects. Interest rate changes, currency fluctuations, commodity prices, and technological standards have always been beyond the control of even the most powerful Third World states. Recently, the intensification of processes of globalization, combined with the emergence or exacerbation of many global policy problems (e.g. environment, AIDS, debt), has increased the direct challenge to the political capacity of all states, including advanced industrial countries (AICs), to determine their own destinies. As Kegley & Wittkopf note: 'present day political reality is a world of borders permeated and transacted by the flow of goods and capital, the passage of people, communication through the airwaves, airborne traffic and space satellites' (Kegley & Wittkopf, 1989, p.511).

In this context it is increasingly difficult for any government either in the North or the South to separate the domestic from the foreign, in that simply to achieve domestic goals, like lower interest rates or cleaner air, may require concerted international co-operation with other states. The permeability of all nation states to transnational forces, combined with growing interdependencies, has eroded the scope of any state to pursue autonomous foreign and domestic policies.

However, for some of the weaker Third World states, these difficulties are almost overwhelming. There is more and more readiness on the part of powerful actors such as Western governments, international NGOs and multinational companies to bypass what may be portrayed as inefficient and corrupt state structures (Figure 13.3).

Figure 13.3   The US-led 'coalition' waged the 1991 Gulf War against Iraq ostensibly to restore Kuwaiti sovereignty. But continued foreign 'interference' is likely to be 'necessary' for some time: (above) pollution from burning Kuwaiti oil wells does not respect borders and needs international expertise to control it; (inset) refugees, such as these displaced by the Gulf conflict, are a major problem in many parts of the world, and international agreements may be necessary to deal with them.

International governance

The fact that no world government exists does not mean that the global system lacks order or processes of governance. As earlier sections of the chapter have indicated, North–South relations are effectively governed by a combination of global market forces and international regimes, such as the GATT regime covering the world trading system or the international telecommunications regime covering the global communications network. One of the consequences of an increasingly complex, diverse and interconnected world system is that conflicts of interests and pressures for international co-ordination and co-operation are amplified. Accordingly, both the scale and the scope of international regulatory activity has increased enormously in the last three decades. One indication of this growth is the expansion in the number of international organizations in the post-war period (Table 13.3).

Regional organizations, such as ASEAN, the OAU, CARICOM, the EACM, and the EC have also become significant forces in managing activities within their jurisdiction, although the authority of these institutions varies enormously. Some, like the EC, have acquired a supranational status (i.e. some Community decision-making takes precedence over national decision-making) whereas others, like EACM, have faded into obscurity. Nevertheless, there is a clear trend towards greater institutionalization of international co-ordination and co-operation between states, both at the regional and global levels (see Table 13.2). However, since governments jealously guard their sovereignty, such institutions and processes of 'governance' are inherently fragile.

13.4 Development prospects in a transitional global order

Of the five dominant characteristics of the contemporary global system outlined above, three in particular will define the development opportunities for the majority of Third World states

Table 13.3 Growth of regional organizations 1981–87

Year	Regional total	Regional NGOs	Regional IGOs
(a) Europe, East and West			
1981	10 965	–	–
1982	15 290	14 539	751
1984	17 104	16 256	848
1987	18 519	17 819	700
(b) Africa			
1981	1949	–	–
1982	3070	2351	719
1984	3731	2898	833
1987	4476	3504	972
(c) Asia			
1981	1839	–	–
1982	2849	2516	333
1984	3199	2831	368
1987	3905	3289	616

Adapted from: Taylor, P. (1990) 'Regionalism: the thought and deed' in Groom, A. J. R. & Taylor, P. (eds), *Frameworks for International Cooperation*, Printer Publishers, London.

throughout the 1990s. The demise of the Cold War, the intensification of processes of globalization, and the broadening as well as deepening of global and regional structures of international 'governance' will have significant ramifications for the Third World. Let us examine each of these features in turn.

The 'Third World' and the end of the Cold War

Although the dissolution of the Cold War has had its most profound consequences in Europe, its effects have been truly global (Figure 13.4). Whereas in the decades of the 1970s and 1980s the Third World experienced unprecedented levels of regional, ethnic and class conflict, fuelled by superpower rivalry, the contemporary era has witnessed the resolution of many regional conflicts, a process encouraged by both superpowers. By 1989 'in some dozen conflicts of Asia, Africa and Latin America, processes of negotiation, encouraged by the great powers, began to take effect: in Cambodia, Afghanistan, the Gulf, the Horn of Africa, Angola, the Sahara, Nicaragua and elsewhere' (Halliday, 1990b). Indeed, one of the dominant themes of bilateral diplomacy between the two superpowers concerned joint action to resolve conflicts and tensions in the Third World. This determination to 'impose' peace and stability on the Third World is driven by the desire to inoculate their newly established relationship against any potential sources of friction emanating from regional disputes around the globe. Although the Third World is no longer an arena for superpower rivalry, it has not ceased to be a focus for political interference by the superpowers.

In the wake of the Iraqi annexation of Kuwait in August 1990, the UN Security Council, with unprecedented unanimity, imposed international sanctions against the aggressor. This decision was enforced by a US-led multilateral military blockade of Iraq and, in Operation Desert Shield, the largest deployment of Western military forces since the Second World War. Significantly, the Soviets not only openly supported the emplacement of over 150 000 US troops in Saudi Arabia but also repeatedly condemned its one-time ally (Iraq) for its annexation of Kuwait. Such a response was indicative of the astounding shift in Soviet foreign policy under Gorbachev. But even more remarkably, when on 15 January 1991 the US-led multilateral force took military action against Iraq to restore Kuwait's sovereignty, the Soviets continued to support the USA. Tacit superpower co-operation over the Iraqi-Kuwait crisis revealed the stirrings of a potential superpower condominium (dominion

Figure 13.4 The 'Third World' and the end of the Cold War: (right) demolishing the Berlin Wall, 1990; (below) what benefits?

together) in the post Cold War era. To some degree this was already evident. As one observer argued: 'The growth of autonomous inter-state and inter-ethnic conflicts in the Third World...suggested that a degree of joint action by the two Great Powers might, in some cases, be preferable to a self-reproducing destructiveness in the Third World itself' (Halliday, 1990b, p.22). Moreover, since neither superpower appears capable of controlling events in Europe or the Third World, the pull of condominium may appear attractive for both the USA and the Soviets, Implicitly acknowledging this, US Secretary of State Baker, in a speech in late 1990, described the new US–Soviet relationship in terms of the construction of 'pathways and mechanisms of co-operation in three broad areas: eliminating the old vestiges of Cold War; addressing the new threats to the post Cold War order; and dealing with transnational dangers to peoples all over the globe' (Baker, 1990).

For the Third World, tacit condominium may imply a more benign but nonetheless intrusive form of superpower involvement in regional and internal disputes. Yet, as many observers have suggested, the US–Soviet relationship is increasingly one of unequals. The drift towards condominium may only conceal (as well as give greater scope for) US intervention in the Third World where fundamental Western interests are threatened – the Gulf War of 1991 being a case in point. Furthermore, the 'collapse' of communist systems in Eastern Europe has reduced the ideological attraction of Soviet development models for the Third World and thereby may have undermined the possibilities for alternatives to capitalist development (see Chapter 12).

Throughout the post-war era the division of the world into two separate socio-economic systems provided Third World nations with two distinctive models of socio-economic development: capitalist and state socialist. Whereas many states, such as Tanzania, attempted to follow autonomous development strategies (involving delinking from the world capitalist system), the rejuvenation of global capitalism in the 1980s and the 'failure' of state socialism have severely constrained the attractiveness and viability of alternative development models. This is not to argue that proponents of unhindered capitalist development have triumphed, for one of the consequences of the ending of the East–West division, according to Denitch (1990, chapter 4), has been to highlight the success of European social democracy and progressive politics as an alternative model to both the USA and the Soviet Union. Nevertheless, in simple practical terms, the West (including Japan) and Western-dominated multilateral institutions have now become the only source of aid, capital and technology for all Third World states. As Ravenhill indicates: 'The North now enjoys a leverage unprecedented since decolonization over the economic strategies pursued in most of Africa, Latin America and South Asia' (Ravenhill, 1990).

An interesting historical parallel (not to be pressed too far) exists between the predicament of the Third World following the dissolution of the Cold War and the situation confronting colonial territories and developing states in the aftermath of the Second World War. For in much the same way as the economic reconstruction of Western Europe in the 1940s and 1950s pre-empted resources for economic development elsewhere, so too in the 1990s is the 'reconstruction' of Eastern Europe drawing resources away from the Third World. The major difference, however, is that the West's inability to fund an equivalent of the Marshall Plan for Eastern Europe makes it inevitable that a vicious struggle for aid and

Figure 13.5

assistance will ensue between the East and the South. There is a disturbing paradox here in that some Third World states, having themselves attempted to emulate socialist trajectories of development, may in the 1990s be forced to bear the burden incurred in redressing the 'failure' of state socialism in Eastern Europe.

In applauding the end of the Cold War, from the relative comfort of affluent Europe, we should beware of making a somewhat ethnocentric and parochial assessment of its benign consequences. That it is to be welcomed is not in doubt since it has tended to bring greater stability to a number of regions. Nevertheless, it must be recognized that this revolution in world affairs has far from transformed the prospects for the majority of Third World nations.

Globalization and the 'Third World'

The intensification of processes of globalization in the 1980s has had profound effects within the Third World. In the economic sphere, the aggressive restructuring of global capitalism, the emergence of a new international division of labour, the integration of financial markets, and the growing power of multinational capital have fostered deeper division and conflict within the Third World while simultaneously widening the North–South 'gap'. In the political and cultural spheres, nationalism, as well as transnational ethnic and religious allegiances, has reasserted itself across the Third World, stimulated by rapid global communication networks, the international migration of labour and processes of uneven development. To some extent, also, the more powerful AICs are consolidating regional spheres of influence, as with the USA in Latin America, Japan with respect to the countries of the Pacific Rim, and France to a lesser extent in North Africa and the Middle East. Globalization must therefore be viewed as essentially uneven in scope, and its effects are highly differentiated. As a result it simultaneously divides and unifies the world but in complex and unpredictable patterns. This is nowhere as evident as in the globalization of production and economic activity (Dicken, 1986).

Already in the 1970s, the globalization of production was seen as bringing about a transformation in the world economy and the emergence of a new international division of labour (NIDL). Embedded in this thesis is the belief that de-industrialization within the AICs, which has brought with it mass unemployment and social dislocation in the West, is part of the same process of global economic restructuring that has also contributed to the emergence of the NICs. As Ernst succinctly summarizes it: 'According to the NIDL theory, a new capitalist world economy has emerged, its main feature being a massive migration of capital from major OECD (Western) countries to low cost production sites in the Third World' (quoted in Gordon, 1988). This restructuring, it is argued, has created a highly integrated global system of production. According to Harris: 'the process of dispersal of manufacturing capacity is a general phenomenon [involving] increasingly complex patterns of changing specialization, interweaving different parts of the world unknowingly in collaborative processes of production' (Harris, 1987, p.116). Overseeing this globalization of production is the multinational or transnational corporation (TNC).

Transnational corporations dominate virtually every sector of global industrial, economic and financial activity. Between them they control a significant proportion of world trade, investment, know-how and employment. They are predominantly of US origin, but the last two decades have witnessed an enormous expansion in the number of Japanese, German and even Third World domiciled TNCs. Since they control considerable resources, their activities and corporate decisions can have profound consequences for communities in disparate parts of the globe. For instance, in late 1990, the giant Phillips corporation announced a complete global reorganization. This involved the shedding of some 45 000 jobs in its subsidiaries across Europe, South-east Asia, Latin America, Africa and North America. Many individuals and local communities in very different parts of the world suffered the consequences of this single boardroom decision taken in Holland. The power of transnational capital is thus considerable. As Dicken concludes:

> "much of the changing shape of the global manufacturing system is sculptured by the TNC through its decisions to invest or not to invest in particular geographical locations... there are few parts of the world in which TNC influence is not important. In some cases...the influence of TNCs on an area's manufacturing fortunes can be overwhelming."
>
> (Dicken, 1986, p.55)

Across the Third World the consequences of this increasing global integration of productive capital have been to sharpen structural inequalities within and between nations. Differentiation occurs between those states increasingly integrated into a global system of production and those which are marginalized: in effect, the NICs versus the Fourth World or LLDCs. Even within the former, social divisions emerge between those communities and groups incorporated into the global production network and those which are bypassed. The globalization of capital thus establishes parallel tendencies towards international integration and national distintegration. These tendencies are acutely evident in the clothing, textile and electronics industries where they also intersect with a growing international sexual division of labour.

Within many of the NICs, the majority of unskilled workers in clothing and electronic sectors are women. In South Korea, for example, some 75% of the workers in these 'new' export industries are women. But what is also fascinating is that this trend is replicated within the AICs. According to Henderson's study of the globalization of the semi-conductor industry: 'In the American owned semi-conductor industry, similar proportions of women in the manual labour force are evident not only in East Asian locations...but also in the United States...and...in Scotland' (Henderson, 1989, p.74). Within the Third World and the AICs the casualization and feminization of the workforce in certain manufacturing sectors is intimately connected to the globalization of capital. As Enloe observes, a new principle

has arisen in the corporate world: 'if you can't move to the Third World, create a feminized Third World in your own backyard' (Enloe, 1989, p.155).

This increasingly tightly spun network of economic connections between North and South, mediated by the multinational corporation, has the important effect of linking together the welfare of otherwise quite unrelated communities in many different locations across the globe. Indeed, some, like Fröbel, argue this is evidence of the complete 'world-wide reorganization of capitalist production' (quoted in Gordon, 1988, p.17). Although this remains a hotly debated thesis (see *Hewitt, Johnson & Wield, 1992, ch. 1* for a critique of the NIDL argument), it is nevertheless widely accepted that the globalization of capital has intensified and has restructured the global economy. In particular, with the rise of NICs, the distinction between North and South has become somewhat blurred since both share the common problem of adjusting to the globalization of production.

The Third World and international 'governance'

As earlier sections of this chapter have argued, the Third World is deeply embedded in a vast array of global institutions and regimes which regulate international activities and relationships. In the economic domain, Third World governments are subject to the disciplines of the IMF, GATT and the World Bank, and a host of organizations, including UNDP, FAO, WHO, UNESCO, UNWRA, and UNICEF, provide specific assistance with development programmes. It is tempting, in view of the origins of the majority of international organizations and institutions, to conceive this architecture of global management as merely another instrument for sustaining Western domination of the Third World. Equally, it is often assumed that most international institutions are controlled by Western governments and interests. These assertions are oversimplifications of a more complex reality. International organizations can be relatively autonomous actors on the world stage and many Third World states (individually and collectively) have exploited this to their advantage.

It would be naïve, however, not to accept that in the economic sphere the primary function of the major international economic institutions is to facilitate the smooth functioning of the global capitalist order. Accordingly, as was described in Chapter 11, throughout the 1980s the IMF and the World Bank imposed 'adjustment' programmes on many Third World states, including Brazil, Argentina, Tanzania and Peru. As a result, these states were persuaded to embrace a neo-liberal agenda of economic reform, involving the privatization of state-owned assets, together with the curtailment of welfarism and state intervention. Yet it is also the case that such 'imposed' policies generated the conditions for collective resistance. For instance, in the early 1980s the debt crisis in the Third World threatened the stability of the whole global financial system. Both the IMF and the World Bank recognized the interdependent nature of the relationship between lenders and debtors. Recognizing a situation of mutual vulnerability significantly influenced the international politics of the debt issue. Such interdependence opened up the real possibility of collective resistance by the debtors to the demands of the USA and others, which in turn stimulated the search for 'global solutions' to the debt crisis through the Baker Plan (1985) and the later Brady Plan (for more detail see *Hewitt et al., 1992*).

The international politics of the debt issue suggests that the global management of North–South relations can no longer be characterized in terms of a simple dominance–dependency relationship. Such a characterization has to be qualified because on many issues the politics of North–South relations is decidedly complex. With respect to trade issues, for instance, the recent GATT round of tariff reduction negotiations in Uruguay witnessed alignments and coalitions cutting across the North–South divide. On agricultural subsidies those Third World states with significant agricultural interests found themselves on the same side as the USA and other industrial states in pressing the EC to reduce barriers to 'free trade' in agricultural produce. The existence of international regimes and institutions facilitates this cross-cutting coalition building, thereby introducing more complex patterns of political interaction into the North–South relationship.

This complexity and interdependence is also reflected in the growing number of global policy problems which cannot be defined so readily in North–South terms. The challenge posed by global environmental degradation and the exploitation of space are two such issues. In both cases, international co-operation and regulation are shaped by a realization that these are issues of the 'global commons' or 'the common heritage of humanity'. Of course, as Chapter 5 has shown, development issues and economic inequalities still play a large part in the diplomacy of environmentalism. The point is that global management of such problems cannot be successfully achieved without some degree of partnership between North and South. UNEP's Executive Director Mostafa Tolba has highlighted the mutual interdependence or vulnerability of the North and South on environmental questions, arguing that: 'the environmental crisis demands nothing less than a revolution in the conduct of international affairs; one that acknowledges the need for global partnership...' (Starke, 1990, p.30).

Recognition of this fact will provide Third World states with a unique opportunity to exert influence across a range of issues central to development prospects in the 1990s. Ravenhill argues that:

> "The changing global agenda...offers a growing number of issues on which there are mutual interests between North and South and on which the co-operation of Southern countries will be necessary if the industrialized countries are to attain their goals. For individual Southern countries there will be opportunities to engage in strategies of issue linkage (between, for example, debt, environmental and market access questions) to improve their bargaining position with the industrialized countries."
>
> (Ravenhill, 1990, p.731)

Across a broad spectrum of issues, the 'old' inter-bloc confrontation between North and South is being replaced by a more complex and fluid international politics of co-operation and resistance. The significance of international regimes and institutions as the primary forums within which this 'new' politics is articulated and nurtured cannot be underestimated. Moreover, the increasing permeability of Third World states and AICs to global forces and problems creates a powerful imperative for strengthening institutions of international, regional and global management. For many Third World states in Asia and Africa, this presents a curious paradox in that the struggle of the 1940s, 1950s and 1960s to achieve sovereign statehood has been replaced in the 1990s by a struggle to protect and preserve it in the face of enormous pressures for greater international co-operation and co-ordination of national policies.

13.5 The end of the 'Third World'?

This chapter commenced with the suggestion that the Third World in the 1990s is confronted by a new set of challenges and opportunities arising from a rapidly changing global order. But whether these 'new' challenges and opportunities are considered historically unique or merely a repackaging of old dilemmas depends upon how convinced one is by the argument that world politics is undergoing a profound historical transformation.

Neo-Marxists, such as Samir Amin, argue that there is still a great measure of continuity with the past: 'The axis of the new world conjuncture is Western capitalist aggression against the Third World peoples with the aim of subordinating their further evolution to the demands of redeployment of transnational capital' (Amin, 1990a, p.48).

In contradistinction, neo-liberals, like Francis Fukuyama, perceive in the historic 'victory' of liberalism over communism, and the global diffusion of Western-led democratization and capitalism, nothing less than the beginnings of a revolutionary transformation of the global order

– the end of history (Fukuyama, 1989). This chapter throughout has stressed elements of both continuity and change in interpreting the current epoch. The implication has been that world politics has entered a turbulent 'transitional' era in which the old order is rapidly decaying but in which the solid architecture of a new global order has failed to materialize. In this respect, then, the Third World is caught between the interstices of a certain past and an uncertain future: it confronts old problems and dilemmas as well as new challenges and opportunities.

Among the most salient of these new challenges is how to protect and promote the interests of the world's poorer nations when there is no longer an effective Third World coalition in world politics. New political strategies are required to exploit the novel political circumstances in which the poorest nations find themselves. This will require a radical rethinking of North–South relations since the traditional frameworks for understanding and explaining this relationship are no longer so convincing. Already the sketchy outlines of new political strategies are emerging with the recognition that inter-bloc confrontation has failed or, as Amin puts it, 'The Bandung era is over' (Amin, 1990a). In its place is a more complex interpretation of North–South relations. This starts from the diversity and disunity on both sides, and from attempts to identify the mutual interests and common problems which may open 'the way for new coalitions that bridge the North–South divide in the 1990s' (Ravenhill, 1990, p.742). What is evident is that international organizations and regimes will play an increasingly significant role in facilitating such coalition-building and issue-linkage. In the 1990s a diplomacy of mutual gain for communities within both the North and the South may thus begin to replace the diplomacy of confrontation.

A distinctive 'new' challenge arises from the political and economic 'fall-out' emanating from the demise of the Cold War. One obvious ramification for the world's poor is that economic reconstruction in Eastern Europe is likely to absorb resources which would otherwise have gone to the South. However, much more critical is the shape of the global power structure in the post Cold War era. Whether the world moves in the direction of US dominance, superpower condominium, or a more polyarchic world order, will have the most profound implications for the military and economic security of all Third World states in the 1990s.

Finally, a recurring dilemma, but with a new twist, concerns the prospects for Third World governments pursuing autonomous development strategies in an increasingly interdependent and internationally 'managed' global political economy. Halliday's view is that: 'The 1980s posed, in perhaps a sharper form than ever before, the question of how far revolutionary transformation, based on a degree of self-reliance or delinking from the world system, can be a viable alternative strategy of development for Third World states...' (Halliday, 1990b, p.37).

Neo-liberals, such as Fukuyama, believe that the collapse of state socialism and the failure of strategies of delinking from the world capitalist economy (e.g. Tanzania) provide a clear answer to this question in that there no longer exists any viable alternative route to development other than that of liberal capitalism (Fukuyama, 1989). Others, like Amin, continue to argue that the more complex, differentiated and pluralistic character of the global political economy in the 1990s is prising open new potential trajectories of development which can combine a necessary degree of integration into the world system with socialist priorities and an attachment to democratic control over national destiny (Amin, 1990b). The 1990s will offer an historic test of these contrasting visions of national development prospects within what is today an increasingly differentiated 'Third World'.

14

THE DEMOCRATIZATION OF THIRD WORLD STATES

DAVID POTTER

There was no mistaking the democratic trend in the early 1990s. This chapter concentrates on Asia but Latin America can perhaps be said to have led the way with the move from authoritarian rule towards more democratic forms during the 1980s in Argentina, Bolivia, Brazil, Chile, Ecuador, Peru and Uruguay, the election of civilian presidents in Guatemala and El Salvador, and at least some political liberalization in Mexico. A similar trend was under way in Africa, a continent where nearly all of the formally democratic regimes left behind by the departing colonial rulers in the 1960s and 1970s had been fairly rapidly replaced by authoritarian forms of one sort or another. Some of the reasons for this were suggested in Chapter 10 (Section 10.4), e.g. African colonial legacies of authoritarianism, use of force, statism, and international economic dependency. By 1990, however, democratic forms were beginning to reappear (or were being strengthened), with multiparty elections in Benin, Botswana, Gabon, Côte d'Ivoire, Namibia and Senegal, moves towards a multiparty system in Mozambique and Nigeria, and serious discussions about such moves in Angola, Cameroon, Ghana, Madagascar, Tanzania and Zaïre. There were pro-democracy movements nearly everywhere in Africa; this time the enemy was not the colonial ruler but authoritarian African leaders. Dr Roger Chongwe, chairman of the African Bar Association, exclaimed, rather wildly: 'there is a [new] wind of change blowing across this continent. It howls for freedom. Whether African leaders like it or not, they cannot stand against it' (*The Guardian*, 11 September 1990, p.19) (Figure 14.1).

Figure 14.1 'Howling for freedom' in South Africa. Jubilant inhabitants of Soweto gather on 11 February 1990 to celebrate the release of Nelson Mandela.

Q Why did this wave of democratization occur in the 1980s and early 1990s?

Q Does democratization help raise the pace of socio-economic development? Does it produce more peaceful and egalitarian social relations?

I shall try in this chapter to address these large questions about the role of the state in development, making use of examples primarily from Asia. Doing so also serves the purpose of setting out important features of Third World states and their relationships to the economy, the society and the global system.

I employ in this chapter a broadly **liberal democratic** definition of democratization and in doing so recognize its limitations. I want to make clear at the outset, however, that I do not accept that the democratization of states (using such a definition) necessarily implies the empowerment of women and men locally to control the decisions and actions that fundamentally affect their lives. Democratic state forms nationally can be comfortable shelters for class rule locally, and voting in national elections may mean little in terms of local empowerment and genuine political participation. Nevertheless, civil and political liberties are much more likely to be maintained and even enhanced over time in democratic state forms than in other forms, and to that extent democratization represents a change in state form in a desirable direction – i.e. it represents political development. But the democratization of national state forms, which this chapter is about, is not the same thing as the achievement of broader political democracy and

> **Liberal democratic:** Such regimes are distinguished from others by (a) *competition* (through elections and multiple parties) for political offices, at regular intervals, excluding the use of force; (b) *participation* of citizens in politics through various forms of collective action at different levels; (c) *accountability* of rulers to the ruled through modes of representation and the rule of law; and (d) *civil and political liberties* sufficient to ensure the integrity of participation, competition and accountability.

egalitarian social relations locally and throughout a society.

Chapter 6 laid out some alternative meanings of development and pointed out several ways in which development is political. It also gave a rough framework for understanding the ensemble of political institutions (coercive, administrative, legal) that can be said to make up the state. It was pointed out that nation states conducting peaceful and warlike relations with each other constitute the basis of the contemporary international political order (Figure 14.2), just as advanced international capitalism is the basis of the world economy. Chapter 10 explains

Figure 14.2   State leaders conduct peaceful relations with each other. Pakistan hosting the four-yearly Islamic Summit in 1974; among those at prayer are: Yassar Arafat of the PLO, Saddam Hussein of Iraq, Zulfiqar Ali Bhutto of Pakistan, Muammar Gaddafi of Libya, King Faisal of Saudi Arabia, and the Shaikh of Kuwait.

some of the likely features of colonial states that may influence the make-up of their post-colonial counterparts.

Whatever its origins, a particular state can relate to development in three basic ways (Chapter 6 again): it can take direct responsibility for development; it can help structure the development activities of others; or it can be the enemy of development for disadvantaged groups and classes. The rest of this chapter uses democratization as a focus for elaborating on the nature of states and development.

14.1 Democratization of states in Asia

Figure 14.3 categorizes 24 major Asian states during 1980–90 in terms of four rough categories: non-democratic, partial democracy, democratization taking place, and semi-democratic or democratic. Before 1980 only six of these 24 states were semi-democratic or democratic (India, Papua New Guinea, Japan, Sri Lanka, Singapore, Malaysia); two others were partial democracies (Thailand and Nepal); the rest were

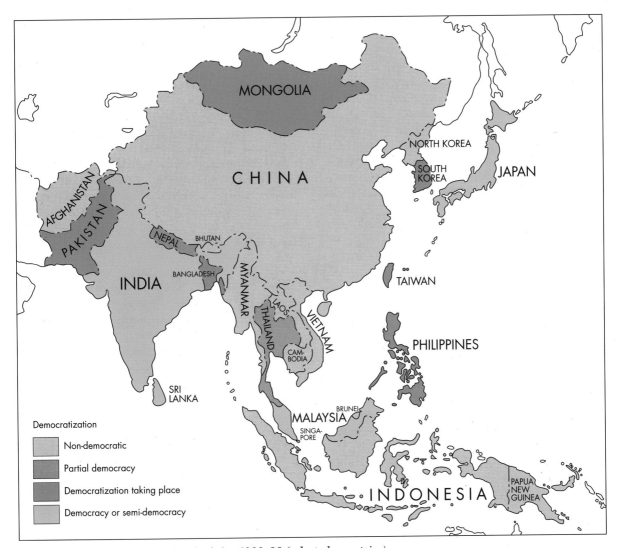

Figure 14.3   Democratization in Asia, 1980–90 (selected countries).

non-democratic. During 1980–90, as the map shows, the six semi-democracies or democracies remained intact, as did the two partial democracies, but six of the non-democracies were moving in a democratic direction (Pakistan, Bangladesh, Mongolia, South Korea, Taiwan, the Philippines). In brief, where there were changes in state forms in Asia during 1980–90, these were in a democratic direction.

As you know, maps like this depend on how the categories used are defined. A word needs to be said about that. Under 'non-democratic', I include two very different state forms. One is a type of regime where the armed forces or a military-controlled civilian party is dominant, initially as a result of a *coup d'état*; officers in the armed forces then place their own people in key posts at various levels within the state apparatus. The military initially provides the political leadership in the regime, to which the state bureaucracy is accountable. Later, military leaders either sponsor a political party or eventually, as citizens, lead one themselves. Whatever the precise mix of military and civilian party, it is characteristic of such regimes that organizations outside the state are initially kept under close control and surveillance. Most such regimes tend to be politically unstable, as they grapple with the twin problems of political legitimacy and bureaucratic control. The two are frequently in conflict. If military control of the bureaucracy is tightened, then political legitimacy amongst the civilian population can slip away; if the military retire to barracks and let their civilian representatives try to build up the legitimacy of the regime, then bureaucratic control is loosened. This dialectic governs the tendency of military regimes to oscillate between tighter military and looser civilian types of political leaderships. The dominant trend in Asia has been for such regimes, following the military *coup,* to move in a civilian direction. Pakistan, Bangladesh, the Philippines, Taiwan and South Korea all moved in this direction between 1980 and 1990 (see Figure 14.3).

A socialist regime, by contrast, tends to have a totalitarian dimension in the sense that it seeks to encompass the entire society in its embrace,

Strong leaderships attempt, through the party and state apparatus, to mobilize the entire society for the purposes of nation-building and social reconstruction in accordance with socialist ideology (see Chapter 12). There have been a number of such regimes in Asia. In 1988, for example, China, North Korea, Mongolia and Vietnam were socialist. Laos, Cambodia (then called Kampuchea) and Afghanistan also belonged in this category, although in these cases the situation was complicated by the presence of foreign troops providing 'stiffening' for the regime (Soviet troops in Afghanistan, Vietnamese troops in Laos and Cambodia). The reformist thrust of *glasnost* (openness) and *perestroika* (reconstruction) in the Soviet Union in the mid-1980s and the dramatic changes in Eastern Europe in 1989–90 were beginning to affect the Asian socialist world as well by 1990. The fundamental nature of these political regimes had not yet altered significantly, although pro-democracy tendencies had surfaced in some of them, notably Mongolia and China.

The other categorizations in Figure 14.3 involve the word 'democracy' (semi-democracy, partial democracy, democratization), and that word, naturally, is a heavily contested concept (Held, 1987). The definition used here and on the map is broadly liberal democratic as defined above.

Most such regimes in Asia have tended to be semi-democratic in the sense that there are competitive elections involving more than one political party, but frequently one 'dominant party' usually wins them and forms governments; there is some political participation through business associations and other interest groups, but such participation locally is fairly closely controlled or monitored by the dominant party and state apparatus; and there are civil and political liberties, but they tend to be qualified to some extent.

This characterization of democracy and semi-democracy would not be accepted everywhere in Asia. Chinese communists, for example, would argue that the procedural guarantees and freedoms in bourgeois definitions are illusory; the workers and peasants in capitalist democracies cannot vote *not* to have capitalism, and capitalism serves the narrow interests of the bourgeoisie. Mao once said: 'I don't trust elections... I was elected from Peking to the National People's Congress but many people have never seen me!' (Nathan, 1986, p.124). Mao's argument was that he represented the people not because he was elected by them but because he stood for their interests. These views reflect the general argument that democracy can be heavily qualified in a class-divided society, and that true (socialist) democracy does not at present exist anywhere but is a goal worth striving for in a classless society.

There is some force in this criticism of the definition of democracy used in Figure 14.3, but the criticism rests too heavily on a critique of elections and other procedural guarantees. The definition used here is actually much broader, involving not only electoral competition but also participation, accountability, and the maintenance of civil and political liberties. Such a broad conception would probably now be widely recognized throughout much of the Third World, including Asia, as a political form quite distinct from other forms, and one towards which many political regimes in the Third World are heading. Why are they heading in this direction? How can this trend in the 1990s be explained?

14.2 Explaining democratization

All political explanations combine elements of *agency* and *structure* (see Chapter 6). Why did Saddam Hussein of Iraq try to annex Kuwait in 1990? Why was there a trend towards democratic forms of state in the Third World in the early 1990s? Why did the Chinese government try to suppress the pro-democracy movement? (Figure 14.4) Part of the answer always involves some agency – the initiatives, choices and actions of persons or groups causing that event or trend. Part of the answer always involves some structure – a set of physical and social constraints, a set of opportunities, a set of norms and values that produce the events or trends and determine the content of the choices and initiatives.

Figure 14.4   Pro-democracy demonstration in Tiananmen Square, 1989.

How agency and structure are combined in an explanation, and the weighting assigned to one or the other, are matters on which there can be fundamental disagreement. Some explanations give more weight to structural factors like the dominant mode of economic production and the class structure related to it. Others emphasize instead the importance of agency factors like political leadership and the demands of particular groups. In this section I summarize examples of 'agency-led' and then 'structure-led' explanations of democratization.

Agency-led explanations

A classic example of an agency-led explanation of democratization is the one by Rustow (1970). His question was: what brings a democracy into existence in the first place? His answer was based on a comparative analysis of Sweden (which moved to democracy between 1890 and 1920) and Turkey (which was moving in that direction when he wrote). Although very simple, his explanation has provided the basis for various subsequent explanations of this type, and can be summarized as follows:

1   A single *background condition* is necessary before a country can start on the road to democracy; that condition is *national unity* within a given territory, i.e. a large majority of persons within the area share a political identity.

2   The road to democracy involves a *preparatory phase*, matched by prolonged and *inconclusive political struggle* between a plurality of relevant élites and groups.

3   There is then a *decision phase*, an historical 'moment' when the parties to the inconclusive political struggle decide to compromise and adopt democratic rules which give each some access to state power.

4   Finally, there is an *habituation phase*, during which those who were party to the compromise decision are gradually replaced by a new generation who have become habituated to democratic rules and who sincerely believe in them.

For Rustow, political identity shared by a plurality of groups provides a sort of structural background from which a country can move towards democracy. The route to democracy is then driven primarily by agency factors, such as inconclusive political struggles between relevant groups and a key moment when the groups, engaged in struggle decide to opt for compromise by adopting a democratic constitution, giving each group some share in state power. Democracy results largely from the agency of political initiatives of the groups involved in the (previous) inconclusive struggle.

A more recent agency-led explanation has been advanced by Diamond, Linz & Lipset (1989), part of a massive four-volume study which appeared between 1988 and 1990 and which purported to explain why democratization has or has not occurred in Latin America, Asia and Africa. For Asia they suggest that 11 variables need to be considered, as follows:

1   *Political leadership*, i.e. 'the actions, values, choices, and skills of both a country's political élite and its one or few top government and party leaders'. Although the authors say that they are not advancing a 'great man' theory of history, they do nonetheless ascribe great importance to the role of political leadership in explaining democratic progress or failure. This is typical of an agency-led explanation. They argue that the success of post-independence democracy in Asia was due in part to leadership qualities of people like Jawaharlal Nehru and his colleagues in India, Tunku Abdul Rahman in Malaysia and Michael Somare (and his successor Julius Chan) in Papua New Guinea, whereas the self-aggrandizing ambitions and authoritarian styles of putatively democratic leaders such as Marcos (Philippines), Rhee (South Korea), Bhutto (Pakistan), Bandaranaike (Sri Lanka), Sukarno (Indonesia), and Indira Gandhi (India), were employed within a democratic process 'to erode or destroy democracy' (Diamond *et al.*, 1989, p.8).

2   *Historical and colonial legacies*. The authors say that the 'democratic prospect' in Asia has varied significantly with the historical and cultural legacies of the colonial experience. For example, they say that India's success with democracy owes something to the British colonial legacy – the establishment of the rule of law as a constraint on arbitrary government, the provision of a system of representation and election during the 1930s that gave educated élites some experience of government, an educational system that taught and praised British democratic concepts and values. Six decades of Australian colonial rule in Papua New Guinea had similar effects; like the British, the Australians introduced electoral competition (three general elections before independence) and encouraged the democratic features of traditional Melanesian culture. By contrast, Japanese colonial rule in Korea strengthened the already centralized and autocratic character of traditional authority there, leaving Korea at the end of Second World War with no countervailing institutions able to balance state power.

3   *Political culture*. This concept refers to distinctive ways that people in a society think and feel about politics, power, authority and legitimacy. The evidence from the case studies makes Diamond and his colleagues cautious about drawing any causal link between political culture and democratization, and they reject Pye's (1985) argument that Asian political cultures are generally paternalistic, making democracy an unlikely prospect there. However, they say that some political cultures are more resonant with democracy than others; in India, the political culture tends to be tolerant and accommodating, and there is a pervasive belief in and reliance on arbitration as the appropriate method for the peaceful resolution of conflict (Weiner, 1965, p.214); but in South Korea compromise is regarded as a 'signal of weakness and lack of resolve, not only by one's adversaries but by one's allies as well' (Han, 1989, p.285).

4 *Ethnic cleavage and conflict.* The argument here is that colonialists or others imposed 'artificial' state boundaries resulting in political regimes with either centralized or dispersed ethnic systems. A centralized ethnic system, with only a few large groups whose interaction is a constant theme of politics nationally, can impede interethnic co-operation which can stifle democratization or lead to democratic instability. An example here is Sri Lanka, where the Sinhalese and the Tamils constitute most of the population. Dispersed ethnic systems, with a multiplicity of small groups scattered to many different points nationally, may contribute to interethnic co-operation and may be a force for political pluralism and democracy. India and Papua New Guinea are the examples given here.

5 *State and society.* Diamond and his colleagues argue essentially that democracy requires a rough 'balance between the 'output' institutions of the state...and the 'input' institutions (political parties, interest groups, associations) that are competing for state control, attention and resources'. The supremacy of the state in relation to other institutions and 'civil society' poses a difficulty for democracy in many Asian countries. In Thailand, for example, the dominant and centralized military and bureaucracy have repeatedly checked the development of autonomous parties, interest groups, and village associations. The challenges to democratization in South Korea and Indonesia are also traced to the formidable power of the state and its coercive apparatus. Again, democratization is easier in India and Papua New Guinea where there is a rich array of voluntary associations making demands on the state.

6 *Political parties and party systems.* The authors argue that democratization is difficult where there are fragmented and weak party systems. There were, for example, more than 100 parties in Indonesia before the collapse of democracy in the late 1950s, and 143 parties put in an appearance in Thailand between 1946 and 1981. In such countries, 'the construction of broad-based, coherent parties...mobilizing and incorporating emerging popular interests, organized effectively down to the local level, and penetrating particularly through the countryside...looms as one of the pre-eminent challenges of democratization' (Diamond *et al.*, 1989, p.29). In India, Sri Lanka and Malaysia, for example, the tendency has been for one or two well-organized parties to dominate nationally (Figure 14.5).

Figure 14.5 Political parties mobilize the vote in the countryside: polling booth queue in South India.

279

7 *Political institutions.* In a similar manner, the authors argue that vigorous legislatures and judiciaries, well established over a period of years, are important for the continued viability of democratic politics.

8 *Decentralization.* They also argue that highly centralized structures of power are not very suitable vehicles for democratization. Democracy faltered in India, for example, when Indira Gandhi undermined federal politics and local autonomy.

9 *Socio-economic development.* The authors make four important, and not entirely consistent, points about the relationships between democracy and socio-economic development. In doing so they refer to the sort of data set out in Table 14.1.

> (a) The Asian examples show no correlation between degree of democracy (as measured by Gastil's (1988) summary measure of civil and political liberties) and *level* of socio-economic development.

> (b) There is no association between degree of democracy and rate of socio-economic *growth*.

> (c) High rates of socio-economic growth are part of a historical transformation that *can hasten* the transition to democracy.

> (d) Rapid socio-economic development *can also threaten* the transition to a durable democracy.

On points (c) and (d), Han (1989) argues that the rapid economic growth rates in non-democratic South Korea brought social changes that facilitated later democratization. Such changes included an increase in the size and political consciousness of the 'middle class'; the growth of a more pluralistic, organized, and autonomous civil society; increasing circulation of people, information, and ideas; greater economic involvements with the industrialized democracies, and with it, the recognition that 'democratization is the necessary ticket for membership in the club of advanced nations' (Han, 1989, p.294). Such rapid development, however, can also loosen accepted patterns of authority and generate a huge agenda of pent-up demands

and conflicts that a fledgling democratic state will have difficulty coping with.

10 *Economic performance and legitimacy.* Diamond and his colleagues suggest that if a democratic state is seen to deliver the goods of development at an acceptable rate, it helps to fuel the people's continuing acceptance of the legitimacy of the regime; but during economic crises such legitimacy can begin to fall away and all states, including democratic ones, will have greater difficulty maintaining their rule. The authors suggest on the basis of the Asian evidence that these relationships can be more complex than that, e.g. successful development can also alter many social relationships and traditional patterns of authority and legitimacy, and can create expectations that the democratic state will have difficulty meeting. But the general relationship between economic performance and regime legitimacy seems to hold in many Asian cases. Das Gupta (1989) argues that India's democracy has been sustained partly because of the unspectacular but steady 'combined development' since independence from colonial rule. On the other hand, Jackson (1989) argues that development stagnation in the Philippines in the 1960s and early 1970s helped to undermine the legitimacy of the democratic state there and provided the context for the imposition of dictatorial rule by Ferdinand Marcos in 1972. Marcos 'delivered the goods' in the early years of his dictatorship, which helped to legitimize his rule, but economic crisis in the 1980s helped to bring him down, leading to the commencement of semi-democratization under Corazon Aquino.

11 *The international environment.* The authors make three main points regarding the importance of international factors on the democratic prospect.

> (a) The larger the country, the smaller the international effects on democracy. The fortunes of India's huge democracy, for example, have had little to do with international factors. Such factors, however, can be an important determinant of democratization in smaller countries. The effects can be positive or negative. Negative, for example, in that Sri Lanka's

Table 14.1 Selected development indicators for some Asian countries, 1960–88

	Civil and political liberties, 1987 (1975)[a]	GNP per capita, 1988[b] (US$)	Average annual growth rate, GNP per capita, 1965–1988 (%)	Average annual rate of inflation, 1980–88 (1965–80) (%)	Infant mortality rate per 1000 infant births, 1988 (1965)	Life expectancy at birth, 1988	Adult literacy rate, 1985 (1960) (%)
India	5 (5)	340	1.8	7.4 (7.5)	97 (150)	58	43 (28)
Papua New Guinea	4 (5)	810	0.5	4.7 (8.1)	61 (143)	54	45 (29)
Malaysia	8 (6)	1940	4.0	1.3 (4.9)	23 (55)	70	73 (53)
Philippines	4 (10)	630	1.6	15.6 (11.7)	44 (72)	64	86 (72)
Pakistan	9 (8)	350	2.5	6.5 (10.3)	107 (149)	55	30 (15)
South Korea	8 (11)	3600	6.8	5.0 (18.7)	24 (62)	70	93 [1980] [71]
Indonesia	11 (10)	440	4.3	8.5 (34.2)	68 (128)	61	74 (39)

[a] Combined score of civil and political liberties, each rated on a 1 to 7 scale with 1 being free and 7 least free. A score of 5 or less (with a 2 on political rights) is regarded as 'free', 6 to 11 as partly free, and 12 to 14 as 'not free'.

[b] 1988 GNP per capita is expressed in constant 1980 US dollars.

Sources: World Bank (1990, 1986) *World Development Report 1990, 1986,* Oxford University Press, Oxford; Diamond, L., Linz, J. J. & Lipset, S. M. (eds) *Democracy in Developing Countries,* pp. 1–52, Adamantine Press, London; *Freedom at Issue,* January–February 1976; Gastil, R. D. (1988) *Freedom in the World: political rights and civil liberties, 1987–88,* University Press of America, Lanham, M.D.

democracy may have become somewhat precarious during the 1980s, due at least in part to deteriorating economic performance (and its effects on political legitimacy) resulting from its economic dependence on international supporters. Positive, in that Lipset (1989) argues that Papua New Guinea's democracy (and economic development) may have benefited from international support – in this case the unconditional aid from Australia (in the late 1980s such aid formed nearly a third of the country's annual budget).

(b) A special category of a negative international impact on democracy in Asia relates to threats (real or perceived) to national security. As Diamond points out, 'a perception of serious threat…legitimates the augmentation and centralization of state power, the militarization of society, and the restriction of civil and political liberties' (Diamond in Diamond et al., 1989, p.40). South Korea's political history, in which military tension was acute from the time of the Korean war until perhaps the latter part of the 1980s, is a striking example, and democratization began in South Korea as such tensions eased. Other examples in Asia where the security factor has affected the democratic prospect include Thailand, Indonesia, and Pakistan.

(c) International influence can be pro-democratic when powerful external agents like the USA explicitly promote it or when there is a general global or regional trend (as in Asia in the late 1980s) towards democracy.

This influential attempt to provide an agency-led explanation of why democratization has and has not occurred in the Third World is not, I think, very coherent. The 11 factors resemble a shopping list of variables. Each one appears to contribute to democratization and at the same time not to contribute. Also, there is little attempt to show how a combination of these variables explains democratization. However, the 11 variables certainly provide a valuable checklist of factors to be borne in mind when trying to grapple with detailed questions about democracy at different levels within Third World countries.

Structure-led explanations

Social Origins of Dictatorship and Democracy by Barrington Moore (1966) is perhaps the classic example of a structure-led explanation of democratization, which deals in part with Asian examples. Moore's focus was the changing relationships between social classes and the state during the historical transformations that occurred in eight countries between the seventeenth century and, roughly, the 1940s. At the centre of such transformations, which Moore called 'modernization', were capitalist industrialization and the commercialization of agriculture. The three main social classes were the landed upper classes, the peasantry, and the urban bourgeoisie. The eight countries whose histories Moore compared were England, France, the USA, China, Japan and India (with some attention given also to Germany and Russia). Moore asked essentially four main questions:

1 Were there similar patterns of relationships between lords, peasants, urban bourgeoisie and the state during the process of modernization which led towards democracy in England, France and the USA?

2 If so, did a different pattern of relationships lead to fascism in Japan (Figure 14.6) (and Germany)?

3 Did still another pattern lead to communist revolution in China (and Russia)?

4 And how does one explain the curious (for Moore) emergence of democracy in India?

Moore's answers can, I think, be expressed in terms of different historical 'routes' to the modern world. Here is my summary of Moore's (1966) route to *democracy*.

- The surplus produced by the peasantry during the process of modernization was extracted by lords and other dominant classes and directed towards industrial growth in the towns.

- The peasantry was gradually transformed by the commercialization of agriculture and/or eventually eliminated as a political factor of consequence.

Figure 14.6   Lords, peasants and the surplus in Japan. Ashikaga Tokaligi presiding at a council. Utagawa Kunigoshi (1798–1861).

- The landed upper classes turned increasingly towards commercial agriculture while setting the peasants 'free'.

- A rough balance of power emerged between the landed upper classes and the state, a balance maintained during the process of modernization.

- A vigorous bourgeoisie with its own economic base emerged in opposition to the state, and eventually went on to become the dominant class in society.

- An inconclusive political struggle developed between the bourgeoisie and the older landed classes.

- Important sections of the landed upper classes were able to develop bourgeois economic habits at a fairly early stage, and were able to maintain a fairly firm economic footing during modernization.

- The process of modernization involved a revolutionary break from the past led by the bourgeoisie.

- The route to democracy was marked by violence and human suffering.

Moore's democratization route can be encapsulated in the form of a general proposition: democracy emerged in conditions where a strong and independent bourgeoisie came into being in opposition to the past regime and managed to exert its control over national policy – while at the same time a rough balance was maintained between the landed upper classes and the state – and where the influence of the peasantry was negligible or non-existent because they were transformed or destroyed by lords and others engaged in the commercialization of agriculture.

The different historical route that led to fascism (as in Japan in the 1930s) can be summarized as follows: fascism emerged in conditions where the urban bourgeoisie was comparatively weak and relied on the dominant landed upper classes to sponsor the commercialization of agriculture through their domination of the state, which enforced labour discipline among the peasantry.

Another route led towards communist revolution (as in China by the 1940s): communist revolution occurred in conditions where the urban bourgeoisie was weak and dominated by the state, the link between the peasantry and the landlords was weak, the landlords failed to commercialize agriculture, and the peasantry was cohesive and found allies with organizational skills.

India, for Moore, was a rather special case. Modernization was a comparatively weak impulse there by the 1940s because the colonial state had protected the landed upper classes and enabled them to pocket much of the economic surplus generated by the peasants rather than directing that surplus towards industrial growth (see Chapter 10). The urban bourgeoisie was therefore weak, the landed upper classes saw no incentive to commercialize agriculture, the peasantry lacked the cohesion and leadership for political actions, and none of these classes had made a revolutionary break from the past. For Moore, this was an unpromising structural framework for democratization, given his 'route to democracy' indicated above, yet democracy did emerge. Why?

Here is Moore's answer:

"British rule rested mainly on the Indian upper classes in the countryside, native princes and larger landowners in many, but not all, parts of the country...Some major political consequences of the tendency to rely on the upper strata in the countryside deserve to be noticed right away...This tendency alienated the commercial and professional classes, the new Indian bourgeoisie, as it slowly put in an appearance during the course of the nineteenth century. By splitting the landed upper classes from the weak and rising urban leaders, the English presence prevented the formation of the characteristic reactionary coalition on the German or Japanese models. This may be judged a decisive contribution towards the eventual establishment of parliamentary democracy on Indian soil, at least as important as the osmosis of English ideas through Indian professional classes. Without at least some favourable structural conditions, the ideas could scarcely have been more than literary playthings. Finally, the British presence drove the Indian bourgeoisie to an accommodation with the peasantry in order to obtain a massive base..."

(Moore, 1966, p.354)

Three main reasons are given. First, the colonial state established a structural condition favourable to parliamentary democracy by splitting the landed upper classes from the rising urban commercial and professional classes. Secondly, English political ideas, including the idea of representative democracy, seeped into the consciousness of these commercial and professional classes during their education. Thirdly, as these commercial and professional classes began to take these ideas more seriously, and even mount a political (nationalist) movement to achieve political objectives, they were driven to an accommodation with the peasantry to increase their political effectiveness. Some of the conditions in the more general route to democracy were there in India, but others were not (for example, no commercialization of agriculture or transformation of the peasantry, no vigorous and dominant bourgeoisie). That is why Moore believed that India's new democracy was rather precarious.

The heart of Moore's general explanation of democratization is that, arising out of the historical development of classes and class alignments, 'favourable structural conditions' emerged which made a 'decisive contribution towards the eventual establishment of parliamentary democracy'. There were political agents involved who were moved by democratic ideas, but their choices and actions were profoundly shaped by class structures, and the democratic ideas would have been no more than 'literary playthings' without those favourable structural conditions. Moore's is a clear example of a structure-led explanation of why some countries in Asia (notably India) became democracies and others (notably China) did not.

Another important, and rather different, structure-led explanation of democratization concentrates on the role of capital. Therborn's (1978) argument is in two parts.

First, certain inherent tendencies within capitalism create conditions which can lead to the rise of democracy. For example:

- Capitalism creates 'free' labour markets, that is, workers emancipated from feudal ties, buying and selling their labour power for a wage. This inherent tendency, along with other features of capitalism, lays the basis for a working-class struggle; 'the labour movement has itself played a vital role in the struggle for democracy'.

- Capitalist relations of production inherently tend to create an 'internally competing, peacefully disunited ruling class', divided into several 'fractions' – mercantile, banking, industrial, agrarian, etc. In the absence of a single centre of power, 'some kind of elective, deliberative and representative political machinery became necessary' for the upper classes. This laid the basis for propertied republics or parliamentary

monarchies (as in Britain) in the early stages of capitalism. From such beginnings, arising out of inherent tendencies within capitalism, eventually came the legislative institutions of parliamentary democracy.

Second, these inherent tendencies do lead to democracy if they are linked to one or all of three main factors which have a decisive influence on when a capitalist country moves to democracy:

- *Military defeat by a foreign power*. A foreign power defeats a country in war and then imposes democracy, e.g. Austria, Japan.

- *National mobilization* of peoples in the face of external threat. This has two aspects: (a) democratization as a means towards bringing about national mobilization for war, e.g. The Canadian War Times Elections Act, 1917; (b) democratization as an effect of popular mobilization for a war effort, e.g. women's suffrage in the USA, 1919.

- *Purely internal developments*, e.g. particular divisions within the ruling class.

The precise focus of Therborn's explanation is somewhat different from Moore's although both are structure led. It is to explain how and why a democracy comes into being *when* it does. The interest in the precise timing is part of Therborn's general argument that democracy came into being *after* the structure of capitalism was well established. In sum, *the rule of capital is paramount; democracy is contingent – it comes or goes within that paramount frame.* Japan's move from fascism to democracy is explained in this context: there were certain inherent tendencies within the structure of Japanese capitalism that were favourable to democracy; when the Japanese were defeated by the Americans in the Second World War and had a form of democracy imposed on them, the structure of capitalism was there to enable that imposition to persist. Both the *agency* of international (and national) initiatives and the *structure* of capitalism were important to the democratic outcome, but the structural feature was paramount.

A number of similar explanations emphasizing inherent tendencies within capitalism and contingent influences have been advanced to account for the more recent democratization trends in South Korea, Taiwan, Singapore, the Philippines and elsewhere. Each society has had its own unique historical trajectory, but Robison (1988) suggests that many of these shared at least some common features:

1 The rapid development of (late) industrial capitalism in Asia created new and initially fragile capital-owning classes; and these developments were nurtured by authoritarian (frequently military-dominated) states.

2 Such authoritarian states were appropriate agents for undertaking a number of vital political and economic functions essential to the growth of (late) capitalism: these included the suppression of reformist or revolutionary threats from workers and peasants, supplying essential investment for heavy industry and economic infrastructure, providing cheap credit and tariff protection for capitalists, enforcing trade monopolies, making lucrative contracts with capitalist entrepreneurs for supply and construction work needed by the state, helping to provide skilled (through public education) and disciplined low-wage labour, and so on.

3 There developed what amounted to a 'pact of domination' between leading capitalists and the military–bureaucratic leaders of the state apparatus based upon a 'complex conjuncture' of common interests. Together, state and class drove forward late capitalist development.

4 As capitalist industrialization proceeded, however, the authoritarian state gradually came under increasing pressure from the growing power of the capital-owning and middle classes, who found the economic and social controls of the authoritarian state increasingly irksome. This weakened the pact of domination, set up complex contradictions and conflicts between the dominant classes and the state, and weakened the social basis of authoritarian rule.

5 Those structural transformations domestically were reinforced by international pressures as the country became more integrated into the international division of labour and moved into a phase of export-oriented industrialization. The freer movement of goods, labour and capital tended to be easier in democratic or semi-democratic regimes, and it was in the interest of transnational capitalist enterprises to advocate some form of democracy as a way of 'opening up' the markets of exclusionary authoritarian regimes to foreign goods and services. Democratization was increasingly seen by important people in the country as a necessary ticket for membership in the 'advanced' international club to which they aspired, and this also provided a strong incentive for economic and political liberalization.

6 A contingent factor also in the 1980s was the weakening of authoritarian states due to growing economic crises. Growing balance of payments deficits, debt, inflation and fiscal crises encouraged such states to hand over economic responsibility for development to the urban bourgeoisie (for example, in Malaysia, Singapore, Indonesia). Democratization was a way of spreading and sharing responsibilities, and defusing the mobilization of discontent.

7 There is nothing automatic about (late) capitalist development of this kind. A number of countries in Asia appeared to be moving along this route in the early 1990s, but as Robison (1988, p.57) points out, 'while political systems may be conditioned by the international context in which they operate at the stage of industrialization over which they preside, the precise nature of regimes is the outcome of a specific history of social and political conflict.' Changing structural conditions are profoundly important, but they do not completely determine the changing forms of states.

It is important to emphasize that little is said in this explanation of democratization about the roles of workers and peasants. The focus is on the structures produced by the economic and political interests of dominant classes, both international and domestic, and the state. Workers and peasants appear to be largely victims of these processes, or if they do attempt to mobilize, are ruthlessly suppressed in order that (late) capitalist development can proceed. In the later stages of democratization and capitalist development, however, working class struggles can become more effective. For example, the democratizing South Korean regime agreed to major wage increases demanded by a large number of workers on strike in the late 1980s, whereas in the past such demands would have met a much tougher response.

The route being described here is also only about rather partial democracy. Greater political participation by workers and peasants in important decisions that affect their lives and the development of more egalitarian social and economic relations are not what the dominant classes, the state and transnational forces have in mind. For example, Cummings (1989, p.34) remarked that in the late 1980s the American Embassy was 'proffering to the men who rule South Korea' a model of democracy with 'one-party rule, with a legitimate but impotent opposition – a labour party of some sort to accommodate the urban working class and render it politically docile'. A form of democracy consistent with the definition used in this chapter is much further down the democratization route for these countries.

One of the principal features that distinguishes such structure-led explanations from agency-led explanations is the grounding of the explanation in the dominant mode of economic production. Moore starts with the particular way the economic surplus is generated by the peasantry and the way social classes are structured in relation to that surplus. Therborn's interest centres on inherent tendencies within capitalism. Robison's sketch of the democratization route is grounded in the class and state dynamics of (late) capitalist development. Such explanations of democratization based on historical economic processes and their structural consequences strike me as more coherent than agency-led explanations that emphasize more voluntarist factors like the democratic proclivities of great leaders. Such explanations, however, do tend to be pitched at a general level and can be less useful for those trying to find answers to detailed questions about democratic politics in a particular country or locality.

14.3 Democratization and development

We have explored the question 'why democratization in the Third World?' by surveying some of the differing explanations that have been advanced in the literature. You will be aware that those surveys throw some light on the other two questions with which we began this chapter: Does democratization raise the pace of economic development? Does it produce peaceful and egalitarian social relations?

As for the latter question, it is noteworthy that the explanations tend to argue that the process of democratization is marked by struggles and conflict between contending groups and classes in society, not by peaceful social relations. For Rustow, the preparatory phase on the route to democracy has as its principal feature inconclusive political struggle between major social groups. It is inconclusive because no one group is able to conclude the struggle by completely overwhelming the others. The groups then agree to a democratic constitution as a way of coping with their conflicts. Moore's route to democracy is similar in this respect: one thing that distinguished countries on the route to democracy, as distinct from those heading towards fascism or communist revolution, was the continuing struggle between the landed upper classes and the urban bourgeoisie, neither of which managed to overwhelm or destroy the other. Therborn also emphasizes the importance of inconclusive political struggle within (between fractions of) the ruling class, enabling the labour movement to find powerful allies during its struggle for a place in the polity. Diamond and his colleagues stress the importance of a plurality of competing groups and associations, as a favourable condition for democratization.

Such inconclusive political struggle on the road to democracy involved intense conflict, sometimes violence. Rustow speaks of a 'hot family feud' (Figure 14.7), Moore draws pointed attention to 'violence and human suffering', and so on. Violence has at certain times in history broken the resistance of a king, or dictator, or dominant

Figure 14.7 'A hot family feud' on the route to democracy in Britain? Police and citizens clash at the bottom of Parliament Street on 'Bloody Sunday', 13 November 1887 after Commissioner of Police Sir Charles Warren's attempts to ban the use of Trafalgar Square for open air meetings.

social class, and thereby prepared the ground for democratization later. As Moore (1969, p.86) once remarked: 'It is simply impossible to put violence, dictatorship and fanaticism in one category; freedom, constitutionalism, and civil liberties in another. The first has played a part in the development of the second.'

It is hardly surprising that most explanations of democratization converge on this point. For one thing, what distinguishes democratic states from other state forms is the degree of conflict allowed – hence the various procedures of debate, majority rule, judicial review and so on which are designed to cope with conflict. In this way, democratic states reveal their origins.

If the process of democratization is not peaceful, neither is it noted for reducing social inequalities. The explanations of democratization that we have considered in this chapter say almost nothing about inequalities, and to gather comparative evidence on changing patterns of inequality would be an enormous undertaking. But enough has been said to make one cautious about exaggerated claims regarding the immediate and beneficial impact of democratization on inequalities. Barrington Moore's argument implies that one of the main victims of the democratization route has been the peasantry, who were gradually transformed by the commercialization of agriculture or eventually eliminated as a political factor of consequence. Behind that rather bland statement lies a history of violence and human suffering on a massive scale, suggesting that the vast difference between the life patterns of ordinary people and of the wealthy probably changed rather little during democratization.

Does democratization hasten economic development? The explanations surveyed in this chapter appear to suggest that there is no necessary association. Neither Moore nor Therborn nor Robison found a causal connection between democratization and capitalist industrialization, nor did they find that such industrial development necessarily encouraged democratization. Capitalist development preceded democratization in Europe and led not only to democracy but also to fascism (Moore). Certain 'inherent tendencies within capitalism' (Therborn) *can* be propitious for democratization only if other contingent factors occur. Late capitalist industrialization in Asia was not preceded by democratization, and the quickest such development took place in non-democratic circumstances (Robison), although as such industrialization matures and the capital-owning classes became more powerful and confident, democratization tendencies can begin to develop. Similarly, democratization did not hasten the economic developments accompanying the commercialization of agriculture, and although this historical transformation in the countryside did in some cases contribute to democratization, in others it led to fascism (Moore). Also, as noted above, Diamond and his colleagues found no association between democratization and rates of economic growth in Asia, nor between democracy and levels of economic development. They also argue that rapid economic growth can help to legitimize democratizing states, but such growth can also undermine them. A similar lack of consistent association between democratization and economic development has been found by Karl (1990), based on her comparative analysis of the recent histories of the countries of Latin America.

You may find this lack of consistent association between democratization and economic development not very encouraging. After all, is not democracy, which aims to empower all women and men rather than just the already powerful, preferable to more authoritarian forms of rule? You may agree with Winston Churchill, as I do, that democracy is the worst political form devised by human beings, except for all those other political forms which are even worse. So, democracy is preferable and yet there is no consistent association between democratization and economic development. I want to say four things about that combination by way of conclusion to this chapter.

First, the comparative evidence from Asia and elsewhere in the Third World does not indicate that democratization and socio-economic development *never* reinforce each other. Sometimes they do, sometimes they do not. What is being said is that the one does not *necessarily* promote the other.

Second, it is important to emphasize that the concept of democratization refers to processes of *moving towards* democracy (or semi-democracy), not *being* a democracy within the meaning of the definition used in this chapter. Indeed, probably no democracies in the Third World (or elsewhere) are fully democratic within that meaning. (As you know, most democracies in Asia are actually semi-democracies or partial democracies.) Whether or not democratization and socio-economic development reinforce each other is a separate issue from whether or not a fully developed democracy is the best possible political framework for sustained socio-economic development and well-being – another large subject not considered in this chapter. The argument here has been, in effect, that Third World people should be wary of assuming that as soon as they get on the road towards democracy their economic and social problems will immediately begin to be overcome.

Third, if you believe that democracy is preferable to other state forms, premised as it is on libertarian and egalitarian values, then it follows that democratization is *intrinsically desirable*, a process of political development. It may be desirable but it may not necessarily immediately trigger other development processes: its realization may be profoundly influenced by the structural alignment of class forces but it also involves the agency of women and men mobilized to continuous struggling against forces hostile to the democratic prospect.

Finally, these three considerations underline the important point that 'development' is not a unified concept. There can be in the same period in the same country positive economic development (e.g. economic growth) and negative political development (e.g. increased repression and destruction of civil and political liberties), or vice versa. Democratization is intrinsically part of the development process, but may have, at least in the early phases of the struggle, a logic separate from other processes of development.

Summary

There are a number of complex and competing explanations to the question: why does democratization take place? Some, like the ones by Rustow and Diamond *et al.*, emphasize agency factors like the importance of political leadership and the initiatives of groups during the push towards democratization. Other explanations, like the ones by Moore, Therborn and Robison, emphasize the importance of economic and social structures which can be more or less favourable to the development of democracy. These two broad types of explanation have been set out here with reference to the literature related to Asian states, but agency-led and structure-led approaches to democratization are also found in works trying to explain similar political processes in Africa and Latin America.

Democratization may be intrinsically desirable, part of political development; but evidence from Asia, Africa and Latin America suggests that democratization, at least in the shorter term, does not necessarily produce peaceful and more egalitarian social relations nor does it necessarily hasten economic development. Economic development may indeed be quicker in more authoritarian political regimes. Sometimes democratization and socio-economic development reinforce each other, and sometimes they do not.

15

GENDER MATTERS IN DEVELOPMENT

RUTH PEARSON

In recent years there has been an increasing awareness that as development has proceeded, or not proceeded, in Third World countries, the impact on men and women has been different. In fact, there is substantial evidence that women have consistently lost out in the process.

Some of the inequalities between men and women had their basis in colonial rule. Although social relations between men and women, as between other groups, were by no means egalitarian in many pre-colonial settings, there is no doubt that colonial capture and the introduction of exploitative labour regimes led to a marked deterioration in the social and economic status of women relative to that of men. The development of a world economy, and the spread of wage labour in both agricultural and industrial production, assumed very different roles for women and men in the economy, sometimes excluding women from wage employment while relying on their unpaid work on family farms or on low-paid work within the informal sector. Even since the Second World War, when national governments and multilateral and bilateral aid agencies have initiated development programmes and projects, there is overwhelming evidence that such assistance has generally bypassed women or sometimes made women worse off. Such was the concern of development professionals that 1975 was declared International Women's Year, followed by the Decade for the Advancement of Women 1976–85. But even after these initiatives there is considerable concern about the lack of understanding of gender relations and the fact that development policies and projects are still in the main gender blind if not actually biased against women.

Q Why has development affected women and men differently? Are women too weak and powerless to take advantage of the challenges and opportunities that development offers? Is it tradition or backwardness that keeps women in a secondary position in Third World societies?

Q Do planners and policy makers discriminate against women because they make mistaken assumptions about women's roles and involvement in production and reproduction? Or, more fundamentally, is the whole process of development, both as historical social change, and in terms of development policies and projects, deeply rooted in the unequal relations between men and women so that it is necessary totally to rethink the objectives and strategies of development?

This chapter will help address these questions by presenting a framework for thinking about gender relations. First, some basic analytical concepts are introduced to help in understanding the social processes that give rise to the disadvantaged position of women. Then I examine some ways in which women are subordinated. These concepts are then used to look at some development initiatives and examine how they have affected men and women differently. Finally, I review some attempts to make women's issues and interests more central in the development field, and argue that the effects of more recent economic policies indicate that there is still a long way to go.

But – be warned – the conclusion of this chapter is that we can't think in terms of analysing development and then looking at its effect on women. On the contrary, whether it is recognized or not, development process or policy inevitably affects and is affected by the relations between the genders in any society. All policies, however technical or neutral they may appear to be, will have gendered implications. One task is to understand how different aspects of social and economic organization are already based on a system of gender relations and thus to clarify how assumptions about the roles of men and women form the backdrop against which policies are formulated, even if these assumptions remain unspoken and unacknowledged.

15.1 Conceptualizing gender and understanding gender relations

'Gender' rather than 'sex' is the key concept here because we are concerned with the social roles and interactions of men and women rather than their biological characteristics. Gender relations are social relations, referring to the ways in which the social categories of men and women, male and female, relate over the whole range of social organization, not just to interactions between individual men and women in the sphere of personal relationships, or in terms of biological reproduction. In all aspects of social activity, including access to resources for production, rewards or remuneration for work, distribution of consumption, income or goods, exercise of authority and power, and participation in cultural and religious activity, gender is important in establishing people's behaviour and the outcome of any social interaction.

As well as interactions between individual men and women, gender relations describe the social

meaning of male and female, and thus what is considered appropriate behaviour or activity for men and women. What is considered as male or female work, or male or female attributes, behaviour or characteristics, varies considerably between different societies and different historical periods. But it is also important to realize that notions of gender identity, and thus what is fitting for men and women to do or be, have a strong ideological content. For example, in Britain, with nearly 50% of women of working age in the labour force, Patrick Jenkin, Minister of State for social services claimed in 1979: 'Quite frankly, I don't think mothers have the same right to work as fathers do. If the Good Lord had intended us to have equal rights to go out to work he wouldn't have created men and women. *These are biological facts*' (emphasis added). In rural sectors of the Third World both men and women often report that women don't do any agricultural work, or that they are just involved as family helpers, or carry out only domestic work. In fact, many of their waking hours are spent in activities such as weeding and harvesting, or collecting animal fodder or fuel wood, which have a direct effect on the productivity of agriculture whether the output is used for self-provisioning or processed and sold on the market.

Because notions of gender roles and activities have such a strong ideological content, policy often reflects normative or prescriptive versions of female and male roles rather than activities actually practised by women and men. But, as will become clear later, it is essential to understand the precise nature of what women and men actually do, and their real contribution to production and reproduction, if development policy is to cease being biased against women.

Using women as an analytical category

Although most people accept that women and men have different social positions, it is of course true that other social divisions such as class, race, ethnicity and age will also affect people's life chances. However, in common with much feminist literature, this chapter 'starts from the premise that all women share a common experience of oppression and subordination' (Young, 1988, p.4).

It is possible to gain some notion of women's common interests, even though the forms of women's oppression and subordination vary widely, both historically and across or within specific societies.

For example, all women are potentially vulnerable to violence from men. Many women in Britain avoid travelling or walking alone at night. However, if a woman is rich enough she can avoid some of the constraints this places on her life by driving her own car, or using expensive forms of transport such as taxis. But her vulnerability to male violence stems from the fact that she is a woman, not from her economic circumstances. Her economic power may be deployed to mitigate the constraints and threats that male violence imposes on her as a woman, but it cannot abolish them.

Other situations can look different. In January 1991 rich women in Riyadh sought to challenge the very strict code of conduct that gender relations in contemporary Saudi Arabia imposed on them. They organized a drive-in to the centre of the city, dismissed their chauffeurs and drove their expensive cars themselves. The authorities retaliated by dismissing many of them from their employment in the University. Again, in this situation the constraints on their action stemmed directly from their gender, which was the basis of the social and political control on their autonomy. Being rich gave them big cars and chauffeurs, but it could not allow them the freedom to drive (Figure 15.1).

Many examples given in this chapter detail the work burden of poor women in the Third World, which systematically exceeds that of men in the same households (and therefore, the same class position). Without focusing on the social relations which produce and perpetuate this situation, it is impossible to go beyond a mere description of the conditions of women in different situations. But if we are concerned to devise policy initiatives and promote social changes which will improve the material and social condition of women, we need to understand how profound and pervasive the structures of women's subordination really are.

Figure 15.1   Gender differences in mobility: women office workers in India travel separately.

How gender makes a difference

One way of measuring how people's gender makes a difference to their share of the benefits of society is to examine macro-level indicators which are differentiated by sex. Look back at Tables 2.6 and 3.3 for literacy rates, educational enrolments and activity rates. Tables 15.1 and 15.2 show some further differences (wage rates and political participation) for selected countries and regions. On all the indicators shown, it is clear that women have an unequal share in some of the 'benefits of development'. Although national average figures hide an enormous amount of variation – by class, occupation and so on, they cannot hide the importance of gender in determining people's access to resources or opportunities.

Competing concepts: subordination or patriarchy

We can go further than talking about *inequality* with respect to gender relations: we can say that men are dominant and women are **subordinated**.

> **Subordination of women:** A phrase used to describe the generalized situation whereby men as a group have more social and economic power than women, including power over women. As a result, women come off worse in most measurable indices of the outcome of social and economic processes. In short, the way the two genders relate to each other is that the male gender is dominant and the female gender is subordinate.

Another term which is often used to describe the unequal relations between men and women is *patriarchy*. This was originally an anthropological term which described the kind of social systems in which authority is vested in the male head of the household (the patriarch) and other male elders within the kinship group. Older men were entitled to exercise socially sanctioned authority over other members of the household or kinship group, both women and younger men.

Some writers prefer to use the term patriarchy to describe the general situation of male dominance over women, in order to develop an analysis of the structures of patriarchy within specific societies (Walby, 1990). However, the form this relationship of domination/subordination takes differs strikingly between different societies and is constantly undergoing change and negotiation. I therefore prefer to use the term *subordination* as it has no connection with any specific system of social organization and does not rest on the basis of a specific structure of male authority over women (Pearson *et al.*, 1984).

Subordination of women

Although the subordination of women is universal, this does not mean that gender relations are the same in every society. In every case, the social meaning of being female or male will be the result of the history of that society, influenced by the nature of the local economy that evolved over time, religious beliefs and political systems. It is necessary to analyse each situation separately in order to understand the precise nature of gender relations, the different factors which influence them, and how gender relations themselves have an impact on social and economic development. It would be a mistake, for example, to pick out one independent variable such as religion and assume that the gender ideologies and related gender roles implied by that religion have a uniform, or even a determining, impact on the specific relationship between men and women. Box 15.1 gives an example.

15.2 Structures of subordination

How and where are women subordinated? The mechanisms of distribution of resources and exercise of power between women and men vary widely, but one can expect to find structures of subordination in areas of life such as property relations, divisions of labour, in law and the state, and in how households are organized. Although I look at these areas separately, there are links between them which will become apparent (for example, between state policy and property relations, or sexual divisions of labour and household organization).

Table 15.1 Wages of females as a percentage of wages of males

	1977	1986
Developing countries		
Cyprus	49.6	56.1
El Salvador	80.8	81.5[a]
Hong Kong	–	77.9
Kenya	55.6	75.6[a]
South Korea	44.7	48.5
Singapore	–	63.4[a]
Sri Lanka	–	75.5
Developed countries		
Czechoslovakia	67.4	67.9
France	75.8	79.5
West Germany	72.3	72.9
Greece	68.8	76.9
Japan	46.0	42.5
New Zealand	73.3	71.8
Sweden	87.4	90.4
United Kingdom	70.8	67.9

Source: UN (1989) Report on the World Social Situation, United Nations, Geneva, p.13.
[a] Based on 1985.

Table 15.2 Percentage of women in parliament

Region	Lower chamber of bicameral assembly		Upper chamber of bicameral assembly	
	1975	Latest year	1975	Latest year
Africa	4.1	6.3	–	4.9
Asia	13.2	12.8	6.9	6.9
Latin America and Caribbean	3.4	10.6	4.2	6.5
North America	3.6	7.0	2.9	7.4
Oceania	1.9	4.9	9.4	22.4
Europe	13.2	17.6	6.4	8.2
Soviet Union	32.1	34.5	30.5	31.1

Source: UN (1989) Report on the World Social Situation, United Nations, Geneva, p.13.

Box 15.1 Gender and 'culture' in the Middle East

That women's legal status and social positions are worse in Muslim countries than anywhere else is a common view. The prescribed role of women in Islamic theology and law is often argued to be a major determinant of women's status. Women are viewed as wives and mothers, and gender segregation is customary if not always legally required. Economic provision is the responsibility of men, and women must marry and reproduce to earn status. Men but not women have the unilateral right of divorce; a woman can work and travel only with the written permission of her male guardian; family honour and good reputation, or the negative consequence of shame, rest most heavily on the conduct of women. Muslim societies are characterized by higher than average fertility, higher than average mortality, and rapid rates of population growth. Age at marriage affects fertility. An average of 34% of all brides in Muslim countries in recent years have been under 20 years of age, and the average level of childbearing is 6 children per woman. The Moroccan sociologist Fatima Mernissi has explained this in terms of Islamic fear of *fitna*: social and moral disturbance caused by single unmarried women. Early marriage and childbearing, therefore, may be regarded as a form of social control.

The Muslim countries of the Middle East and South Asia also have a distinct gender disparity in literacy and education, and in Bangladesh low rates of women in the labour force. High fertility, low literacy and low labour force participation are linked to the low status of women, which in turn is often attributed to the prevalence of Islamic law and norms in these societies. It is said that because of the continuing importance of values such as family honour and modesty, women's participation in non-agricultural or paid labour carries with it a social stigma, and gainful employment is not perceived as part of their role.

However, these conceptions are too facile. In the first instance, the view of woman as wife and mother is present in other religious and symbolic systems. The Orthodox Jewish law of personal status bears many similarities to the fundamentals of Islamic law, especially with respect to marriage and divorce. Secondly, the demographic patterns mentioned above are not unique to Muslim countries; high fertility rates are found in sub-Saharan African countries today, and were common in Western countries at the first stage of the demographic transition. Thirdly, high maternal mortality and an inverse sex ratio exist in non-Muslim areas as well. [The sex ratio measures the number of women to men in the population. Normally the ratios have a higher number of women to men, reflecting greater vulnerability of male infants, adult males and women's longer life expectancy. In a few countries the normal ratio is reversed.] In north India and rural China, female infanticide has been documented. In the most patriarchal regions of West and South Asia (including India) there are marked gender disparities in delivery of health care and access to food, resulting in an excessive mortality rate for women. Thus, women's disadvantaged position in the Middle East cannot be attributed solely to Islam. Religious and cultural specificities do shape gender systems, but they are not the most significant determinants and are themselves subject to change.

There is also considerable variation in gender codes in Muslim countries, as measured by differences in women's legal status, educational levels, fertility trends, and employment patterns. For example, sex segregation in public is the norm and the law in Saudi Arabia, but not so in Syria, Iraq or Morocco. Following the Iranian Revolution, the new authorities prohibited abortion and contraception, and lowered the age of consent to 13 for girls. But in Tunisia contraceptive use is widespread and the average age of marriage is 24. Women's employment levels are dissimilar in such countries as Turkey, Saudi Arabia, Iran, South Yemen and Algeria. Changes and variations across the region and within a society are linked to state economic and legal policies. It is important not to regard 'culture' as a constant; it is variable, and the extent of its impact depends on other factors, notably on the depth and scope of development, on state policy, and on class and social structure.

(Extracted and adapted from
Moghadam, 1991, pp.4–5)

Property relations: ownership and access to land

Systems of land tenure and inheritance vary widely. In most cases women have use rights rather than outright ownership rights to land. Where full rights to ownership exist in traditional legal codes, a woman's share is not usually equal to a man's. And even where traditionally women are entitled to inherit land, new laws often reduce women's rights to succession. For example, the Colombian land reform law confined land redistribution to married males over 18, and deprived women of rights they had in the pre-reform system (Whitehead, 1985, p.53).

Even where women have customary use rights to land, these are rarely exercised in their own right, but only through kin relationships with individual men – brothers, fathers, other male relatives or, most commonly, husbands. In many African systems which prescribe the payment of bridewealth to the husband's family, women very often can only farm residual land after the rest of the kin group have had their allocation, even though, as wives, they are required to work on their husband's land. Without direct ownership or right to farm land, a woman has little say in the allocation of the income or crops. Since, as we detail below, men and women have different responsibilities within a household, lack of access to land can mean not just dependence on the head of the household, but real difficulties in meeting her own and her children's needs for food and clothing.

Even in quite different systems where property is combined in marriage, which is common in many Asian societies, a woman rarely receives direct proceeds from the land, but depends on her husband to support her and her children. Although property relations are determined at the level of social and legal institutions, women's subordination in this realm has important consequences reaching into the allocation of resources, income and product within the household (Whitehead, 1985, pp.55–6).

The sexual division of labour

Another important aspect of women's subordination is the allocation of different tasks and responsibilities to men and women. Strictly this should be called the 'gender division of labour'; however, although it refers to social, not biological, categories, the term *sexual division of labour* is generally used. One division often made is that between productive work and domestic labour. Women are generally regarded as responsible for the latter, although men may also play a role (Table 15.3). However, it does remain true that women carry out the bulk of what has been called reproductive work – including biological, generational and daily reproduction. In Britain in 1988, 72% of women claimed total responsibility for domestic work; and in Cuba, where by law, men are supposed to share domestic tasks, 82% of women in Havana and 96% of the women in the countryside have sole responsibility for domestic chores (Momsen, 1991).

In many Third World societies, in the urban informal sector and in the rural economy, it is extremely difficult in practice to distinguish between domestic labour and productive work. Many

Table 15.3 Gender divisions of time use in Nepal

	Time input into village and domestic work (%)	
	Men	Women
Cooking and serving	10	90
Cleaning	5	95
Maintenance	7	93
Laundry	10	90
Shopping	54	46
Other domestic	22	78
Childcare	16	84
Animal husbandry	55	45
Family farm enterprise	45	55
Gathering and hunting	60	40
Fuel gathering	34	66
House construction	72	28
Food processing	13	87
Water collection	8	92
Outside earning activity	69	31
Local market economy	43	57

Source: adapted from Joss, S. (1990) 'Your husband is your god', GADU Pack no.12, OXFAM, Oxford.

apparently domestic tasks are part of maintaining production. But allocation of time to different activities including so-called domestic labour in rural production systems does vary to a certain extent according to the seasonal and other demands of production. Table 15.4 gives a seasonal breakdown of time spent on different tasks by men and women. The sexual division of labour within both agricultural and household tasks is fairly rigid. During peak times, women's time allocated to domestic work is reduced – though only by 13%. Men's domestic tasks are usually of a different nature (house maintenance, fencing, preparation of land for gardens etc.). In this example, women have much longer work time overall and consequently less time for leisure or sleep in both peak and slack seasons.

The sexual division of labour therefore operates not just between domestic and productive work but within each category of work as well. Also, what is considered to be a male or female occupation varies considerably between different societies, and in some cases between different groups and classes within the same society. A number of activities which are 'male' in one place (petty trade in North India, for example) are predominantly female tasks in others, such as Jamaica or Nigeria. On the other hand, the vast majority of domestic servants are women in South American cities, whereas in Lagos they are men; most typists in Jamaica are women, but not so in Madras (Figure 15.2).

The sexual division of labour may be modified to respond to changes in economic circumstances which alter the supply of men or women to carry out 'male' or 'female' tasks. For example, when the plough was introduced earlier this century in Zimbabwe, women were displaced from tasks associated with the preparation of ground for farming. However, in recent years, with large-scale male out-migration from rural areas, women have largely taken over ploughing and land preparation. In the northern province of neighbouring Zambia, the response to male out-migration has been different. Women have responded by hiring in casual labour to carry out traditional male tasks such as tree cutting and ploughing.

Gender roles have their own dynamic, evolving and adapting over many generations, exhibiting considerable resistance to changes in economic circumstances, or even to market forces. Often strategies which avoid changing the sexual division of labour are pursued, as in the Zambian case. Another example occurred when many women, often mothers of young children, migrated from Sri Lanka and other South Asian countries to the high-income Gulf states during the 1980s to work

Table 15.4 Gender divisions of time use (hours per month) in the dry zone of Sri Lanka

| | Peak season | | Slack season | |
	Male	Female	Male	Female
Agricultural production	298	299	245	235
Household tasks	90	199	60	220
Fetching water and firewood	30	50	30	60
Social and religious duties	8	12	15	15
Total work hours	426	560	350	530
Leisure/sleep	294	160	370	190

Source: Momsen, J. (1991) *Women and Development in the Third World*, Routledge, London, table 4.3.

Figure 15.2   Women's jobs differ from place to place: (left) street trading in Lagos; (above) working on a building site in north India.

as domestic servants. Most often their child-care and domestic work at home devolved on other female family members, rather than on the fathers or male kin. This persistence of a strict sexual division of labour in domestic work is common all over the world (possibly more so than in household production) and should warn us not to assume too readily that gender relations can be modified or changed in response to change in the demand for or availability of male or female labour.

Sexual divisions of labour are neither constant in any particular society nor paralleled between different societies. But in every society there are clear demarcations which define what is appropriate, or possible, for women and men in that society – demarcations which may be transgressed by individuals without necessarily challenging the principle of allocation of tasks according to gender. Box 15.2 gives an example of how sexual divisions of labour are established in childhood and are carried through into adult life.

Box 15.2 Your husband is your God

Rural women are the backbone of Nepal's sub-sistence economy. Their work starts in the early morning and goes on throughout the day with little time for rest. Their endless toil is taken for granted. Their major role is to be a wife and a mother carrying out most of the domestic tasks in the household. No one questions the division of roles. In the research conducted by Tribhuvan University Centre for Economic Development and Administration into the 'Status of Women in Nepal', it was calculated that on average women work 10.81 hours each day compared with men's 7.51 hours. Also, 86% of all domestic activities, 64% of the village work and 62% of the agricultural activities are done by women.

The disadvantaged life of a woman starts at birth when everyone is disappointed that the baby is not a boy. Girls cost a lot to marry (i.e. in dowry) and although by the age of 7 years they are contributing to the household income, they are considered a burden. In rural Nepal relatively few girls go to school, although numbers are increasing in primary schools. Education is now seen as a plus for a better marriage. Any difficulty at home and the girl is the one to take time off lessons, particularly if it is to look after younger siblings. But she is lucky to go to school at all; most girls have to stay at home helping their mothers in the fields and with the never-ending domestic chores. It is the family's dependence on girls' labour that keeps them out of school. Female literacy rate is 18%, whereas the male rate is 52%.

In comparison, everyone makes a fuss of the boys. They are important and soon learn that they can order their sisters around to serve them.

They wash first, they eat first and never have to clear up afterwards. Their early life is one of playing games with other boys in the village. Sometimes they might have to look after cattle but many jobs like cooking, cleaning, laundry, food processing and water collection are never theirs. A boy's education at school is important.

When girls start menstruating they are hidden away in a dark room for 9 days. The second month it is less time and from then onwards the young woman knows that each month she is unclean for 4 days. She must not enter the kitchen, touch anyone else's food but her own and must eat and sleep separately. At the end of the 4 days she must wash all the clothes she has worn as well as the bedclothes. If it were not for the feeling of being ostracized, she could enjoy her days 'off'.

Legal marriage has a very high position in society. If the parents have arranged the marriage they will do everything to protect it. In marriage it is the Nepali woman who has to adapt – in fact, one of a bride's virtues is the ability to adjust to her new home. She is the one who moves house, she is the one in a new family coping with a mother-in-law whom she must please to make her life bearable. She has a lot of work to do in the home, up at 4 a.m. to start preparing *dal bhat* (main meal eaten morning and evening) for the whole family and finishing her day about 8 p.m. Then she has to get to know a total stranger, who is now her husband. Having previously been very inhibited about showing any part of her body, she is now thrown into living with a man.

In modern families there is a terrible dilemma that mothers face. If they have learnt some independence with the agreement of their husband, it would seem right to teach their daughters the same. But this could 'jeopardize' their chances of a 'good' marriage, so the cultural traditions keep the status quo. It is often thought that a woman with 'too much' education might fail to make a 'good' wife as she could well not be used to housework and might even consider herself above it. But times are changing, particularly in the urban areas, like Kathmandu; families allow love marriages and allow sons' families to live separately from the extended family home much more often. Also, work can necessitate living away from home. In contrast, in rural areas parents often marry their children off before they are 14 years old.

Under Nepali law men and women are equal. Women do not inherit property or land: only sons benefit. The one exception is an unmarried woman over 35 years of age, since she has the right to some of her father's land. But unmarried women are rare. Divorce is exceptional and there is a small percentage of separations. Widows are considered unlucky since, for some reason, they must have caused their husband's death. Men are able to remarry, whereas women cannot, and, although illegal, some have two wives. If the wife fails to produce a male child it is considered very inauspicious and the husband is within his rights to find another wife.

The women undertake 62% of all subsistence agricultural activities. The backbreaking weeding, irrigation, harvesting, muck spreading of crops is done by women. Men take a major role in land preparation, terrace upkeep, crop protection and spreading of chemical fertilizer. If there is anything needing technology or 'modern' understanding men insist it is their responsibility. So if any new development is brought to the village it is assumed that it will be taught and discussed with the men. The 'Status of Women in Nepal' did find that women's high input into agriculture production is not reflected by a commensurate level of participation in farm management decisions, although often choice was made according to tradition, e.g. crop choice, seed selection.

Outside the immediate household men dominate any economic activity. Of all earnings outside the home, 69% is earned by men, and they make most of the decisions on how the money is spent. Wives are consulted from time to time regarding household durables and clothing. Transactions concerning land or animals or major household loans are primarily decided by men. This outside earning capacity, meeting others in market places, bargaining and selling, etc., contributes to men gaining confidence to dominate decision-making. A number of men in Nepal, particularly in the far west, migrate into India for work. Since the extended family prevails, women are still subject to decisions made by men other than their husbands, in the family. Women do have a voice in how the house is run, particularly as they get older; also by giving guidance to their daughters and daughters-in-law.

(Extracted from Joss, 1990, pp.1–3)

The state, law and gender relations

The state influences how women are viewed in society in various ways: through legislation, in how public institutions are staffed and run, and through social policies. These mechanisms can affect whether women are seen as autonomous individuals or as dependents of men. In some countries, women's civil rights (in divorce, custody of children, ownership of property, autonomy in matters of employment, financial contracts) are discriminatory. Entitlement to benefits such as health insurance, pensions and welfare payments are often linked to women's relationship with men, rather than granted in their own right. In many countries the police and law courts treat physical and sexual assaults of women within the family quite differently from other forms of assault.

The state also appoints public officials in the judiciary, the civil service, industrial planning and regulatory boards, etc. Where such appointments are not adequately shared between men and women it reinforces the notion that primarily it is men who have the authority and competence to represent the whole population in serious matters. While many discriminatory laws have been repealed, particularly in the socialist Third World countries, in many places the extensive powers and practices of the state continue to legitimate unequal rights and unequal treatment of men and women.

Gender relations within households

In developing countries, most households are the site of many different kinds of activity: residence and consumption, subsistence and market production, distribution of resources and output, socialization of children, care of the elderly, and so on. These activities may take place in a single unit or in several connected ones. Different members of households may have different rights (over property, labour and income) and obligations (particular productive tasks, food provision, domestic work, etc.), often depending on age and marital status, as well as sexual divisions of labour.

Development policies are often directed to an idealized model of households where production, income and consumption are shared: they are seen to have convergent interests in which all members contribute according to their ability and share the benefits according to their needs. However, empirical studies have shown that far from being a unit in which all resources and benefits are pooled equitably, the use of resources and labour, and the distribution of income and output, have constantly to be negotiated, and intra-household relations are often conflictive (Dwyer & Bruce, 1988). In different household systems, men and women have different responsibilities in terms of work in domestic or other activities, obligations to provide food and other resources (water and cooking fuel), and cash income to meet family needs such as school fees, clothing, agricultural inputs, etc. In some households, far from a pooling situation, women trade grain they have grown themselves for cash to buy clothing and other necessities. In others, women's labour in family farms and enterprises give them no direct rights to cash, no 'owned resources' to exchange for cash, and little influence in deciding how household income is to be allocated between competing demands. In each of these cases, intra-household relations can be seen as a form of *co-operative conflict* (Sen, 1987): although household members may depend on each other in terms of labour and output, there is considerable negotiation and conflict about how

such inputs and benefits should be distributed.

As well as these intra-household divisions, the distribution of inputs and benefits can also systematically disadvantage women. For many households, especially those on the margins of subsistence, women contribute a major proportion of their labour time, and their total income, to the needs of the household, particularly to their children. In contrast, men tend to have considerable autonomy over how they spend their income:

> "A general finding is that women's income is almost exclusively used to meet collective household needs, whereas men tend to retain a considerable portion of their income for personal spending. A key source of male bargaining power within the household is determining what proportion of their income to pass on to other household members. Rather than personal spending money for male adults being residual after other household needs have been met, it is often a priority and the amount allocated to household expenditure is residually determined by the difference between male income and personal spending. The good husband is the one who strictly limits his personal spending, but his prerogative to enjoy such spending remains."
>
> (Elson, 1991, pp.3–4)

Box 15.3 is an illustration of how household divisions are linked to access to land, sexual divisions of labour and changes brought by commoditization.

15.3 Using gender to analyse development

In the previous sections I have mapped out some basic concepts necessary to analyse the changing social relations of gender. I have stressed that the precise nature of gender relations differs according to specific historical contexts. We will now see how these concepts can be applied by analysing briefly two examples of change in developing countries.

Box 15.3 The impact of cash cropping in Ghana in the 1920s

In the earlier period, peasant households largely concentrated on food-crop production. The principal food crop then was yam, although vegetables were also grown. Both men and women cultivated the same plot. The men generally cleared the land on their own, but the women shared the work of weeding, planting and harvesting. However, when cocoa cultivation became profitable, a new sexual division of labour emerged. Men took over the production of the cash crop and women were left to produce the food crops for subsistence. Land that earlier had been cultivated jointly by both sexes was gradually taken over by the men to cultivate cocoa, and the women were left with small, infertile plots. Men now devoted their labour almost exclusively to cocoa and stopped helping with the food crops. The money from the sale of cocoa did not contribute to the family's subsistence; rather it was used for larger investments either in the cocoa crop, in house-building, and occasionally on the children's education. Often the money was also spent by the men on their own clothes and on drink.

Where women managed to produce a surplus for sale, the cash they obtained could not be spent by them but was controlled by the male heads of households and used for the family's subsistence needs. In the absence of help from the men, the women found it increasingly necessary to shift out of yam and vegetable cultivation into the cultivation of less labour-intensive crops such as maize and cassava. These crops, however, had a higher starch content than yam and were less nutritious.

The desire to have control over some cash income, as well as the takeover by men of land previously used for food crops, caused the women to try and establish separate farms. Women's access to independent plots, however, was severely constrained by a number of factors. An important one was the increased privatization of communal land, and the confining of inheritance rights of this land to men. Hence, where previously women had the same rights to communal property as men, even though the social structure was patrilineal, now they had to ask permission from male relatives for the use of the latter's fallow land. With an intensification in the competition for land this became increasingly difficult. Another problem related to the fact that women had no direct access to labour other than their own or that of their children, and hired labour was difficult to get (quite apart from problems of payment); hence they were constrained in the type of land they could cultivate. For example, they could not use forest land, which would need manual clearing, even though such land was usually the most fertile; rather they had to opt for a less fertile plot, but one that was locationally accessible and that they could prepare on their own or plough with a hired tractor (hiring a tractor was cheaper than hiring labour).

The difficulties of access to land and labour as well as to extension services and credit meant that survival became a battle for many women and their children. In short, the introduction of cash cropping in Ghana may be seen to have (a) significantly increased women's work burden; (b) led to a fall in the nutritional standard of their diets because it necessitated a shift from yam to cassava and, for some, even a fall in their absolute levels of consumption; and (c) brought no gain in terms of their cash, since the scale of their independent activities was usually so small as to make it very difficult to produce enough even for subsistence needs.

(Agarwal, 1985a, pp.106–7)

Women and the Green Revolution

The 'Green Revolution' is regarded as an important example of successful development. A technological package based on new seeds, but with biochemical, mechanical and institutional aspects (i.e. new forms of credit and training) has been responsible for substantial increases in production of wheat and rice, particularly in Asia.

The Green Revolution started in the mid-1960s and continues today. It is probably the first systematic attempt to increase worldwide food production. Agricultural production is increased in three ways: average yields are increased, the area devoted to agriculture is extended, and the number of crops grown in a year is increased.

The major technical elements are:

1 Mechanization – the introduction of tractors, and other machinery for harvesting and irrigation etc. These may be used to extend the cropped area; often machines are used to replace labour.

2 Biochemical innovations – such as new seed varieties and new inputs such as fertilizers, pesticides and weed killers. These can all increase crop yields.

Most assessments of the impact of the Green Revolution in India recognize that as well as increasing overall productivity, it has also increased socio-economic stratification. This has resulted in worsening of income distribution between rich farmers on the one hand and poor farmers on the other, many of whom have joined the ranks of landless labourers.

The economist, Bina Agarwal (1985a), has used gender to analyse this particular development process within the three major rural groups or classes in India, identified as:

1 agricultural labour households (which have no land or insufficient land for household subsistence);

2 small cultivator households (which have sufficient land to provide for household needs using family labour);

3 large cultivator households (which have sufficient land and other resources to farm using hired, i.e. paid, labour, and which do not use the women of the household for field work).

For women in the landless or land-short households the impact of Green Revolution technology was complex. The numbers of households in this group increased (including a large number of female headed households) and there was a rise in the number of women who had to hire themselves out as casual wage workers because many families were living on lower material resources than before the changes. Women had to increase their time and energy devoted to agricultural work for

wages as well as maintain their subsistence production and domestic labour to ensure their household's survival. Women were squeezed between conflicting demands – the need to earn income, and the need to increase time devoted to domestic activities.

While there was a rise in the total amount of hired labour required by larger farm households, the increase in employment has not been gender neutral. The new technology package requires additional labour for tasks such as transplanting and weeding (normally women's work) but it is not clear that this has kept pace with the increase in the numbers of women seeking such work. Other activities, particularly harvesting, threshing and grain-processing, have been subject to substantial mechanization, reducing the employment opportunities for poor women. The widespread adoption of grain processing mills has vastly reduced women's employment in dehusking of rice, for example (Agarwal, 1985a).

The Green Revolution has at best had a mixed impact on the employment opportunities and incomes of poor women. Statistics show increases in days worked and in wage rates, but the latter have been offset by price increases (Agarwal, 1985a, p.97). Even where women's incomes have improved, it is not clear that they experience the benefits: women in landless and land-short households continue to suffer chronic deficits in food intake, both in absolute terms and in relation to men.

Women working as casual wage labour also face health hazards from standing in water containing fertilizers and pesticides to do weeding and transplanting (Figure 15.3). There are reports of women field workers suffering intestinal and parasitical infections, leech bites, skin complaints, rheumatic joints, arthritis and gynaecological infections (Agarwal, 1985b, p.329).

Women from small cultivator households are in a different position. Although in general they are not required to work outside family farms for wages, their work within the family holdings has intensified. New seed varieties shorten the gestation period and make possible two harvests a year. Since most of women's agricultural tasks are not

Figure 15.3   Standing in water: Indian women transplanting rice.

mechanized, their work burden has increased. In addition, they still carry the burden of domestic work. They are of course subject to the same health hazards described above.

Total household incomes of this group have in general increased, but decisions on the spending of this income remain biased against women. For instance, substantial shares of the extra income have been spent on tractors which save male labour time – but few households in this group have invested in home-grinding mills which would reduce women's domestic work burden in food processing.

The impact on women from large cultivator households is more closely linked to the considerable rise in household incomes. Although male bias in intra-household distribution remains, women have generally shared in the new prosperity. But even prosperity has implications for women. With increasing wealth, women are withdrawn from field production (a common way of demonstrating prosperity), but this does not necessarily signal an increase in women's autonomy. First, there may be an increase in domestic labour as women are responsible for preparing food for the increased numbers of hired labourers, and for other farm management tasks. Secondly, women's ability to bargain over allocation of household income may be further constrained. Withdrawal from field work can often lead to total seclusion and an increase in women's dependence on men.

The effect of the Green Revolution therefore cannot be measured just in terms of increased production or rises in aggregate household incomes. Consideration should also be given to the increases in women's workloads and to the fact that women often fail to receive an equitable share in the extra income earned, even though their labour is often central to achieving higher levels of productivity. Increased demand for casual waged work for women from the poorest households has not necessarily increased household income where, as has often been the case, real wage levels have declined. Since women are in danger of losing out (of working harder, for less reward, with added health risks), this policy for improvement in agricultural productivity cannot be considered an unqualified success.

Development projects: the dangers of gender blindness

Many development initiatives have been introduced in areas of the Third World in the form of large-scale projects financed by external aid and organized by technical experts. One well-known example was the attempt in the 1970s to introduce irrigated rice production in an area in The Gambia where, traditionally, men had farmed groundnuts as a cash crop (with the aid of family labour) and women had cultivated rice for household consumption on unirrigated wetlands (Figure 15.4). The irrigation project was unsuccessful largely because the planners failed to understand the specific dynamics of gender relations within the local household-based farming system (Dey, 1982; Carney, 1988).

"The lack of success stemmed in part from [a] 'male dominant', 'domestic sharing' model of the household which shaped the project. An initial assumption was that the men were the rice growers with full control over the necessary resources. Incentive packages included cheap credits, inputs and assured markets offered to male farmers. But it was women who traditionally grew rice for household consumption and exchange, within [a] kind of complex set of rights and obligations...The scheme proposed to develop irrigated rice production on common lands to which women had secured use rights. Backed by project officials, men established exclusive rights to these common lands, pushing the women out to inferior scattered plots to continue cultivating traditional rice varieties. All access to inputs, labour, and finance was mediated through husbands, and women became notably reluctant to participate in their planned role as family labour. For what work the women did do on the irrigated rice fields husbands had to pay their wives... the disappointingly low levels of improved rice production arose substantially from these misunderstandings."

(Whitehead, 1990, pp.63–4)

As well as assumptions about the structure and operation of households, the planners took other things for granted. First, they assumed that women were 'free' labour – that their labour would not have to be recompensed in any way, and thus involved no resource cost. Secondly, they assumed that women's labour had no existing productive use and therefore was available for intensive application to irrigated rice production. In fact, women were busy with their own rice lands in the rainy season, during which men did not have access to their labour and, therefore, did not grow a further rice crop as planners had expected. Men only grew rice in the dry season when their women were free to work for them (for wages) and when they themselves earned higher wages elsewhere. In addition, ignoring that women were farmers in their own right, planners neither took advantage of women's expertise in swamp land rice nor helped women to improve their own output.

Figure 15.4   Gambian women have expertise in rice production: Mandinka women hoeing weeds in the rice fields.

The Gambian example illustrates two further points. Women are not necessarily the passive victims of circumstances with no possibility of resisting. By refusing to make their labour available at no cost and to the convenience of the male farmers involved in irrigated rice production, the Gambian women demonstrated their determination to defend their existing gender position as separate producers, albeit with various obligations as well as rights within the household. The example demonstrates that in households where men and women hold separate purses with no pooling of cash income, and have responsibility for different items of household expenditure and food production, as well as different access to resources for production, it cannot be assumed that increasing household production and incomes will of necessity improve the living standards of all members.

15.4   Putting gender on the development agenda

The concern with the effects of development on women goes back some 10–20 years. The UN's Decade for the Advancement of Women (1976–85), which was preceded by the International Women's Year in 1975, was the culmination of successful pressure by women activists and academics (Pielta & Vickers, 1991). The argument was that development policies, co-ordinated by the leading multinational agencies, such as the World Bank, UNIDO and FAO, as well as many bilateral development agencies and NGOs, had ignored the needs of poor women in Third World countries.

It was not just in economic terms that the marginalization of women needed to be addressed.

There was concern that women were being 'left out' in four critical areas: political rights; legal rights; access to education and training; and their working lives (Boyd, 1988).

'Women in Development' policies

The new emphasis led many agencies and experts to commit themselves to Women in Development policies in order to enable women both to enjoy the fruits of development, and to make an appropriate and necessary contribution to the development process. Many development agencies financed projects aimed at helping poor women, particularly setting up income-generating projects ranging from poultry and fishing co-operatives to handicrafts and school uniform sewing, with various inputs of credit, technical expertise and training provided. Of course, women's labour was unpaid, since the income depended on the sale of the project's output.

A parallel development was the establishment of Ministries for Women's Affairs or Women's Bureaux, particularly in Commonwealth countries. These were charged with negotiating and co-ordinating the activities of the various international agencies keen to finance projects for women. These agencies also helped to raise the profile of women's issues, and divert some aid resources to targeted women's groups. Unsurprisingly, they were chronically under-staffed and under-resourced and most have found it extremely difficult to meet their broad range of objectives (Gordon, 1985).

A major criticism of the Women in Development approach is that its basic assumptions are flawed: it starts from the premise that women have been excluded from development. But as our previous discussion has clearly illustrated, women's time, energy, work and skills are involved in every aspect of the development process; it is the inequality of gender relations and the continuing subordination of women that ensure that women's contribution is not matched by recognition and remuneration in social, political and economic terms.

The shortcomings of the Women in Development approach, which focuses on women rather than on gender relations, is illustrated by research which has assessed various projects aimed at improving the situation of poor women (Buvinic, 1986). Most projects focused on one aspect of women's problems (often lack of access to cash income) without considering the implications of such initiatives. For example, income-generating projects do not take into consideration how women's time budgets are already overstretched (see Chapter 3). Furthermore, many projects were staffed by unpaid leaders, with little commercial or technical expertise available, rather than professional salaried workers. Many projects are based on what are assumed to be women's traditional skills (embroidery, handicrafts etc.), though in many cases the women have no direct experience of these activities. Few have identified any attainable market and as a result are continually dependent on further aid for survival.

Increasing visibility of women?

One clear advance from the UN Decade has been an enhanced understanding of the nature and variety of women's work in the Third World. This has been accompanied by an increasing recognition that conventional methods of collecting statistics consistently underestimate the extent of women's work (Waring, 1989; see also Chapter 3). But even a more comprehensive recording of women's paid work would not represent the real extent to which all economies rely on women, both in productive activities and in the range of reproductive activities discussed above.

"Women often contribute to family earnings in a number of ways, making it important that all income-earning activities be reported. For example, a woman may help on the family farm or in a family enterprise. During the off-season she may take in laundry, produce handicrafts, brew beer for sale, or work on someone else's farm. Especially if she is poor, she may scrape together a living by engaging in several such activities. Unless there is an inquiry into time use or a sequence of questions on multiple jobs, secondary occupations may well be omitted...Unpaid family work on farms and in small non-farm enterprises is an important part of women's work in developing countries...A special effort is often necessary to make certain that it will not be overlooked."

(Mueller, 1983, pp.276–7)

In spite of the increase in research and awareness of the specific circumstances of women, there was still no evidence at the beginning of the 1990s that the implications of taking gender seriously are understood or acted upon at the level of macroeconomic planning.

15.5 Structural adjustment policies: lessons from the past

The 1980s saw widespread implementation of structural adjustment policies (see Chapter 11), which were intended to transform Third World economies into efficient and competitive producers for international markets.

What have we learned about the nature of gender relations that might be relevant to the analysis of those market-oriented policies? It should be clear that economies do not only work through market relations. Given the complexity of gender relations within smallholder households, the invisibility of women's work both in the rural and urban sectors, and the multiple and complex roles women actually perform, market-only policies such as structural adjustment are likely to have specific, and probably negative, effects on women.

For example, women agricultural producers may grow marketable crops, but their energies are often directed to producing food for household consumption. If prices are allowed to rise as an incentive to increasing marketed production, women will often be in a situation where they are under pressure from husbands and other male kin to work harder in order to produce more cash crops at the expense of household food. Women also often find that their workload is increased because new crops and technology intensify the need

for weeding and other female agricultural tasks. Even so, they will not necessarily be compensated by payment for their labour or increased cash income from the sale of the new crops. Furthermore, there is no certainty that cash earned will be used to purchase food to compensate for the reduction in their own provisioning, or that the money earned will be sufficient to purchase equivalent food products on the open market.

These problems are borne out by a nutritional study carried out in Zambia in 1988, after an adjustment programme had been in place. It was found that 'in the predominantly maize growing areas it is the children of the farmers growing maize for sale who are most likely to suffer from nutritional stress' (Feldman, 1989, p.17). In this case, it was also suspected that the extra burden of work imposed on women by producing for the market gave them less time for childcare, infant feeding and food preparation. Thus, as well as underestimating the centrality of women's work in agricultural production, policies based on market models also ignore women's domestic or reproductive roles. Not only is time taken away from non-market production, it can also squeeze the time required to carry out essential domestic tasks. Women will always attempt to stretch themselves in order to meet their multiple obligations, with the result that their total work burden is increased, but their ability to ensure the well-being of their families is jeopardized, to say nothing of their own health and welfare.

Raising prices as an incentive to increase agricultural production for sale, plus the reduction of state control of food markets has meant sharply rising prices for consumers. Together with increased costs of housing, water, school fees and medicines, households can only survive by acquiring several incomes. Women especially have been forced to find alternative income sources, mainly in the informal sector of the economy (Figure 15.5). A recent survey in Dar Es Salaam showed that women of all classes were doing extra work for cash (Feldman, 1989, p.25). Young girls contributed by selling cooked food before and after school. In addition, 58% of households reported that they had reduced the number of meals taken from three to two.

Figure 15.5   Women earn low wages making matchboxes in Bangladesh.

The other plank of structural adjustment policies – that of reducing government spending – has had specific and direct effects on women, who, as teachers, health workers and public sector office workers, have often taken the brunt of cuts in public sector employment. Many women have had to turn to the informal sector for work. Cuts in public services such as health and education have also taken their toll on the poor. In some places, health budgets have been reduced by 50% or more, and a lack of foreign exchange has led to shortage of medical supplies within state-run hospitals and health centres. In Tanzania, mortality of children under 5 has increased since 1980 from 193 per thousand to 309 in 1987. In Zambia there has been a sharp increase in childhood malnutrition, and a decline in the child immunization programmes. The uptake of antenatal care has also fallen off. In countries such as Jamaica, where the government has introduced charges for previously free public health care, many women, who face declining incomes and increased prices for basic needs, have simply been unable to afford medical attention for themselves or their children (Antrobus, 1988).

Reduction in other areas of state expenditure also affects women in ways which are less obvious. Housing, sanitation, clean water programmes, rural road projects – all these have been the victims as governments seek to reduce 'unnecessary' or 'unproductive' expenditure. But all these services affect the ability of women to manage their conflicting roles. Without rural roads, women are unable to travel to urban centres for health care, for education, or to market their own crops. Without sanitation and clean water, women use much time and energy fetching water from insanitary and distant points; intestinal and other diseases afflict children and adults alike, adding care of the sick to women's already overburdened tasks.

In some countries, governments have also introduced or increased school fees as a means of raising revenues. This has contributed to the marked fall-off of enrolment of girls in schools, for several reasons. Additional expenditure cannot be justified where household incomes are falling. The possibility of women obtaining a job where literacy is necessary is slight. Girls have been withdrawn from school in order to care for younger children as women have increasingly had to take up income-generating work. The girls themselves are also required to contribute to household income either by assisting relatives, or working as domestic servants (Pittin, 1990).

Adjustment policies which favour deregulation of the market are premised on ideas about 'getting the prices right' – that is, eliminating what are seen to be distortions in the market. But within a conceptualization of the economy which ignores any analysis of gender, women's time and energy have no price: they become further stretched. As earnings of other members of the household decline, women extend the range of income-generating activities, seeking new sources of cash and investing their time and resources in new activities – hence the increase within the urban economy of activities such as vegetable growing, petty trading, beer brewing, poultry keeping and tailoring (Figure 15.6). Common strategies to reduce household expenditure initially focus on purchasing cheaper commodities, either by seeking cheaper sources of supply or by purchasing more basic goods which require home processing – both of which involve time and work undertaken by women. As the crisis deepens, items of expenditure are eliminated. As the total amount of goods and services the household can afford to buy falls, the share of women and girls of these shrinking resources also tends to decline. For example, it is common for male family members' food consumption to be protected as food intake of women and young girls is sacrificed.

A more market-oriented economy will not have negative consequences for all women in Third World economies. Increased market-based economic relations tend to cause increased stratification in the economy, with some groups doing better and being in a position to accumulate, as was the case in the Green Revolution. Inevitably, groups who have weaker claims on resources lose out. For many women in both rural and urban Third World communities, adjustment has meant increasing workloads and falling living standards as they have struggled, often unsuccessfully, to maintain their standard of living and to ensure the basic survival of their households.

Figure 15.6 Gendered development? Women learn to build water tanks in Kenya.

15.6 A challenge for the 1990s

The 1990s will present enormous challenges to all those who are concerned with development. However, all development issues require analysis from the perspective of gender relations if they are to be successful and serve the needs of women as well as men. There is some evidence that the World Bank is drawing back from its 1980s policy agenda and prioritizing areas of social investment which are crucial to women's survival. But other key issues such as environmental conservation, political stabilization and democracy all need to be subjected to a similar scrutiny. Efforts to halt soil degradation will not succeed if they rely on (unlimited and unrewarded) applications of women's scarce time and energy resources; democracy will be meaningless unless women are represented and have full access to political office and power. The whole of the development agenda will need to be reformulated if the message that gender matters is to be taken seriously.

Summary

1   Gender relations are social relations, not biological or natural.

2   Like all other social relations, gender relations will be affected by, and will affect, how societies and economies change over time.

3   Therefore, gender *matters* – in development analysis as in any other sphere of social analysis and policies. Without understanding gender relations policies can go wrong.

4   The key to redressing the inequality of gender relations and the mismatch between women's contributions to society and their rewards goes much further than directing aid programmes to women.

5   The importance of analysing gender applies at all levels of development policy – planning, implementation and evaluation.

TECHNOLOGY IN DEVELOPMENT

GORDON WILSON

Technology is crucial for Third World development. Think of attempts to beat drought and you think of dams. Think of efforts to improve agricultural production and you think of new seeds, fertilizers, pesticides and herbicides. Think of improving manufacturing industry performance and you think of new machinery. It is difficult to think of any aspect of development that does not involve either a tool, a machine, or some other piece of hardware or 'thing'.

Q How should the role of technology in development be understood?

Previous chapters have developed ways of assessing development from different perspectives. Similarly, the aim of this chapter is to provide more tools for understanding Third World development – this time, tools for thinking about and beginning to analyse the contribution of technology to development.

16.1 Technology as social process

I have, by implication, given you a commonly held perception of technology – 'things' with a human purpose. I have also implied that the intention is for technology to solve problems. Dams solve the problem of drought and so on.

Is technology, therefore, simply 'things' that solve problems? Consider the following quotation from an article about a dam project in the Indian state of Gujarat (Figure 16.1) which appeared in the *New Scientist* of 27 May 1989:

"The Sardar Sarovar dam is huge. It is Gujarat's answer to the equally huge problem of recurring drought. Like large parts of India, the state has had no monsoon for the past four years. Rain was so scarce last year that the worst-hit areas received water for only 15 minutes every two days.

…The Gujarat government estimates that the stored water will irrigate 1.8 million hectares, supplying 3.5 billion litres of drinking water every day and producing 1450 megawatts of hydroelectricity. To do all that, and much more, says the government, the dam project will cost more than $3.5 billion and will take 22 years to complete."

Yes, here we have a problem and a massive piece of hardware designed to solve it. Nor is the hardware confined to a huge vessel for trapping and holding water:

"…the entire length of the main canal will be lined with concrete to prevent water from seeping into the soil and waterlogging it. Only two other irrigation projects in the

Figure 16.1   A massive piece of hardware: the Sardar Sarovar dam under construction.

world employ computers to control the delivery of water. Gates every 10 to 15 kilometres along the main canal will open and close by remote control. The operators will similarly control the delivery of water to 37 main branches of the canal.

Piezometers in wells dotted around the command area will record fluctuations in the water table and will alert the operators to the risk of waterlogging. If the water table rises too far the operators will pump water out of the wells and into the main canal."

Computer-controlled gates every 10–15 km along a main canal with 37 branches; piezometers, wells and pumps to control the water table: obviously these and many other components require a high level of technical co-ordination if the dam as a whole is to do what it is supposed to do. In this sense it might be more useful to think in terms of a dam *system*, just as we might recognize that different components are co-ordinated in, for example, a railway *system* or a telecommunications *system*.

The seeds, fertilizers, pesticides and herbicides that I mentioned in the opening paragraph also together form a system in that the sowing of the seeds and the application of the chemical inputs has to be co-ordinated. You don't simply buy and sow a new variety of seed and hope for the best – at least you don't when your livelihood depends on it. You must also buy the rest of the package of chemical inputs and apply them correctly to ensure that the seed germinates well and provides good crops. For the system to work, all the components have to be present and working together. This is what technologists call the *technical imperatives* of the system.

The seeds and chemical inputs system actually forms part of a wider agricultural system. Similarly, the railways are part of the wider transport system and a dam is part of a water system with three distinct purposes: to supply drinking water; water for agriculture; and water for hydro-electric power.

The Sardar Sarovar project is intended to fulfil all three of these aims, but if it is ever completed (as I write in April 1991 there are doubts) it will have involved rather more than a problem to be solved and a complex technical system of hardware. These aspects only form the beginning and end of it. *People* have also been involved, and not just technical people. Politicians in the national and affected state governments were involved in protracted negotiations about the options and could not agree. This resulted in recourse to a legal tribunal, but even after the go-ahead was given, the project had to await investment clearance from the Planning Commission. Meanwhile, funding was being negotiated with the World Bank and the Overseas Economic Co-operative Fund of Japan.

All the above happened just to get the project started. Implementation would then require recruitment of a team to design, manage and build the dam. Elsewhere in its article, the *New Scientist* refers to 'thousands of engineers and construction workers'.

Another aspect is the vast amount of knowledge that has been necessary and, again, not just engineering knowledge. The knowledge of economists, politicians, lawyers and planners has been important. The project claims to be concerned with its social and environmental impact and the *New Scientist* quotes Sanat Mehta, chairman of the project company: 'We are not taking this just as an engineering project. It is a multidisciplinary development project.'

Finally, it is fairly obvious that the project could not have started, nor could it continue, without the *social organization* of those politicians, lawyers, planners, economists, engineers, scientists and construction workers, so that their efforts are co-ordinated towards a common goal.

Between problem beginning and hardware end, therefore, we have a long, complicated *process*, and because organizations and people have such a big part to play, it is a *social* process. I have chosen to dissect a dam project, but the elements I have identified (problem to be solved, people, different kinds of knowledge, social organization and 'things') can be applied to any technology. Understanding this social process is essential for understanding the role of technology in development, and is the focus of the following sections.

The process starts with the *choosing* of the technology. Implementation follows and generally requires considerable technical expertise – in the case of the Sardar Sarovar project, for designing and building the dam. This raises the question of how that expertise or *capability* has been (or will be) acquired. Finally, the *consequences* of implementation, especially those consequences that are not part of the original aims, have to be considered – in the case of the Sardar Sarovar dam the national government gave specific environmental clearance to the project. These three factors raise a common issue – that of *control* of technology.

16.2 Choosing technology

Technology choice is not straightforward. It is not only a matter of considering the technical factors, but also of economics, and moral and political choices. Moreover, the factors that influence our choice may vary over time. Today we may take into account environmental concerns, for example, that would have been unheard of 25 years ago.

A major issue in Chapter 2 concerned technology choice in the control of diseases: good diet, clean water and effective sanitation. In that chapter I discussed Cuba's excellent health statistics which have generally been acknowledged as the result more of a long-standing primary health care programme involving basic drugs, improved diet, clean water and health education than of any hi-tech exploits. Yet in the 1980s there was a discernible shift in policy. Alongside a continuing commitment to primary health care arose the intention of becoming a technological 'medical power', investing heavily in biological research, genetic engineering and new diagnostic techniques.

Box 16.1 comprises extracts from a 1989 *Financial Times* article (also see Figure 16.2). Read it and try to list the factors you think might have encouraged Cuba to develop this particular type of medical technology.

This example shows well how diverse factors interact when choosing particular technological directions. One way of looking at it is to say that the political commitment to meeting the health needs of the Cuban people plus the economically attractive possibility of opening up export markets constitute a pressure to develop medical technology. At the same time medical research is providing pressure in the same direction and this is underpinned by the prestige of Cuba becoming a medical power. The prestige is associated with underlying beliefs and values concerning the superiority of scientific and technological progress. In this concrete case, the pressures seem to be funnelling the development of medical technology in a particular direction.

Box 16.1 Extracts from 'Cuba: a picture of health'

Public health was given very high priority by the post-1959 Government, with the result that Cuba has achieved levels of health care more normal in a Western than a developing country…

The Castro Government has placed strong emphasis on preventive medicine, which entails decentralization of facilities throughout the island and free access to services. Other social policies reinforce the benefits of this restructuring, notably improvements in housing and sanitation and the raising of nutritional levels. The eradication of illiteracy has made active health education campaigns possible…

Cuba has also for some years been investing heavily in high-tech medicine – biological research, genetic engineering, new diagnostic techniques and new treatments and research. The Government has said its objective is to become a 'medical power' capable of providing expertise and technological transfer to Latin American and other developing countries.

Cuban scientists have developed a new vaccine against the killer Group B strain of meningococcal meningitis at the Centre for Biological Research. This is likely to be the first such vaccine to complete efficacy trials anywhere in the world. The patent has so far been requested by 19 countries.

Another achievement is the development of an ultra-microanalytic immunoassay system, Suma, originally developed to test for congenital disorders such as Down's Syndrome, hypothyroidism and allergies. It is now available to all pregnant women.

Cuban specialists claim that Suma is much cheaper to operate than similar Western projects as it uses as little as 5 per cent the quantity of reagents and is much faster. This makes it potentially attractive for mass testing in Third World conditions. Brazil is reportedly interested in purchasing it.

Suma is being used in mass testing for AIDS: the Health Ministry planned to carry out 7m tests during the course of last year…

The main hospitals are well equipped with the latest in medical technology…. Inside Ameijeiras you will find computerized axial tomography (CAT) scanners, micro-surgery and even, perhaps, a brain tissue transplant…

In 1983 Cuba held its first biennial 'Health for All' medical technology fair. The 1987 fair lasted seven days and was attended by 282 companies, including many of the largest European and Japanese companies in the field. The organizers hope to attract enough overseas buyers to turn the event into a show-case for the Caribbean and Latin America.

(Gareth Jenkins, *Financial Times*, 17 February 1989)

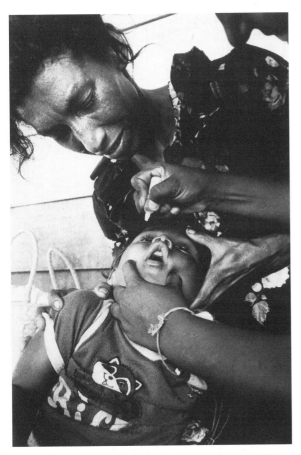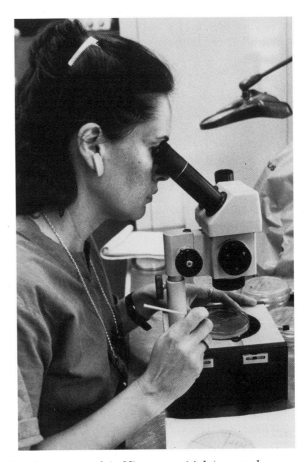

Figure 16.2   Cuba as a 'medical power': (above left) a Cuban doctor at work in Nicaragua; (right) research under way on meningococcal meningitis at Cuba's Centre for Biological Research.

A similar analysis of the pressures and interests shaping technology choice can be applied to the Sardar Sarovar project. One pressure is political (the need to provide the population with water). Another is economic (the loss in agricultural production alone during the latest drought was estimated by the Gujarat government to be 50 billion rupees). These combine with the pressure of the latest scientific ideas in dam technology and the ideology of the technocrats in the Gujarat state government to create the project.

I have just described one model of the process of *technological innovation*. The complex interaction between pressures and interests shapes the direction of science and technology, resulting in new processes and products.

Like all models, however, this one is a simplification of reality and it is necessary to go beyond it. The basic principles underlying dam technology have been around a long time. Yet dam projects have not always been satisfactory. Waterlogging and salination, for example, have long been problems with big projects. Such problems are acknowledged and have fed back into the innovation process so that, as we have seen, the Sardar Sarovar project employs sophisticated computer technology in an attempt to overcome them. Performance reality always feeds back and creates modifications.

Technology is often viewed in a static sense. We have a problem, we choose a technology that will solve it to a greater or lesser extent, and that is that. However, I have introduced a dynamic element where technology is in a constant state of flux, forever being adapted and feeding back into the process of innovation.

Technology is also dynamic in terms of the knowledge which it both requires and generates. In manufacturing technology, for example, the choice may be between a capital-intensive and short-term economically efficient system and a labour-intensive system that is relatively inefficient in the same terms. The latter would generally only be chosen if there was a strong, political commitment to conscious, visible job creation which outweighed the short-term efficiency argument. But the debate does not stop at this point. As Kaplinsky has argued: 'there may be cases in which economically inefficient techniques may be highly desirable because they empower small-scale producers or provide valuable industrial experience which facilitates alternative technological trajectories in the future' (Kaplinsky, 1990, p.24). Kaplinsky also quotes Riskin who argued that, despite the fact that many of the small-scale techniques employed in China during the Great Leap Forward (1958–60) were inefficient from both economic and engineering viewpoints:

"many of the thriving regional industries of today had their origins in a primitive workshop established in 1958...

Even where the shops established during the great leap were forced to close, however,

the initial experience with industrial methods they had afforded the peasants and the lessons, both positive and negative, to which they gave rise, proved invaluable later when local industrialization was again pushed vigorously."

(Kaplinsky, 1990)

Technology, therefore, cannot be viewed statically, but dynamically in terms of what else it generates (Figure 16.3).

Figure 16.3 Small-scale manufacturing, though inefficient in the short term, may increase skills and capacities in the longer term:
(left) small firm making clock parts in Maharashtra financed by a loan from the Industrial Development Bank of India;
(above) locally designed melting furnace at the Intermediate Technology Transfer Unit, Tema, Ghana, which produces cornmill plates to make sprockets for motorcycles.

16.3   Technological capability

So far in my discussion of innovation I have begged two important questions. How is basic scientific research converted into the new processes and products that are being demanded politically or economically? How are modifications to existing processes and products effected so that they do their jobs better?

The missing link that answers both questions is *technological capability*. Its acquisition is crucial in any country, but particularly Third World countries where often the demand for a process or product outstrips the technological capability to supply. We saw in Chapter 2, for example, that in many Third World countries there is a high demand for vaccines to help combat infectious diseases, but frequently in these countries both the technical capability for manufacture and the basic scientific study of the diseases lags behind this demand. This lag is linked in turn to a lack of demand for these vaccines in the industrialized countries.

A major question in Third World development, therefore, is: how is the required technological capability to be acquired?

One way is to *transfer* the relevant technology from the industrialized countries to the Third World countries, but four problems stand out.

First, the industrialized countries and the multinational enterprises that operate from them will not readily give away the technologies they own. One solution is to purchase the technologies and associated expertise required. This is a regular occurrence with any transactions between Third World based and First World based companies. Success depends on having enough technological capacity to 'run' the technology purchased. Other 'solutions' are to 'rent' the technology; in other words, operate it on licence, or to invite the multinational corporations that 'own' the technology needed to make an investment. These last two solutions raise questions about lack of control of the technology, a subject to which I will return in the last section.

Secondly, organizations in industrialized countries may not wish to transfer their latest technologies to the Third World. Instead, they may sell 'yesterday's' technologies. These may also be 'dirty' technologies no longer allowed in the industrialized countries because of health and safety and/or environmental considerations. It is interesting to note, however, that current environmental concern in the industrialized nations does give the developing countries some bargaining power. We noted in Chapter 5 how countries such as India and China are demanding that, if they are to cut down on their industrial emissions of CFCs, for example, the requisite alternative technologies must be transferred to them.

Thirdly, simply transferring technology in the form of machines is not enough. Skills also have to be transferred, not only for the operation of the technology, but for maintenance too. On top of this the material resources necessary for operation and maintenance need to be available.

Fourthly, it is now widely accepted that technologies are not value free or value neutral. They reflect the 'social values, institutional forms and culture' of the transferring party (Anderson, 1985, p.57), or, to put it another way: 'Technology can be considered to resemble genetic material which carries the code of the society which conceived and nurtured it and which, given a favourable milieu, tries to replicate that society' (Reddy, 1975). This transfer of values embedded in a technology can have profound consequences for the recipient, something I will consider further in the next section.

Technology transfer may occasionally be seen as an end in itself, but more often it is an initial requirement for developing a new *indigenous* technological capability (Figure 16.4). The important question to ask about technology transfer, therefore, is whether it assists or hinders the development of such capability.

It is useful in this context to see how Japan developed initially. The first stage involved technology transfer by copying technologies of the industrialized countries. Far from simply 'reinventing the wheel', this process of copying can

Figure 16.4   Transferring old technology: Georgina Degbor with a small lathe given her by a customer of her former employer, which enabled her to start her own engineering business making parts for textile machinery at the Intermediate Technology Transfer Unit at Tema, Ghana.

build up indigenous expertise in the technology, and, if successful, the expertise can be used to develop the technology further, to adapt it; in other words, to participate in the innovation process.

Writing about the acquisition of hi-tech, flexible steel-making technology by Third World countries, Hyung Sup Choi of the Korea National Academy of Sciences states that 'technology development in the developing countries naturally starts with the importation of advanced foreign technology.' But this must be accompanied by the capability to modify it and improve it for domestic applications, and Choi adds:

"To achieve viable results from technology transfer, we have found every reason to conclude that it is fundamental that a corresponding effort be made to digest and adapt foreign technologies. In this respect, it is desirable to adopt what might be called a three-pronged approach, emphasizing accelerated introduction of foreign technologies, creation of basic capabilities including manpower development, and stimulation of basic R&D activities."

(Choi, 1988, p.297)

Choi is less strong about how developing countries might actually acquire the advanced foreign technology, which he states is the first step, referring to the possibility of 'collaboration with countries where there are enlightened governments and people', but his basic thesis of the necessity for technology transfer to be accompanied by an indigenous adaptive capability is echoed by other writers on technology in the Third World. Arnold Pacey, for example, in his book, *Technology in World Civilization*, writes with a rather different emphasis. His examples are mainly agricultural or concerned with small-scale 'appropriate' technology. He suggests:

"when new technologies are introduced into a country, their success almost always depends on the local innovations which they stimulate. These are usually of an *adaptive* nature, and help to match the transferred technology to local conditions."

(Pacey, 1990, p.192)

Pacey cites the Green Revolution in India, where high-yielding cereals developed under US auspices in Mexico were transferred to the subcontinent. As a result, India became self-sufficient in grain and Pacey states:

"Much of the adaptive work necessary to support India's green revolution was carried out by Indian plant breeders at local agricultural research stations. But these institutions were situated on good land, and it was experiment and innovation by ordinary farmers which ensured the success of the new crops on the poorer soils. Sometimes, indeed, the farmers chose to use crop varieties which had been rejected by official research. One rice variety developed by the plant breeders and then discarded was known as IR 24. Farmers who got hold of this liked it and multiplied the seed themselves, selecting plants that did well in local conditions. The result has been rice which is much more resistant to a local insect pest than are any of the varieties approved by the scientists."

(Pacey, 1990, pp.192–3)

We see here two stages in the technology transfer. The first is from the USA–Mexico programme to India's agricultural research stations, which results in some adaptations. The second stage is the transfer from the research station to the farm, which results in further adaptations. The first stage is transfer between countries: the second stage transfer within the country.

The essential point, however, is that adaptation accompanied transfer. First the research station and then the farmer was able to modify and hence guide the technology to suit local conditions. The process of adaptation is in itself an integral part of acquiring and developing capability and expertise.

This discussion brings us back to the dynamic aspects of technology. It is forever changing and being changed, to changing circumstances, to changing needs.

16.4 Technology in the driving seat?

The intention of technology is to solve problems, as I stated at the beginning of the chapter. Others, however, view technology itself as part of the problem, e.g. the *New Scientist* on the Sardar Sarovar dam again:

"The Sardar Sarovar Project will destroy the homes and lands of 70 000 people in Gujarat, Maharashtra and Madhya Pradesh. It will drown almost 14 000 hectares of forest land and, inevitably, much of the wildlife that inhabits it. The dam will create pockets of stagnant water that will increase the prevalence of malaria in the vicinity of the dam. The reservoir will also destroy important temples and shrines.

Indian environmentalists claim that irrigation on such a large scale is bound to cause waterlogging and a build-up of salt in some of the soils in the area to be served by the dam..."

(*New Scientist*, 27 May 1989)

So, in this contrary view, think of a dam and you think of waterlogging and salination of the soil, of stagnant pools breeding malaria mosquitos, of destruction of the environment.

Similarly, think of new manufacturing machinery and you think of job loss, exploitation and alienation. Think of new seeds, fertilizers, pesticides and herbicides and you think of ecological disruption and transfer of land resources from poor farmers who cannot afford these inputs to rich farmers who can, or from women members of a household who have traditionally grown crops to the male members.

'Out of control' technology usually manifests itself in what some writers have called 'unforeseen and unintended consequences'. The designers of the first dams in the Third World did not *intend* them to be breeding grounds for malaria also. One implication of our social process view of technology is the complex interconnectedness between technology and society. Technological choices cannot be made without setting off a chain of consequences that go far beyond the original intention. Some of these unanticipated consequences may be beneficial, but others will inevitably pose problems. As Winner has commented: 'The uncertainty and uncontrollability of the outcomes of action stand as a major problem for all technological planning. If one does not know the full range of results that can spring from innovation, then the idea of technical rationality – the accommodation of means to ends – becomes entirely problematic' (Winner, 1977, p.96).

An example of what Winner is getting at can be found in the Indian Green Revolution, mentioned earlier. This was originally conceived as benefiting rich and poor farmers equally because of the scale neutrality of the technical inputs (seeds, fertilizer, pesticide and water). But, the sinking of tubewells for irrigation water requires considerable investment, the introduction of multiple cropping (i.e. more than one crop per year) benefits those with access to resources for mechanized equipment for harvesting and other activities, and the fact that the inputs have to be bought disadvantages those farmers not already integrated into the money market. In short, the Green Revolution has been associated with the rise of the rich peasantry in India, so that today they form a dominant rural class. The original conception has been largely lost.

These arguments are variants of technological determinism (Box 16.2). At its most extreme this implies that technology dominates language and the way we behave: 'people come to accept the norms and standards of technical processes as central to their lives as a whole,' states Winner. Citing technical efficiency, the quest for which is of paramount importance in technical systems, he adds:

"But now efficiency takes on a more general value and becomes a universal maxim for all intelligent conduct. Is the most product being obtained for the resources and effort expended? The question is no longer applied to such things as assembly-line production. It becomes equally applicable to matters of pleasure, leisure, learning, every instance of human communication, and every kind of activity, whatever its ostensive purpose."

(Winner, 1977, p.229)

If technology can create a poor deal for the human race, it can also undo that deal and create a good one. This more optimistic view was taken by some of the advocates of 'appropriate technology', who came into prominence in the 1970s. Reviewing their argument, Kaplinsky states:

"it is argued that because capital-intensive technologies are generally imported, their utilization worsens the savings and foreign exchange problems faced by developing

---

Box 16.2 Technological determinism

This is the idea that technology ultimately controls or determines the nature of society, which is why we often define societies by their dominant technology. An article in *The Guardian* newspaper of 22 June 1989 described a seven-point national plan to turn Singapore into 'the world's first information society'.

In industry, advocates of technological determinism claim that states throughout the world converge in their industrial policies, driven by the imperative of the dominant technologies. This was the view expressed by Galbraith, when he wrote: 'there is a broad convergence between industrial systems. The imperatives of technology, not the images of ideology, are what determine the shape of economic society' (Galbraith, 1972).

However, viewing technology as a social process means seeing industrial systems as constantly changing outcomes of that process. Technology reflects the society that produces it. Although in turn it is one influence in shaping that society, it is only one influence among several.

---

countries; it also necessitates location in urban areas and this leads to concentration of social services (such as schools and health) and infrastructure (such as electricity) and the consequent neglect of rural areas."

(Kaplinsky, 1990, p.28)

The solution to inappropriate technologies, according to this argument, is to choose small-scale, labour-intensive (appropriate) technologies that can be situated in rural areas (Figure 16.5).

The general argument, therefore, is that if a technology has bad consequences for society, the solution is to replace it with a technology that has good consequences. This is the essence of the *technical fix*. Frankenstein's monster is not a monster after all.

Figure 16.5 'Appropriate' alternatives to big dam projects? (Left) small hydro-electric plant in Nepal; (below) building a 'bund' or small dam in Maharasthra, India.

The image that technological determinism conjures up (and what I have just written shows that it need not necessarily be a bad image) is that of 'things' or a system of 'things' controlling our lives. The corollary of this is that human beings cannot control their own destiny – which is a depressing view of human beings. We are in danger, however, of falling into the trap here of considering technology only as 'things', rather than as a social process which results in these 'things' carrying with them a value-laden baggage. It is important to recognize that the 'determining' aspect of technology arises from it *not* being value free or value neutral. In other words, technology *reflects* the nature of the society in which it originates and is nurtured, which has inevitable consequences when it is transferred from one society to another.

This is well exemplified when we consider the gender consequences that arise from the adoption of new technologies. Reviewing the implication of technology transfer for women in Third World countries, Anderson refers to a study conducted for the United Nations on the impact of scientific and technological progress on employment and work conditions in various trades:

"In every case where machinery was introduced in activities traditionally done by women, men either completely replaced women or the activity became subdivided and men took over the tasks that used the technology and required greater skill while women were relegated to less skilled, menial tasks. These shifts were accompanied by loss of income-earning opportunities or marginalization and lower income for women...

In Java, when rice mills were introduced, women who had traditionally earned their only monetary income from hand milling were displaced as men assumed the positions in the factories. In Korea, when the government installed rice mills, men in the mills did jobs previously done by women."

(Anderson, 1985, p.61)

Underlying these processes are assumed beliefs that men work with technology and women do not, or that men do 'modern' work while women only work in the subsistence sectors. But the transferred technologies are also 'blind' to values of the recipient societies which may determine where women may work and where they may not.

Later in her chapter, Anderson talks of this gender blindness in relation to technologies intended to make life easier for women, two examples being solar stoves that obviate the need for collecting firewood, and pipes from distant water pumps which obviate the need for carrying water:

"Women have traditionally used the time of long walks to fetch water and firewood for social organization, conversation and interchange. Technologies which alter these functions eliminate these opportunities, so that other social forms have to be found."

(ibid., p.66)

On the other hand, the consequences of technology transfer are not necessarily negative, precisely because they do disrupt cultural norms (Figure 16.6):

"Technologies which gather women in certain areas, such as grain mills, can facilitate social activity and opportunities for education. For example, women in Asia have received literacy training while they wait for their rice to be ground at mills and women in Africa have received nutrition training while waiting in line at clinics."

(ibid., p.66)

Figure 16.6 Mothers in rural Uganda, waiting for their children to be immunized, are given a talk on health education.

The essential lesson of this discussion is that when we consider (or attempt to predict even) the consequences of technological activity, we have to take into account societal values.

16.5 The quest for technological control

Choosing technology, acquiring technological capability and technological determinism raise the common issue of control. This is a matter of politics in the sense developed in Chapter 6: it is a process of guidance in a framework set by power relations, interests and value preferences.

Kaplinsky claims that power relations between interest groups are central to innovation. In the generation of agricultural technology he cites the bias of new seed production towards varieties that use chemicals intensively, rather than seeds that are less dependent on these inputs. The reason he cites is that the large chemical companies own most of the major seed companies (Kaplinsky, 1990, p.21).

Similarly, Kaplinsky argues that a foreign investor who decides to build a large-scale, capital-intensive plant in a developing country may run into only poorly articulated opposition, partly because some interest groups operating within the country will gain from the investment as suppliers, customers and receivers of tax revenues. But this investment may damage the operations of small-scale indigenous operators and may rule out future domestic investors.

Many interest groups operate within a state, each having more or less power. Often the power has an economic base, but this is not exclusively the case. I have previously mentioned those who articulate the ideological supremacy of scientific and technological progress. These are the *technocrats* or professional experts. Often operating in quite narrow fields, their job is to propose the technical solutions to economic and political problems, but frequently it is also to rationalize and legitimize political decisions as technical (and hence politically neutral) choices. Technocrats are to be found

in government civil services and in influential positions in large enterprises. They may not be responsible for the final decisions but their advice is important. This can endow them with considerable power in choice of technology, so that technologies may be chosen because they are technically 'sweet', as well as for their appropriateness to the job in hand.

The power invested in technocrats is power associated with ownership of knowledge: scientific and technical knowledge that is allotted an important ideological position. Indigenous, local knowledge, on the other hand, is less visible, is less recognized and carries far less power. This unequal relationship between the two forms of knowledge can create blockages to the successful use and adaptation of new technologies on the ground. The ultimate casualty is the development of technological capability itself for, as we noted in Section 16.3, the assimilation of indigenous knowledge and the consequent ability to adapt transferred technology is an essential feature of such development.

Appropriate technology

'Appropriateness' or, conversely, 'inappropriateness', are words that have cropped up more than once during this chapter. To the extent that technology succeeds in its aim of solving problems, then what technology is 'appropriate' for development depends on what is the problem to be solved, which in turn depends on who has the power to define what are the problems of development, or on which view of development is taken. However, in practice the notion of appropriate technology is a particular response to issues of technological control. In terms of the views of development put forward in Chapter 6, appropriate technology fits with a populist view. It sees as appropriate that which gives control to individuals and communities at a local level, rather than to the technocrats of either states or capitalist firms.

Box 16.3 gives a list of 'criteria of a technology's appropriateness' for a developing country developed in 1975 – with two additional notes. If we adopt these criteria, the Sardar Sarovar dam project, for example, is clearly inappropriate.

Box 16.3   The criteria of a technology's appropriateness

Appropriate technology should be compatible with local cultural and economic conditions, i.e. the human, material and cultural resources of the community.

The tools and processes should be under the maintenance and operational control of the population.

Appropriate technology, wherever possible, should use locally available resources.

If imported resources and technology are used, some control must be made available to the community.

Appropriate technology should wherever possible use local energy sources.

It should be ecologically and environmentally sound.

It should minimize cultural disruptions.

It should be flexible in order that a community should not lock itself into systems which later prove inefficient and unsuitable.

Research and policy action should be integrated and locally operated wherever possible in order to ensure the relevance of the research to the welfare of the local population, the maximization of local creativity, the participation of local inhabitants in technological developments and the synchronization of research with field activities.

(Brace Research Institute, quoted in Lawand *et al.*, 1976, p.132)

Notes:

1   Appropriateness for women and taking account of gender relations would now be added to this list.

2   A weakness in the list is the poorly defined notion of 'community'.

Look at Box 16.3 and note how these criteria, particularly those emphasizing compatibility with local cultural and economic conditions, go beyond the technical fix model of appropriate technology mentioned in Section 16.4 above on technological determinism. More importantly, note also how many of the criteria relate, directly or indirectly, to maintaining local control – and not only in the obvious sense of equating 'control' with 'power'. Appropriate technology enthusiasts also insist that control is about *guiding* technology towards the goals of meeting human and social needs.

Appropriate technology applied according to these criteria, therefore, is seen as empowering, and it should come as no surprise to note that primarily it has been aimed at those who lack power – such as the rural poor and women.

'If it's not appropriate for women, it's not appropriate' was the major slogan at 'Tech and tools: an appropriate technology event for women' in Nairobi in 1985. Out of this event arose *The Tech and*

Tools Book: a guide to technologies women are using world-wide, which has this to say in its introduction:

"The question is how to involve women in determining which technologies they need for their own development and that of their families, their communities and their countries. How can women have better access to, use of, and control of such technologies?…the main consideration must be, what do women really want? This includes technical as well as social, cultural and political considerations of a given society or country. For example, the family structure and its balance of power and authority must be considered, as well as any changes that the new or improved technologies might bring. It is not that such changes are undesirable or that they should be avoided, but that they have to be anticipated and strategies developed to cope with them."

(Sandhu & Sandler, 1986)

The Tech and Tools Book gives basic information sheets for 57 technologies under the headings of agriculture, communications, energy, food processing, health and sanitation, and income generation. Each information sheet assesses the technology in line with the considerations outlined in the introduction.

One example, the cassava grater, is given in Box 16.4. Study this box and try to answer the following questions:

1 What was the problem that needed solving?

2 What was the pressure for a technological solution?

3 Has the introduction of the grater involved technology transfer, development of indigenous capacity, or both?

4 Has the cassava grater solved the problem? (In other words, has it achieved its aims?)

5 What have been the consequences beyond these original aims?

Box 16.4   Cassava grater

The cassava-grating machine is used for grating peeled cassava. It consists of a motor placed in a wooden box and blades for chopping and grinding cassava tubers; the machine runs on diesel fuel. The cassava tuber is inserted and passes through different sections of the machine where it is chopped and ground. The machine can process approximately 3000 lbs [1361 kg] of cassava daily. It can be operated by one person and can also be used to grate coconut.

Strengths
1 Quicker and less physically demanding than the traditional method.

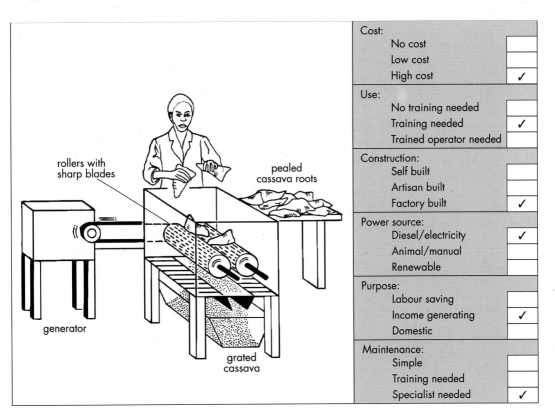

Cost:		
No cost		
Low cost		
High cost		✓
Use:		
No training needed		
Training needed		✓
Trained operator needed		
Construction:		
Self built		
Artisan built		
Factory built		✓
Power source:		
Diesel/electricity		✓
Animal/manual		
Renewable		
Purpose:		
Labour saving		
Income generating		✓
Domestic		
Maintenance:		
Simple		
Training needed		
Specialist needed		✓

rollers with sharp blades

pealed cassava roots

generator

grated cassava

2 Processes larger quantities of cassava than the traditional method.

3 Easy to learn how to use the machine.

Weaknesses

1 Disrupts cultural traditions of grating cassava since it involves less people in the process; detracts from the social value of grating cassava collectively.

2 Dependent on technicians for maintenance; thus production stops when the machine breaks down.

3 Requires large capital investment initially.

How it has been used

The Lu Fuluri Dangriga Women's Group in Belize has approximately 13 active members between the ages of 50 and 70. They had been grating cassava all their lives and decided to start a cassava bread-making project to produce income. They started the project by using the traditional, manual methods of grating, but found that it was too slow when they wanted to expand production outside the community.

The women found that it was inefficient to use a commercial grating machine service as it meant walking long distances and waiting their turns in queues. They persuaded a funding agency to provide about BZ$500 for purchase of a machine (with casing around it), and for the training of several members in its operation.

The machine changed the way the group worked together: (1) it has forced the group to be more cost conscious and business-like; (2) they are now more production orientated; (3) there is a greater division of labour; (4) they are planting much more cassava since they have the capacity to process more. The women are well known in the country and the community as they are the only women's group that owns a machine and produces collectively.

The group has had the machine for six months and is generating an income from selling the cassava bread and from hiring out the machine.

(Sandhu & Sadler, 1986, pp.68–9)

My answers are as follows:

1 The problem was that the traditional, manual method of grating cassava was slow and physically demanding.

2 The pressure for a cassava grating machine was economic. The women had decided to start a cassava bread-making project to produce income.

3 The technology was certainly transferred to the women but we are not told whether it is manufactured using the country's indigenous capability or not. The fact that the grater is factory built, runs on diesel and a specialist is needed to maintain it suggests that the input of indigenous knowledge from the women may be low. Training in use will add to their knowledge, however.

4 Yes, the cassava grater apparently achieved its aims. It is faster and less physically tiring than the traditional method, which means that larger quantities of cassava can be processed. It is also easy to learn how to use the machine.

5 The consequences are mixed: (a) The women are planting more cassava (the information given does not specify whether or not this is instead of other crops); (b) they have become cost conscious, business-like and production orientated with a greater division of labour which has detracted from the social value of grating cassava collectively; (c) fewer people are required to grate cassava which, together with the previous consequence, has led to a disruption of cultural traditions; (d) the women may have less overall control than when they grated manually, being dependent on outside technicians for maintenance and breakdown of the machine. However, the income generated may increase the women's control over other aspects of their lives, and there may be a group empowerment aspect also.

Perhaps the only universal point about appropriate technology is its specificity. What may be appropriate at one site in one part of the world at one time may not be appropriate at another site or at another time. Hence, the ambiguity over whether a particular example like the cassava grater is fully appropriate. There are questions to be asked about the technical and physical appropriateness to a particular location, and the social appropriateness. We might, especially, ask: appropriate to whom? Owners of the technology, the people who will work with it, and (if it is for consumption) consumers will each judge it by different criteria.

Appropriate technology accords with the populist vision of small-scale producers having a large measure of control of the technology they operate (Figure 16.7). Critics acknowledge the attractiveness of such visions, but argue that the potential contribution of small-scale 'appropriate' technology to raising productive capacities to ensure economic development is extremely limited, and without economic development there can be no general, permanent improvement in well-being.

Figure 16.7 The Indian government has long had a policy of promoting or protecting village industries. This blacksmith's stall at Muirpur village market sells iron implements for specific local agricultural and domestic uses.

Others, like Kaplinsky, argue that appropriate technology cannot be viewed in such narrow, static terms. The acquisition of industrial experience and skills via small-scale industry, to which I referred earlier, is a case in point.

Kaplinsky brings us back to the central issue of power relations, however, when he writes of the problems that surround the general adoption of appropriate technology. The obstacles to its widespread diffusion, he writes, lie with 'the strength of vested social and political interests' which the appropriate technology movement has often had to confront head-on (Kaplinsky, 1990, p.3).

Perhaps the main contribution of the notion of appropriate technology is to point out questions to be asked of all technology. If a technology does not meet criteria of appropriateness based on values of local needs and empowerment, it nevertheless is likely to be 'appropriate' for those who control the technology and thus who guide or inhibit development.

Summary

1   Most aspects of Third World development involve technology as an essential component.

2   Technology should not be viewed simply as 'things'. These are only the outcome of a social process involving people, different forms of knowledge and the social organization of those people and their knowledge.

3   Technology choice involves a complex interaction of technical and social factors. Because of this technologies are never value free or value neutral.

4   A major issue in Third World development is the acquisition of technological capability, via technology transfer from (mainly) the developed countries and/or via the development of an indigenous capacity. However it is acquired, the ability to adapt technology to local conditions is crucial for the continuing development of capability.

5   Technology is intended to help solve problems, such as those of drought or ill-health. Often, however, it also has adverse consequences, many of which may have been unforeseen, due to its complex interconnectedness with other social relations.

6   *Who controls technology?* is a fundamental question concerning technology choice, acquisition and use in the Third World. Appropriate technology, a populist notion, cannot provide a mechanism for transforming society, but does offer a useful critical perspective on the role of technology in development.

17

TAKING CULTURE SERIOUSLY

TIM ALLEN

This chapter is about why it is important to take culture seriously when investigating issues of development. Unfortunately this often does not happen. Sometimes culture is treated as something relatively unimportant, and sometimes culture is taken into account but misunderstood. Both result in highly misleading analyses.

Q What are the pitfalls in thinking about culture in Third World development?

By the end of the chapter you should be able to recognize these pitfalls, and will be equipped to make your own judgement as to whether other chapters in this book adequately avoid them.

I must make a few comments at the outset about what I mean by the word 'culture'. This is no easy task, because a straightforward definition is impossible. If you ask most British people what they think culture is, they might tell you that it is literature, art and music, and they would perhaps be thinking particularly of Victorian novels, paintings by Turner and music by Elgar. These are the things 'cultured' people enjoy. Those who are not 'cultured' eat fish and chips, go to football matches or bingo, drink beer, watch soap operas on television, and read the tabloid newspapers. On the other hand, and here is where everything starts to become muddled, these 'uncultured' folk might be classified as enjoying 'popular culture', something that may be thought of as better than the 'high-

brow' culture of those who go to the opera. Additional complications arise if you ask a question like: 'What is the difference between French culture and English culture?' Your answer might then include some of the following: international languages, frogs' legs, snails, garlic, making love, sense of humour, the Royal Family, tweed jackets, crimes of passion, tea, 'stiff upper lips', and the Battle of Waterloo.

In fact 'culture' is one of those words which is used without any precise shared understanding of what it means, and social scientists are no exception. In 1952, two anthropologists, Kluckhohn and Kroeber, were able to survey over 150 different definitions, and subsequently many others have been attempted, some of them couched in the kind of obscure jargon that is incomprehensible to the uninitiated. Part of the confusion is that culture cannot be thought of in isolation. It relates to other elusive notions, like 'society', 'civilization', 'art', 'ideology' and possibly even 'Truth'; and how the term ends up being used is inevitably bound up with the way human activity is to be generally comprehended.

There is no space here for a review of all these different usages, nor to explore the complex debates about culture within the fields of sociobiology, social psychology, philosophy, art history and literary theory. I will make brief remarks about some anthropological perspectives (in

Section 17.3), since these have had a strong influence on 'Western' perceptions of life in formerly colonized countries. Also, I will outline a general categorization of approaches in the final section of the chapter. But at this point let it suffice to say that I am using culture in the sense of a *shared set of values*. As you read on it will become apparent what I mean by this, and what are the ramifications for the study of development when shared values are set aside (Section 17.1) or conceived of in simplistic terms (Section 17.2).

17.1 Setting culture aside

In this section I discuss the first of the pitfalls I have mentioned: the danger of treating culture as something relatively unimportant. I do so with reference to developments in Iran during the late 1970s and early 1980s. I try to explain why a political economist writing just before the revolution underestimated religion, and I indicate ways in which the specific nature of Iranian Islam had a fundamentally important bearing on what happened in the country. It is not my intention to suggest that political economy is misleading compared with other approaches. My point is more general: the assumption that society operates according to economic principles is a partial perspective, and one that can readily lead to flawed interpretations.

The following extracts are taken from a scholarly book on Iran by Fred Halliday. They refer to the activities of religious scholars, known as '*mollahs*', and their leaders, known as '*ayatollahs*'. The book was published in 1979 just before the Ayatollah Khomeini swept to power, and established an Islamic state (Figure 17.1).

> "The role of the *mollahs* and their associates, the merchants of the bazaar, is of especial interest, given their apparent obscurity in the years prior to 1978 and then their major role in the events of that year. There has no doubt been a revival of religious sentiment in Iran in recent years, but it would be misleading to analyse the events of 1978 in purely religious terms or to accept the

Figure 17.1   'No one expected the Hidden Imam to arrive in a Jumbo Jet': Khomeini returns to Iran from exile, 1979.

concept of an 'Islamic revolution' as propounded by the religious leaders. Both this characterization, and the Shah's references to his opponents as 'religious fanatics', obscure the deep material, i.e. social and economic factors, underlying this movement and the specific nature of the class alliance present within it...The *ayatollahs* and *mollahs* on their own can probably not sustain and channel the popular upsurge...It is quite possible that before too long the Iranian people will chase the Pahlavi dictator and his associates from power, will surmount the obstacles in its way, and build a prosperous and socialist Iran."

(Halliday, 1979, pp.216, 299, 309)

Three short quotes of this kind cannot do justice to a detailed study, which has much that is insightful to say about Iranian politics. Furthermore, what Halliday tells us here and in other parts of his book about the Islamic element in the resistance movement seemed reasonable to many people at the time he was writing. Certainly it would have been a mistake to analyse events in 'purely religious terms'. There was undoubtedly much more going on. In Iran, for example, *mollahs* have the right to collect taxes on trading profits (the *khums*) and on wealth (the *zakat*). Thus their association with the merchants of the bazaar is underpinned by an economic relationship, and religious rhetoric has often been used to protect traders' incomes. In the 1890s, the *mollahs* succeeded in mobilizing a national protest against the tobacco monopoly granted by the government to an Englishman, and in 1905–06 similar pressures forced through constitutional reforms aimed at protecting the business community from foreign competition. The *mollahs* and merchants can thus be seen as powerful elements in a faction which has persistently sought to promote its own class interests, and which became increasingly antagonistic to the Shah's government from the mid-1970s, because attempts to introduce a one-party state led to attacks on the economic privileges of the religious establishment and to an 'anti-profiteering' campaign.

However, the fact that there were class interests of this kind at work in Iran does not explain how and why religion could so readily be harnessed as the language of political dissent. By setting up as a straw target the idea that Islam was the primary influence on what was happening, Halliday ends up deflecting attention away from religion as a key factor. Notice how he makes a distinction between 'religious sentiment' and 'deep material, i.e. social and economic factors'. He thus implies that a study of changing Iran can largely exclude certain kinds of beliefs and values. A consequence of this approach is that the book as a whole persistently sets aside or devalues the role of Islam in Iranian politics. Khomeini's influence is characterized as ill-defined and ambiguous, and treated only in passing. Far more attention is focused on the secular resistance movements, which for Halliday

appear to be crucially significant for the country's future. In the light of subsequent events, this seems remarkably misconceived.

It is always easy to be wise after the event, and it should be stressed that Halliday was not the only one to underestimate the religious element. Khomeini would never have come to power had he not been a shrewd and manipulative politician. He carefully avoided direct criticism of the Left, and argued for the restoration of the 1906 constitution right up until he came to power. He would probably have failed to establish an Islamic state had it not been for the active support of a broad range of opposition groups during the crucial months of takeover, including secular-minded intellectuals and the Marxist guerrillas (the *Fedayeen*). Many viewed Khomeini as no more than a useful figurehead, and would never have supported him had they recognized his ambitions and his capacity to achieve them. In retrospect it seems as though there were a lot of people with their heads in the sand.

Yet for those with the eyes to see, the writing had been on the wall for some time. In 1972 Hamid Algar had remarked:

> "Iranian national consciousness still remains wedded to Shi'i Islam, and when the integrity of the nation is held to be threatened by internal autocracy and foreign hegemony, protest in religious terms will continue to be voiced, and the appeals of men such as the Ayatullah Khumayni to be widely heeded."
>
> (Algar, 1972, p.255)

There was good reason for thinking this. Khomeini's own views had been clearly laid out in 1970, when he gave a series of lectures to young *mollahs* in Iraq, which were then published as *The Islamic Government* (1971). He explained that *mollahs* would be the nucleus of the revolutionary vanguard under the leadership of the *ayatollahs*, with the 'virtuous faqih' (himself) at the head. These same clerics would then rule the post-revolutionary Iranian state. Halliday and others seem to have dismissed this manifesto as the deluded musings of a frustrated old man. But there were characteristics of Iranian Islam and political philosophy which might have prompted a closer investigation of Khomeini's assertions. Indeed, one author has gone so far as to claim that the main lines of development for the Iranian revolution had been laid down in the sixteenth century (Ruthven, 1984, p.219).

The population of Iran is predominantly made up of adherents of the main sect of the Shi'ite branch of Islam. They are sometimes called 'Twelvers', because they venerate twelve descendants of the Prophet Muhammad as his spiritual heirs. These descendants are known as the imams, the first being Ali, the cousin and son-in-law of Muhammad. The Twelfth Imam was Muhammad al Muntazar, who is believed not to have died, but to have disappeared in 940. He has become the Hidden Imam, and will eventually return as the *Mahdi* (Messiah), 'to fill the earth with equity and justice just as it was filled with oppression and tyranny' (Ruthven, 1984, p.204).

Iranian Shi'ites, together with other Shi'ite groups, comprise about 10% of all Muslims in the world. In most respects, their practices and traditions are similar to those of the majority of Muslims, known as Sunnis, but in interpretations of history and in political philosophy there are some important differences.

The schism between Sunni and Shi'ite Islam is linked to events in the mid seventh century. For the Shi'ites, everything went wrong when Ali's rightful claim to the caliphate of the Arab empire was usurped by Uthman of the Umayyad clan. They curse Uthman as the perpetrator of evil, and Ali's eventual assassination by a group of former supporters in 661 is seen as the tragic consequence of Uthman's actions. In contrast, for the Sunnis, 'Uthman was one of the venerated, 'right-minded' caliphs of the early Arab empire. Khomeini's refusal to recognize this was a reason why it proved difficult to export the Iranian revolution to the Arab world and to other Sunni Moslem countries during the 1980s (Sivan, 1985, p.186).

The death of Ali is bewailed by Shi'a as an appalling catastrophe, but even more significant in Shi'ite mythology is the killing of the Third Imam, Husayn, by the Umayyads in 680. In several respects the event is recognized as a redeeming sacrifice akin to that of Christ in Christian belief, and the date, known as Ashura, is the most important Shi'ite holiday. On this day, the circumstances leading to Husayn's death are recounted in great detail by specialist narrators at huge gatherings of mourners. As the tale unfolds, and one terrible event follows another, the audience expresses its grief with unrestrained wailing and self-flagellation. Something of the charged atmosphere of these occasions is illustrated by the following extract from a narration. It was taped in the bazaar of Tehran by Gustav Thaiss, an anthropologist who researched there during the 1960s.

> "With his six month old son Ali Asghar in his arms, the Imam cried out to the enemy that as this innocent babe had defiled none, at least he should be spared and a little water given to allay his thirst but the reply was an arrow shot at the child's neck which pinned it to his father's arm. After returning the cruelly murdered child to its sorrowing mother's arms (who then sang a mournful lament over her dead child) the Imam returned to pay the last of the sacrifice with his own blood. Arrow after arrow followed, piercing his body into a sieve unti, when the aged Imam fell from his horse, his body did not touch the ground but was held off the ground by the arrows which were sticking out of his body..."

> (Thaiss, 1972, p.356)

This dwelling upon Husayn's last agonies has similarities with the fascination of the Catholic Church in the details of the crucifixion and the sufferings of the saints. But the dramatization of the Husayn story has more explicit political resonances than the Christian Easter. The Imam's martyrdom symbolizes the wickedness of all corrupt governments, and the flourishing of evil since the golden age of the Prophet. One informant explained to Thaiss:

> "If we cry for Imam Husain, it is because he is the one who was sent by God to rise up and bring justice for human beings. So right now, if we want to have justice among human

beings we should bring Imam Husain back to life… Imam Husain still asks for help. He says 'Come to the battlefield and help me.' What should I do? I have to follow him? How? By rising against tyranny and oppression…"

(Thaiss, 1972, p.359)

It is not surprising, therefore, that the gatherings of Ashura served as landmarks in the development of the revolutionary movement in Iran, from the public assemblies of the 1963 Ashura, which signalled the rise of the massive protest movement against the Shah and his enterprise of 'modernization', to the more organized gatherings of December 1978, when revolutionary ferment and fervour resulted in hundreds of thousands demonstrating against the Shah in the streets of Tehran (Shivan, 1985, pp.187–8) (Figure 17.2). *Mollahs* persistently emphasized contemporary meanings in the old myths, even providing updated versions of Husayn's own words, as in a revealing extract from a sermon recorded by Thaiss during the 1963 disturbances.

"Husain said, 'This is not a type of revolt that you can co-operate with by giving wealth or giving speeches or giving religious magazines or newspapers to people, the only way you can co-operate with this revolt is with martyrdom and with self-sacrifice.' Husain, in his last sentence, said, 'I am not asking help from merchants or from powerful writers. The only help I can get is from those sincere devotees, and heroes, who are willing to sacrifice themselves, who are truly willing to sacrifice their blood.' We, the prophet's descendants, deserve more to be your ruler, to be the leader of your region of the world."

(Thaiss, 1972, p.360)

The suggestion here that the clergy ('we') should be thought of as the Prophet's heirs and ought to take over political authority, would seem very odd to Sunnis. Sunni fundamentalists have found it difficult to establish a religious basis for political opposition to an existing Muslim regime, providing

Figure 17.2   Anti-Shah demonstrators, September 1978.

the ruler does not publicly repudiate the religion. This is because Sunni myths, traditions, and laws evolved in large part under governmental auspices, and therefore tend to legitimize governmental powers. According to Sunni theorists, even a bad Muslim ruler who tramples upon some Islamic principles is preferable to chaos. The believers therefore have no recognized right to rebel (Sivan, 1985, p.190).

But this is not so for Shi'ites. Indeed, according to Shi'ite teachings, a state that is not ruled by the descendants of Ali cannot be legitimate. In practice, the Shi'ites developed positions of accommodation with ruling governments, on the grounds that this prevented anarchy, or was necessary to defend the state against foreign invaders. Nevertheless, the accommodation was generally recognized as pragmatic. In the Shi'ite mythological timeframe, it is only a temporary measure to provide some social order for the faithful from the time of the 'disappearance' of the Twelfth Imam, to the future date of the Hidden Imam's return as the *Mahdi*. In Iran, these messianic ideas were bound up with the very formation of a national identity, as the following potted history reveals.

Iran was once the heartland of the Persian empire, which had fallen to Arab armies in the seventh century. The Persian religion, Zoroastrianism, was replaced by Islam, but the Persians retained their language and arabization was generally resisted. It was partly this sense of separateness, and the continued awareness of being under foreign domination which made Imami Shi'ism appealing. It was perhaps also significant that the Fourth Imam, Zain al'Abidin, was the son of Husayn by the daughter of Yazdigird, the last of the kings of ancient Persia. Thus several of the Twelve Imams were related to the Persian royal dynasty.

By the end of the fifteenth century there were significant Shi'ite groups in Iran, but it was the rise of the Safavid dynasty from 1501 which turned the sect into the official, Iranian, state ideology. With the help of a forged genealogy, Isma'il I claimed descent from the Seventh Imam, asserted his right to rule on behalf of the Hidden Imam, and empowered Shi'ite *mollahs* to promote religious

conformity. In due course, ideological uniformity cemented territorial unity. The modern Iranian nation came into being (Ruthven, 1984, p.222).

When Safavid fortunes declined during the eighteenth century, the Iranian *mollahs* began to draw upon the anti-statist connotations of Shi'ite political philosophy to assert an independent authority. With the end of the dynasty in 1722, some of them argued that, in the absence of the Hidden Imam from whom their authority came, they had the right to make individual interpretations of law on his behalf. This view was eventually accepted, largely because none of Iran's later rulers was able to lay claim to spiritual authority, and tended to rely heavily on the *mollahs* to provide some ideological underpinning of their rule. By the nineteenth century, the *mollahs* were in a strong enough position to resist attempts to abolish their rights, notably their right to give sanctuary in holy places even to the Shah's enemies. It had become firmly established in the popular, national consciousness that the only truly legitimate rule was that of the Hidden Imam. Governments were to be tolerated until the Hidden Imam's return, but the clergy, as the Hidden Imam's representatives, were the protectors of the faithful.

It took no great leap of imagination on Khomeini's behalf to tap into these notions. He could claim that a pragmatic accommodation with the Pahlavi state was impossible, because the Shah had relinquished part of Iranian sovereignty to foreigners, had barred the influence of Islam on public life, and had even made use of pagan symbols (derived from ancient Persia), like the title 'Light of the Aryans'. It was also quite natural for the clergy to be called upon to be the vanguard of resistance. They had already played this part on several occasions in the past. Furthermore, while Khomeini did try to emphasize that he was not imbued with the Hidden Imam's spiritual powers, he nevertheless allowed followers to call him Imam (and adopted the title 'Vicar of the Hidden Imam' after coming to power). Comparison was therefore invited between Khomeini and the *Mahdi*, and the huge demonstration at the time of his triumphant return highlighted how widely messianic expectations had been aroused.

It is worth repeating that the characteristics of Iranian Islam do not in themselves explain Khomeini's success. Nevertheless, they were undoubtedly of considerable importance. So why did Halliday largely overlook them?

Part of the answer is probably that, like many others who were concerned about Iran's future in the 1970s, he had a tendency to indulge in 'wish fulfilment'. But there was an additional theoretical reason. Halliday did not deal with culture seriously because he thought it was relatively unimportant. His book is an example of taking materialism too far. Materialism is a way of thinking about the world that asserts the primacy of material things (as opposed to ideas or beliefs) in determining human behaviour. It is a premise of much of the scholarly writing on development issues. Unfortunately, it has become common for it to be applied simplistically (or 'mechanistically'), thereby giving an excessive emphasis to narrowly defined notions of politics and economics. Some writers have called this tendency 'economism' (Sahlins, 1976; Geertz, 1984; Gramsci, 1988), not because it is necessarily a fault of economics, but because it assumes the theoretical separation of economic factors and forces from other aspects of human life, and in effect reduces these other aspects to economic causes.

Many writers influenced by Marxist ideas (like Halliday) emphasize social stratification in terms of class structures at the expense of almost everything else. Indeed, it is not rare to find academics and political activists stating that social forms are determined by class relations as if there was a straightforward correlation. Meanwhile, some of those of more conventional persuasions (such as neo-liberal economists) habitually utilize an even more reductive methodology. Suspect statistics purporting to represent levels of economic 'growth' in diverse parts of the world are compared using sophisticated mathematical models. Both modes of analysis end up treating culture as a residual factor. Shared values, which conflict with what the author perceives as objective realities, are readily interpreted as 'false consciousness', and phenomena such as religion dismissed as the consequence of superstition, ignorance and poverty. Or alternatively, culture becomes a set of variables which impinge upon 'rational' economic behaviour, and therefore have to be allowed for in equations. Either way, culture is not seen as fundamental, and is treated as something separate from the business of production. As the anthropologist Clifford Geertz has pointed out, the implicit assumption is that society runs on the energies of want.

In a review of the debates about social change in rural Java, Geertz takes gleeful pleasure in demolishing the arguments of scholars, obsessed with numbers, who divorce agricultural activities from the rest of daily life, and is equally mocking of those Marxist scholars who have vainly tried to explain inconvenient findings like the absence of an established landlord class. He concludes with the following observation:

> "Whatever one may think of omega point models of social change, in which everyone ends up as a class warrior or a utility maximizer (and I, obviously, think very little of them), there is no chance of analysing change effectively if one pushes aside as so much incidental music what it is that in fact is changing: the moral substance of a sort of existence. The Renaissance, the Reformation, the Enlightenment and the Romantic Reaction made the modern world as much as trade, science, bureaucracy and the Industrial Revolution; and, indeed, vast changes of social mind, they made it together. Whatever happens in Asia, Africa, and Latin America…it will…involve comparable passages, comparably vast."
>
> (Geertz, 1984, p.524)

This ought to be self-evident. The study of development, or indeed the study of human life generally, necessitates the study of shared values of all kinds, and the examination of their multifaceted transformations. Religion and kinship are just as significant as economic transactions and the political life of nation states, and in fact these things are not really separable or comparable. They do not exist in isolation from one another. Yet 'economism' has a wide currency among development analysts. Why is this so?

Reasons include such things as ideological dogmatism, academic overspecialization, and the constraints of the aid business (in which numerical economics often dominates development planning). But there is another factor which needs to be discussed at some length. It is the difficulty of writing about peoples' beliefs and moral codes in such a way as to make comparative generalizations possible. A tendency among scholars who have attempted it is to resort to simplistic models of human behaviour which lay undue emphasis on different ways of thinking. They replace 'economism' or simplistic materialism with its opposite, idealism (the view that mental phenomena, or 'ideas', are the basis of what happens in the world). This approach is dangerous in development studies because it leads to the assumption that ways of thinking are what facilitate or hinder social change. The poor may therefore end up being blamed for their own poverty: it is the consequence of the way they see the world, because if they thought 'progressively' they would become more affluent.

Idealism underpinned European, colonial notions of cultural superiority, was a premise for theories of 'modernization' during the 1950s and 1960s (see Chapter 11), and retains a firm hold on the popular imagination in the rich countries, as well as among educated élites elsewhere. It is also implicit in some of the policy prescriptions of international aid agencies, particularly when programmes involve 'educating' a target population. I will discuss idealism in the following section. Here it is important to note that the materialism of much contemporary academic writing on development issues is partly a reaction to idealist explanations of poverty, and to an extent this is quite healthy. The problem is that too often the baby is thrown out with the bath water, and analysts simply ignore crucial issues, like religion and kinship, in the apparent hope that they will just go away. The trick is to find the balance.

17.2 Getting culture wrong

I have noted that in spite of the materialism of many development analysts, idealism is by no means something only to be associated with the European colonial past and Victorian ideas about social evolution. There are writers on the Third World who have continued to assert the primacy of ways of thinking in explaining the occurrence of poverty and affluence.

Criticism has been levelled at some anthropologists on these grounds. It has been pointed out that Geertz's own interpretation of Javan social life treats class relations as little more than instances of religious affiliation. Geertz has asserted that his critics have not properly understood his analysis (due to their 'economism'). Nevertheless, it is certainly the case that he has widely been interpreted as suggesting that the Javanese 'involuted' value system is their main obstacle to progress.

This idea, that culture is the principal barrier to development, is commonly expressed in much less subtle terms by policy makers and by journalists in the Western media. Simplistic conceptions of 'tribalism', or assumptions about the traditional attitudes of peasant farmers are held up as the explanation for what happens. By and large, it would seem that the more peculiar the lifestyle of the people being commented on looks to the outsider, the more likely it is that the way those people think about the world is taken to be an adequate explanation of what they do.

The case of Africa's pastoralists is a good example. Their relationship with their animals is sometimes accepted as the reason why they are materially poor, why their soils become eroded, and why regional or even national economies fail to grow. Alternatively, the relationship may be viewed as something worth preserving, because it indicates that these people are unaffected by the corrupting impact of capitalism and are living in harmony with nature. The logical conclusion of the second line of argument ought to be that development should occur in such a way as to leave the pastoralists alone. However, as often as not the two views are combined (Figure 17.3). The following passage shows what I mean. It is taken from an article by Robert Mister, describing the kinds of problems facing aid workers in Somalia during the early 1980s.

"Somalia is one of the world's poorest countries, with a stagnating economy which is frequently close to collapse. Many of its people live at a level of absolute poverty. Since nomads form the bulk of the population, their pastoral lifestyle pervades almost all aspects of life. The economy is very dependent on the export of nomads' livestock, which accounts for some 80% of export earnings. The nomad way of life is slow, people act individually, outlooks tend to be conservative, and the nomad struggle for survival has tended to produce a society that exploits and damages the long-term environment.

The Somalis are Muslims who adhere to Koranic law when it does not conflict with local customary law. They have strong sentiments of national esteem, but tend to guard the secrets of their culture and only share them on their own terms. And they are proudly independent, aggressive and wary of outside influences. As in many traditional societies, the people of Somalia are also fatalistic…"

(Mister, 1988, p.127)

We are presented with a grim picture of the Somali economy, and it is clear that the author intends us to draw the conclusion that the basic problem is the Somali people themselves. Their culture may be worthy of respect, but it has disastrous consequences. This is a shockingly misinformed view, and reveals a dangerous mixture of romantic and derogatory notions about ostensibly exotic ways of life. The fact that it appears in a book which purports to provide a local-level, grassroots perspective on implementing development projects, reveals how deeply ingrained such attitudes remain. In the rest of this section I show what is wrong with this analysis of Somalia, and point to ways in which the Somali economy and Somali pastoralism might be discussed more insightfully. Whereas in the last

Figure 17.3   Romanticized and abused: Somalia's pastoralists.

section I demonstrated how Iranian politics cannot be properly understood without analysing Iranian culture, in this section I will argue that Somali culture cannot be comprehended without analysing Somali political economy.

There is no doubt that in the early 1980s Somalia was in trouble, but only a decade or so previously it had been self-sufficient in grain production as well a being a significant livestock and livestock produce exporter (it was one of the world's major suppliers of high-quality sheepskins). It may have been the case that the national economy had been weak, but it was a combination of factors during the 1970s which led to the situation Mister confronted. Drought, oil price increases, inflation, war with Ethiopia over the Ogaden (1977–78), an influx of thousands of refugees, and a disastrous attempt by the government to regulate the marketing system as a step towards 'scientific socialism' all contributed. To make matters even worse, in 1983 an embargo was placed on Somali livestock exports due to a rinderpest epidemic. The country's main market in Saudi Arabia was lost, and catastrophe was only avoided by an Egyptian agreement to accept imports.

By this time, economic difficulties and a dissatisfaction with the USSR as an arms and aid supplier had prompted an ideological shift towards the USA and a loan agreement with the IMF. In 1982 the World Bank had put together a 'structural adjustment' package, and although the Ministry of Planning had stated in 1983 that self-sufficiency in food was a remote possibility, Somalia was again in grain surplus by 1989. The apparent improvement was partly due to the enormous quantities of food relief which the country secured from the United Nations' World Food Programme by exaggerating the numbers of refugees, but also due to the introduction of maize production in irrigation schemes. These schemes seem to have benefited the relatively affluent rather than the very poor, and the economic upturn has been reversed in recent years due to the civil war. Nevertheless, there are indications that Somalia is far from being a hopeless case.

The purpose of this brief sketch is to emphasize that the difficulties of Somalia in the 1980s were not something that had been inevitable. Mister himself goes on to explain in his article how the refugee influx played a part in the country's problems, but the economy was also not as stagnant as he thought, and certainly was not in a mess as a consequence of the immutable characteristics he ascribes to the Somali people. It is undoubtedly the case that groups living in Somalia hold cultural values very dear, and over certain issues will resist change. But to exaggerate this conservatism, to suggest that the Somalis 'guard the secrets of their culture' (whatever that means), to accuse them of fatalism, and to claim that 'the nomad struggle for survival...exploits and damages the long-term environment' contradicts the evidence.

It is true that some 55% of the country's population have what might loosely be called a pastoral lifestyle, but 45% do not. Some 20% are almost exclusively engaged in agriculture (bananas are the second most important export), while the remainder live in the towns or work abroad. Moreover, for generations Somalis have been renowned for their entrepreneurial skills and eagerness for paid employment. Back in the 1930s, the British Member of Parliament representing the Tiger Bay area of Cardiff was known as the 'MP for Somaliland' because of the large number of Somali sailors and their families residing in his constituency, and it was not so long ago that there were two Somali cafés in Salford near the docks to serve the Somali seamen. Throughout East Africa, it is more often than not Somalis who drive the long-distance trucking routes, and in Muslim areas it is often Somalis who are the butchers. In addition, there are some 500 000 Somalis working in the oil-producing states in the Gulf, whose repatriated salaries and 'gifts' have been critical to the economy for many years (something which does not appear in official statistics).

But Mister has not just overlooked these things, he is also fundamentally mistaken about pastoralism. The Somali pastoral sector may be thought of as comprising two main sections:

1   various groups living in 'traditional' ways, herding camels, cattle, sheep and goats, who may be termed 'nomadic' in that they move their animals along established routes between water

points and grazing land (but it should be stressed that there is in fact nothing aimless about their movements);

2 sedentary groups of 'nomadic' origin who retain their herds.

Sedentarization of pastoralists is something that has happened for generations. Particularly at times of drought or war, groups from the north and centre of the country have migrated to the arable south, and taken up farming. Often they have tried to rebuild lost herds at the same time, but over the years have given this up and become permanently settled, retaining only a few animals. The number of these sedentary pastoralists in the south has grown rapidly since 1974, when a drought provided the then Soviet-backed government with the opportunity to 'modernize' the rural economy. A spectacular programme of enforced settlement of 'nomads' was launched, modelled on the schemes imposed on the pastoralists of the southern Soviet republics. In 1975, 110 000 'nomads' were forcibly located at three agricultural and three fishery projects, far removed from their home areas. These were organized as production co-operatives, and were subject to rigid state controls. Not surprisingly, the enterprise was a fiasco. Ten years later the projects were still being supplied with relief food by the World Food Programme, and most of the men had left for the towns, the Gulf, or tried to mix farming with herding (something which sometimes leads to an excessive concentration of animals in one place).

Since the 1960s, there has also been increased sedentarization in other parts of the country. This has been due to a shift towards a more commercialized way of life among 'nomadic' pastoralists in the arid zones. From colonial times, governmental posts, new wells providing permanent watering places, and facilities like schools and shops have attracted ever larger settlements. Some family members began to reside permanently in such places, and this process has been accelerated since the 1974–75 drought. Settlement has also been greatly encouraged by the construction of a tarmac road running through the centre and north. It is in these places that the effects of overgrazing are most severe.

Where there is evidence of long-term damage being done to the environment, it is almost always where pastoralists have become settled but have retained some or all of their herds. Even in these instances the evidence is far from conclusive. But the point is that where soil erosion is found, it is not the consequence of 'the nomad struggle for survival'. On the contrary it is precisely because of the enforced or encouraged abandoning of the old ways that the problem has arisen. Far from being a consequence of conservatism, it is the direct result of inappropriate development.

Nevertheless, in spite of almost continuous drought during the 1970s and 80s, repeated interference by the government, and all the upheavals of recent years, there remain a large number of 'nomadic' pastoralists in Somalia. There are perhaps as many as two million, and with good reason. In 1986 a study of the Somali 'nomadic' production system by an adviser to the Ministry of Livestock observed (in the understated language of a semi-official document) that it is 'quite efficient'. Recognizing ecological constraints, and that 'the pastoralists have perfect knowledge of their production system including pasture and water management', the adviser concluded that development in the arid zones of Somalia will involve neither the promotion of agriculture nor the teaching of better stock-raising methods. The existing livestock system 'has proven its efficacy and steadiness under normal climatic and drought conditions over many years', and rearing practices are well adapted to the prevailing climate conditions. With the proviso that Somalia's 'nomadic' pastoralists have additionally demonstrated extraordinary adaptability and deep political skills (they have had to in order to survive), it seems reasonable to agree with the study's conclusion. If the government succeeded in providing the necessary moderate support (i.e. water, animal health, marketing), the pastoralists would probably continue generating a great deal of the country's protein and foreign currency for the indefinite future (Hübi, 1986, pp.62–3, 72).

The wonder is that anyone should be taken aback by these insights. The life of Somali pastoralists is far from being a closed and secretive world. There is a readily available literature on them, much of it published in English (De Lancey *et al.*, 1988). Furthermore, since the early 1980s, the Somali Academy of Sciences and Arts in Mogadishu has published a series of excellent working papers entitled *Camel Forum* which has investigated a wide variety of aspects of camel husbandry in the country. Virtually all these studies concur that livestock production among the pastoralists is both economically efficient and ecologically sound.

Why is it that so many people who really should know better persist in the belief that progress for pastoralists means settling them down in one place and 'educating' them to be farmers? Why do schemes which promote or enforce sedentarization continue to be implemented (and by no means only in Somalia)? Why has there been so little interest in the difficulties which pastoral peoples themselves worry about and try to cope with? What is it about pastoralists that so often causes them to be defined as the problem?

Some of the answers to these questions are fairly obvious. Defining pastoralists as the problem may be cynical or deliberate. It may underpin a policy of integrating pastoral groups into the national economy – to secure resources from them and (ideally) provide them with modern services. Or it may simply be the excuse for imposing governmental controls, particularly if the pastoralists in question have managed to secure automatic rifles (as they have in several parts of north-east Africa). But such political motivation does not explain why even those wishing to assist pastoralists or to comment objectively on development issues may lack a basic understanding of pastoral ways of life. In these instances wrong-headedness seems to have more to do with innocent ignorance, unconscious ethnocentrism (i.e. the belief that ultimately 'our' way of doing things is better than 'theirs'), and, perhaps above all, a muddle about culture.

Mister clearly considers the shared values of the Somali as extremely important, but his conception of how these values relate to behaviour is unhelpful. Notice his use of the expression 'local society and culture'. He conceptualizes a group of people he finds strange as a society with a specific culture. The society is the product of the culture,

and vice versa. In effect, the society and the culture are so interwoven as to amount to the same thing: Somali society is like this...Somali culture is like that...He writes about the Somali as if their thinking, and their corresponding way of behaving, exists within a closed system – one with its own logic and rules. In so doing, he reflects an influence from anthropology.

17.3 Culture in anthropology

Numerous references have been made in this chapter to the work of anthropologists. It is not surprising, since 'culture' is often regarded as the anthropologist's subject. Anthropologists are commonly used as consultants for photojournalism of the kind that appears in *National Geographic* magazine, and in making ethnographic films like the *Disappearing World* series of Granada Television. It is usually assumed that anthropologists hope to protect or preserve cultural forms by recording them. This, however, is by no means always the case. There are numerous schools of thought within anthropology, some of which are far more 'economistic' than most economists. Marvin Harris, for example, has gone so far as to argue that ideology and social organization can be explained in terms of 'adaptive responses to techno-economic condition' (Harris, 1969, p.240). Nevertheless, it is the case that anthropologists have usually been more interested in the significance of shared values than their colleagues from other disciplines, in large part because their methodology requires them to try to think like the people they are studying.

Academic anthropological research involves 'participant observation'. This requires the anthropologist to learn the language of the people he or she is studying, to live with people for a long period, and to try to see things from the point of view of the actors themselves. There are tremendous strengths with this method, not the least of which is that it makes it possible to look at the interconnections between things. Thus, I. M. Lewis has revealed in fascinating detail how kinship, poetry, livestock, sexual desires, lines of authority, attitudes to exchange and markets, and a host of other things are interwoven in the fabric of the daily life of Somalis whom he came to know very well (Lewis, 1965, 1975, 1985). But the methodology also has limitations.

1 It generates extremely complex micro-level data, which are conventionally coped with by conceptualizing the people being studied as comprising a separate, usually small, autonomous unit. The problem is that, like so many other social models, the artificiality of the construct is sometimes forgotten, and ideas and values exchanged between units, as well as wider political and economic linkages, are played down or overlooked. *The world is described as if it is full of socio-cultural islands.*

2 Because anthropologists have wanted to undermine ethnocentric views of exotic ways of life, and have therefore concentrated on the logic of ostensibly odd behaviour and thinking (like witchcraft beliefs), they have tended to explain *how things work, not how things change.* Indeed, in many of the older anthropological studies, change seems to be only accounted for by external interference, and is depicted as something unfortunate, upsetting to the social balance.

3 The tendency not to examine processes of change is also a consequence of basing knowledge on observation and participation. It is impossible to examine long-term processes of change *unless fieldwork is carried out over many years* (something which has rarely been possible).

4 Making a virtue of necessity, many anthropologists have responded by taking an *ahistorical approach to theoretical analysis*, arguing that it is possible to understand aspects of human life by comparing the characteristics of one society with those of another. They may, for example, compare different conceptions of gender, or the symbolic meanings of rituals and myths found in different parts of the world. Often such studies are sophisticated and interesting, but in separating beliefs from their social context, there is a tendency towards idealism which can be amplified when comparative anthropological work is simplified or summarized. In effect, it may be implied that different ways of life are comprehensible as though they are systems of thought.

Anthropologists have themselves recognized these problems, and have tried to grapple with them. The flexibility of ethnic identities and group boundaries has been emphasized, and economic anthropologists have been at the forefront in arguing that rural people in poor countries are not 'tradition bound', but highly adaptive. It has been shown that peasants and pastoralists alike are constantly having to make decisions about the use of productive assets, labour, savings and investment, that innovation and experimentation is going on all the time in livestock management and agricultural techniques, and that 'traditional' institutions and values are not necessarily incompatible with modern capitalist forms of political and economic organization.

But academic anthropologists have not fully resolved the difficulties of their methodology. More importantly, the older anthropological writing continues to be interpreted in such a way as to lend credence to the view that 'other cultures' are relatively static and internally logical – they are imbued with a special set of values which determines social activity in such a manner that acts as an obstacle to development. As I have mentioned, Geertz's discussion of 'agricultural involution' in Java can be (mis)understood like this. So can the anthropological literature on pastoralism in Africa. David Turton, himself an anthropologist, has highlighted the problem:

> "Until recently it was widely assumed among development planners, and probably still is among administrators and politicians, that the chief obstacle to the development of pastoralists lay in their 'irrational' preoccupation with the accumulation of stock as an end in itself and their predilection for a wandering way of life. One of the authorities most frequently cited as responsible for this view is the anthropologist Melville Herskovits who coined the term 'cattle complex' to refer to a collection of cultural traits (such as the identification of individuals with 'favourite' beasts, the use of cattle in sacrifice and bridewealth and preferences for certain skin colours and patterns and horn shapes) which are widely distributed among cattle-keeping people in East Africa.

Neither Herskovits, nor those anthropologists who subsequently wrote detailed monographs about particular pastoral peoples, argued that pastoralists were economically irrational. But, by concentrating on the social and cultural elaboration of cattle at the expense of analysing pastoral production systems…they did help to give an exaggerated impression of the uniqueness of the 'pastoral way of life' and of its isolation from regional and national economic and political processes. They must therefore take some of the blame for the fact that the term 'cattle complex' came to be used by others in a Freudian sense, never intended by Herskovits, to mean a sentimental and obsessive attachment to cattle and for the stereotype of the proud, aggressive and conservative pastoralist which became part of the conventional wisdom of pastoral development."

(Turton, 1988, p.138)

The continued currency of this conventional wisdom is clear enough in Mister's observations on Somalia. He views culture, or at least the exotic culture he is observing, as a set of values which defines behaviour and constrains processes of change. Instead of the economic determinism of some materialist writers, Mister has resorted to cultural determinism. It is hard to decide which is the more unhelpful perspective.

17.4 Approaching culture sensibly

I have used 'a shared set of values' as a working definition of culture, and have commented on what happens when culture in this sense is ignored and what happens when it is misunderstood. Now I am going to break down this catch-all phrase into a general categorization of approaches. I will briefly remark on what I see as seven useful ways of thinking about culture. My categorization is not a universally accepted standard classification, so there is nothing sacrosanct about it. Alternative categorizations could have been proposed. I merely provide it as a frame of reference. It will

also be apparent that one approach does not necessarily preclude another.

The generalized (or generic) usage

This is the oldest and most all-embracing way of employing the word 'culture'. It encompasses the whole range of human activities which are learned and not instinctive, and which are transmitted from generation to generation through various learning processes. Culture is thought about like this in the controversy between those who believe that human beings are 'pre-programmed' by Nature, and those who believe that behaviour is mostly a product of what happens to people after birth. In effect, culture in the generic sense refers to all aspects of human life which are not biological.

The expressive usage

The usage of the word culture to refer to various forms of expression, usually artistic expression, might be loosely termed the expressive approach. Culture began to be thought about in this way in Europe during the late eighteenth century. Underpinning it is the idea that every group (or even every individual) has special ways of reflecting perceptions and experiences which are inherently of great value. Culture in this sense is not usually seen as the motivator of human action, but as something that emanates from it. A study of Italian Renaissance culture, or a study of the culture of the South African townships, might therefore have little to say about the ways of life in which distinctive forms of literature, painting or music emerged. These phenomena might be commented on almost exclusively in terms of themselves, with discussion concentrating on creative techniques or on aesthetic principles. Nevertheless, as Chapter 19 on southern African music demonstrates, the examination of culture as a form of expression can be most revealing of peoples' attitudes and aspirations.

The hierarchical usage

This approach is premised on a stratified conception of culture in which one culture is superior to others. It became the dominant approach of

Victorian times, being reinforced by sociological interpretations of Darwin's theory of evolution. It may be linked with a belief in universal moral and/or aesthetic principles. Alternatively or simultaneously, it may be connected with ideas about historical processes governing human activity. A prevalent form is the juxtaposition of 'high' culture and 'low' culture in discussions of culture as a form of expression. When applied more generally to different sets of values, the result is usually a schema of social or cultural evolution, ending in what the author considers to be the highest stage of human development, such as Protestant Christianity, or Islam, or socialism, or multi-party democracy.

The dangers of a hierarchical approach are élitism, dogmatism and ethnocentrism, and subsequent approaches to culture have tended to react against it. But it is worth bearing in mind that it is common for human groups of all kinds to assert their superiority over others. Somali pastoralists, for example, are certainly proud people, and are likely to be fairly contemptuous of settled agriculturalists who have no animals. Moreover, it is necessary to accept some hierarchical notions in order to avoid excessive cultural relativism. This is the view that ways of life have to be interpreted according to their own specific rationality, and by implication that no way of thinking or acting is better than any other. It is a position which, if taken too far, can render any discussion about how life might be improved almost meaningless.

The superorganic usage

An approach which can be termed superorganic requires that the word 'culture' be used in an analytical rather than a descriptive sense, to refer to a form or pattern which is abstracted from observed behaviour. Thus, culture is treated as something relatively independent of day-to-day life, but as having an overarching influence of which the actors themselves may be unaware. This approach has its origins in the work of the great French sociologist, Emile Durkheim (1858–1917). Often it will mean that the writer attempts to highlight the essential, aggregated characteristics of a particular group, and will look at how these characteristics have a bearing on what

people do. The strength of this mode of inquiry is that it helps us to understand different ways in which groups of people experience things, and to locate a logic behind different ways of behaving. But, as I have shown, there is a danger of slipping into crude idealism if the approach is used in a simplistic manner. Durkheim and his followers rejected psychological explanations of society, yet concepts like 'involution' and 'cattle complex' have sometimes been interpreted as closed ways of thinking which determine collective action.

The holistic usage

The holistic approach to culture, and social analysis generally, is premised on the recognition that social phenomena of all kinds are interconnected. Economics cannot be divorced from religion, or kinship, or conceptions of gender. In practice, it is more of a methodology than a form of interpretation, and has commonly been combined with a superorganic approach. It has been closely associated with the fieldwork techniques of anthropologists since the 1920s, and remains a fundamental tenet of anthropological inquiry. The works by Lewis on Somalia and Geertz on Java are good examples. The advantage of holism is that it avoids reducing human lives to a set of possible responses to particular forces, which the analyst believes are universal. A problem, however, is that it is difficult to systematize the complex data produced, and it is hard to draw a line between what is relevant and what is not. There is a consequent tendency for holistic analysis to concentrate on the inner logic of lifestyles within an artificially bounded unit, and to make a direct correlation between a society and a culture. I have noted how this may lead to the construction of static social models, which end up being compared in misleadingly idealist terms.

The pluralistic usage

The pluralistic approach does not necessarily contradict other ways of looking at culture, but highlights the existence of multiple cultural forms within political or socio-economic units. This phenomenon has probably always been recognized, but has persistently been played down in analysis, and a group is commonly supposed to have an overall cultural coherence, even when discussion centres on such things as class conflict. In contrast, pluralistic analysis may focus on subcultures, like the homeless in London, or the interchange of shared values across group boundaries, or between classes. It avoids making any assumption that there is a common boundary between culture and society, and stresses the fact that shared values are constantly being negotiated and are always likely to change. The approach takes various forms, but a particularly important contribution was a book on highland Burma by E. R. Leach, published in 1954. Here Leach showed how complex were the connections between political action and ideal models of society, when examined regionally and through time. He noted, for example, how people readily adopted different ethnic identities when moving from one village to another. Similarly, the discussion of East African 'Asians' in Chapter 18 demonstrates how they are far from being a discrete cultural unit.

The hegemonic usage

This is an approach to culture which has become increasingly important in recent years. It is strongly influenced by the work of the Italian communist thinker, Antonio Gramsci (1891–1937). The hegemonic approach recognizes that cultural forms are pluralistic, but emphasizes the relationship between culture and power. Social dominance of all kinds, whether political, economic, aesthetic, ethnic, religious or sexual, is not simply maintained by force, but involves the consent of the subordinate. Power is linked to a hegemonic ideology, i.e. an overarching, dominant set of norms (see also Chapter 10). Social institutions, education, rituals, even language itself, reinforce the values and world view of the dominant, such that these become 'common sense' for everyone. Social dissent is thereby usually contained. But dissent may nevertheless be expressed in subcultures, or in limited forms of protest. Although the hegemonic ideology may appear to have a permanence, in fact it is continuously responding to and absorbing partial critiques. The approach is especially useful in that it helps explain why oppressed groups are often so actively involved in preserving their own subordination.

17.5 Taking culture seriously

So now I have come to the final couple of paragraphs, and must nail my own flag to the mast, although I would anticipate that you can guess what my views are. It seems to me that all these approaches to culture have a value, and should, as Peter Worsley has put it, be treated as a 'family' of overlapping meanings. All of them:

> "direct our attention to society as a whole and insist that it cannot be reduced to the economic or the political. This does not imply a new 'culturalism', in which 'culture' becomes more important than political economy: rather, it involves examining the interplay between economic and political institutions and the rest of daily life…What we need to avoid is not only the assumption that the 'cultural' is a separate sphere, but that it is causally secondary…It is in fact, the realm of those crucial institutions in which the ideas we live by are produced and through which they are communicated…"
>
> (Worsley, 1984, p.60)

Thus, taking culture seriously means avoiding simple determinisms of whatever variety, and accepting the complexity of life, even rejoicing in it. Social analysts will always have to construct models, because it is necessary to set some issues aside in order to be able to comment on others. But they should never lose sight of what they are doing and conflate the model with reality, for in fact it is only a tool to get a handle on things. Moreover, no model, or hypothesis, or theory of social change is worth much if it simply omits most aspects of human behaviour, and makes no reference at all to what people think and feel. Clearly, material factors are crucial to the study of development, but not exclusively so. The current prevalence of 'economistic' models among analysts is not an effective antidote to the simplistic cultural assumptions that still infect popular perceptions of Third World poverty. They merely compound the problems of understanding.

18

ETHNICITY AND CLASS: THE CASE OF THE EAST AFRICAN 'ASIANS'

JANET BUJRA

Writing in 1965, Tandon described the colonial societies of East Africa as systems which 'hinged on racial separatism'. More precisely, the 'three principal races (Europeans, Asians and Africans) were divided not only on a cultural and social plane, but also on an economic and political basis: the Europeans [less than 0.5% of the population] occupied the apex of the triangle; the Africans [98%] formed the base; and the Asians [1.5%] occupied the middle' (in Ghai, 1965, p.82, proportions p.79).

At first sight, the story of 'Asians' in East Africa might appear as a convincing illustration of the potency of 'race' or 'ethnicity' as factors in social life. What we will consider in this chapter is why such explanations might be problematic, despite appearances and popular understandings to the contrary. In the first place the notion of discrete and highly distinctive 'races' or human 'types' is now discredited in scientific accounts and avoided by most geneticists. The thesis that 'races' could be ranked in a hierarchy, from the most backward and inferior to the most advanced and refined, is similarly discredited, and more convincingly interpreted as legitimization for the 'civilizing mission' which imperialism claimed to pursue. However, the view that there is a causal link between genetics, racially conceived, and social behaviour is reiterated in more academic formulations such as those of sociobiology (which argues that social

relations are biologically determined), though this view is contested by other sociologists).

The concept of 'ethnicity' is of a different order, since it designates social groups on the basis of a shared cultural, rather than genetic, identity. However, 'culture' is often presented as if it were fixed and immutable, and in popular consciousness there is often an assumption that culture is given by birth, rather than something learned. Anthropological accounts of 'other cultures' (see Chapter 17) have sometimes appeared to confirm such a perception, misrepresenting them as unchanging and homogeneous.

In analysis of both colonial and contemporary Africa, explanations based on 'race' or on 'ethnicity' are very much to the fore, although here the notions of 'tribe' and 'tribalism' generally stand in for the concept of 'ethnicity' (Box 18.1). Journalists commonly seize upon 'tribalism' as an explanation for Africa's ills. It is made to appear as self-evident, for example, that peoples who do not share a common cultural identity will be naturally antagonistic towards each other; in other words, ethnic differences make for social conflict. Violence in South African townships may be presented as a case of Zulu against Xhosa; the Biafran war in Nigeria in the 1960s as an 'Ibo affair'; the problems of Zimbabwe understood as Ndebele against Shona, and so on. And if 'tribalism' is employed as an obvious explanation for social conflict among Africans, the language of 'racial differences' comes into play as a seemingly even more potent mode of instant understanding of the relations between Africans and others. The most prominent example is South Africa, while this chapter focuses on a similar phenomenon in East Africa. From this perspective, 'races', like 'tribes', are seen as discrete, natural social units, founded in a biological distinctiveness which finds cultural expression. And 'races', like 'tribes' are perceived as naturally opposed.

Superficial appearances may appear to confirm the 'race'/ethnicity explanation for human behaviour. For example, the Mau Mau uprising in colonial Kenya was largely carried through by people calling themselves Kikuyu (see Box 18.2 later in this chapter); social divisions in South

Africa do set black against white. But many sociologists challenge this perception. They argue that both 'race' and 'ethnicity' are social constructions, woven out of claims of assumed inherent and immutable differences. Which cultural differences or aspects of physical appearance are held to be markers of social value or discredit varies widely.

It is not 'society' in the abstract, or as a whole, which engages in this project. Processes of social classification, which involve the ranking of 'racial'/ethnic groups, are generally a manifestation of power, a means (though not the only means) whereby dominant groups assert their pre-ordained superiority. This is why some theorists deny the salience of 'race' as an explanatory concept, and instead focus on 'racism' (as ideology and institutionalized practice, i.e. established ways

---

Box 18.1 'Tribes' and 'tribalism'

Africa is no different from other continents in having a population that is ethnically diverse: it contains many peoples, speaking a multiplicity of languages and living a diversity of cultures. These ethnic differences rarely coincide with the boundaries of nation states: the colonial carve-up largely ignored their disposition, and independence has not generally led to a redrawing of national boundaries. Labour migration, forcible resettlements and displacements, the expansion of socially diverse urban areas, the constitution of independent states with the imperative to create new 'national identities', have all conspired to complicate the question of ethnicity, and to throw into question the assumption that 'tribalism', whether defined positively or negatively, offers a definitive understanding of African development.

When colonial officials used the term 'tribes' to distinguish between their African subjects, it generally had a pejorative connotation. To speak of 'tribalism' suggested primordial attachments to primitive and savage ways of life, sharply distinguished from the 'civilized' ways of Europeans. Today it can still be used with this implication by outsiders, which is why it is often

avoided in scholarly discussion. However, it is commonly used by Africans themselves, both to describe their own ethnic origin and that of others. Africans may also use the term 'tribalism' to denote what they perceive as the undue political or economic prominence of a particular ethnic group. In contemporary Kenya, for example, the distinction between a positive and a negative view of 'tribe' is being argued from pulpits around the country, with some clerics denouncing 'state tribalism', and urging employers to 'offer jobs purely on merit and not according to tribal background', while at the same time they insist that it is 'God's design that people are born into various tribes' and that 'all Kenyans should be proud of their tribes' (*Daily Nation*, 15–16 April 1991).

There is thus no undisputed or neutral definition of the terms 'tribe' and 'tribalism': these terms are the very stuff of political struggle and intellectual debate. The same point could be applied to the term 'race', as well as to the apparently more academic and less emotionally charged concept of 'ethnicity'. It is wise therefore always to specify and put in context the meanings you give to these words.

of thinking and doing) as the key concept in understanding such situations. Conversely, where individuals and groups identify with 'racial' or ethnic labels, this can be seen as a form of defensive or offensive political mobilization.

The language in which these power struggles are conducted is one of 'cultural differences', 'origins' or 'blood', but for some authors this is merely a symbolic representation of contention over more material distinctions. Writers influenced by Marxism tend to argue that social divisions founded in the system of production, and especially class differences, offer a more grounded explanation for patterns of 'racial' or 'ethnic' conflict.

For those of you who wish to follow up the theoretical debates around 'race'/ethnicity, introductory and further reading is suggested at the end of this chapter. A critical examination of the fate of 'Asians' in East Africa will allow you to see how the contrasting explanations have been operationalized in one particular case. This example will also raise challenging questions about the influence of the colonial past on contemporary issues of development strategy in Africa.

18.1 Asians in East Africa: the colonial period

When we speak of ethnic minorities in Western Europe we automatically assume them to be among the most disadvantaged in society (the 'lowest of the low', as Wallraff, 1985, has put it in the German context). It would be invalid, however, to assume that social disadvantage and ethnic minority status always go together, or that one explains the other. In colonial East Africa, ethnic minority 'Asians' occupied the middle level of society. The very concept of 'middle' entails not only that there is a dominant category above the group in question, but also that there is a lower element, relative to which those in the middle enjoy certain privileges and material comforts. How did the Asians come to occupy this position?

The people who are designated as 'Asians' in East Africa either originated in, or are the descendants of people from the Indian subcontinent. Arabs are not counted as Asians here. Within East Africa, Asians would not generally so name themselves (except in relation to other similar social categories: 'Europeans' and 'Africans'), but would rather refer to their religious affiliation or sect: Bohoras, Ismailis, Ithnasherias, Shi'a or Sunni Muslims, Hindus or Sikhs. The significance of this will become clear presently; the point I wish to make initially is that the name 'Asians' is an outsider's rather than an insider's term. It is an indicator of their common position in society rather than a statement about shared culture. The 'naming problem' is not peculiar to this case of purported ethnicity. You will be familiar with a key instance in British society, though of a different order, concerning the term 'black' – the controversy as to who is included and who excluded, whether the term embraces 'Asians' or not, whether it denotes a particular political position rather than physical appearance, the history of previous 'acceptable' terminologies, and so on.

What are Asians doing in East Africa? Indians first settled in this area along with Arabs as traders and middlemen in the centuries-old Indian Ocean trading networks. Unlike the Arabs they did not attempt to exert political control in coastal areas of East Africa, nor did they take wives from among the local people. By the nineteenth century they were well established, especially in Zanzibar, which had by that time become a Sultanate ruled by an Omani dynasty. Their role was to provide financial capital and supplies for trading caravans which were sent into the hinterland as far as Uganda, buying slaves and ivory in exchange for cloth, chinaware and weaponry (Mamdani, 1976, pp.66–7).

When the British arrived in the closing years of the nineteenth century, they found Indian traders already established and willing to supply their requirements. Not that they gave them much credit for doing this. Already in 1897, 'the Indian' was being stereotyped in racist terms as: 'crafty, money-making, cunning, intensely polite, his soul bound to his body by the one laudable and religious anxiety of its helping him to turn his coin to better advantage' (quoted by Mangat, 1969, p.22).

The British first annexed the area which was later designated Kenya and Uganda; the Germans appropriated the area which became Tanganyika, only losing it to Britain at the end of the First World War. One of the first actions of the new colonial power was to build a railway up from the Kenya coast to Kampala, the capital of Uganda. The railway was constructed by importing labour from India (at that time under the British *raj* and hence within the British Empire) (Figure 18.1). 'An alien army' of workers (as it was described in 1899, quoted in Mangat, 1969, p.38) was contracted to build and service the railway: skilled craftsmen and semi-professionals as well as manual labourers; from 'assistant surgeons to water carriers'. Of the 32 000 workers who were imported, only half returned home unscarred, 2500 died, 6500 were incapacitated by accident or disease, and 6700 remained voluntarily (Mangat, 1969, p.38).

Although the origins of the East African Asians were quite diverse in socio-economic terms (from wealthy traders down to humble manual labourers or 'coolies', in local parlance of the time), they soon began to converge in their economic position and social role in the new colonial society. The more lowly gravitated upwards, bypassing the African population, while those with professional skills or capital were prevented from rising further by the concrete ceiling of colonial racial separatism, which reserved the topmost positions and ultimate political power for Europeans. Mangat refers to this as the 'tripartite system' of colonial society. It would be simplistic to assume that such a system was simply imposed by the British as part of a conspiracy of 'divide and rule'. Although, ultimately, the system functioned in this way, it was the outcome of protracted political struggles which, for Asians as well as for the African population, centred on land and political rights, and took a slightly different form in each of the three countries.

Kenya

European settlers in Kenya were granted land alienated from Africans, but when Indians also wished to purchase land in the alienated areas, the settlers resisted fiercely. The free market was not left to settle the matter: it was decided by the Colonial Secretary that 'as a matter of administrative convenience, grants in the uplands area should not be made to Indians' (Lord Elgin, 1908, quoted in Mangat, 1969, p.102). From restrictions on the purchase of agricultural land, the colonial state also extended its limitations to Asians in the towns. A form of residential and commercial

Figure 18.1   Building the East African railway.

segregation was established, rationalized as 'the need to maintain proper sanitary standards – especially in view of the information that the Secretary of State had received, that most of the Indians were of low-caste origin and prone to insanitary habits' (Mangat, 1969, p.107).

A contemporary observer noted that European settlers had a particular antipathy to Indians, whom they saw as threatening their bid for political dominance in this new territory; they wanted to 'keep the Indians, not only out of the [European] uplands, but out of the country altogether' (Sadler, 1908, quoted in Mangat, p.102). When Indians demanded political rights, they were rebuffed, eventually having to settle first for nominated members in the Legislative Council, later for a communal electoral roll, which did not put them on a par with Europeans, but gave them more rights than Africans (who did not get even nominated members until much later). Ironically, however, European antagonism to the Indians was generally rationalized in terms of protecting *African* interests.

Effectively excluded from the agricultural sector, Asians in Kenya began to specialize either in commerce (retailing, wholesale, insurance, banking, financial services, import/export) or they shifted from skilled crafts into professional work and public service occupations (especially clerical and lower administrative positions). Here, too, their contribution was derided on the grounds that Africans might legitimately have taken these positions, and done the work for lower wages (Mangat, 1969, p.116).

Uganda

Kenya had the most sharply segregated colonial system, but what happened to the Asians in the other two territories differed only in degree. In all three colonies Asians were prepared to set up small shops in remote areas of the country, where they sold Western manufactured goods to peasant farmers, and bought African agricultural products in return. This role – what Dharam Ghai called 'the extension of the monetary economy into the subsistence areas' (1965, p.101) – was recognized in Uganda as early as 1920. In Uganda there

was no question of European settlement, but Asians were still excluded from owning agricultural land (Mamdani, 1976, p.70); colonial exploitation there consisted of pressing African peasants into producing cotton for the mills of Lancashire (assuring cheap supplies). In this endeavour the Asians began to play a central role as buyers of the crop, later diversifying into processing. Early European attempts to establish cotton ginning factories in the rural areas were undermined by Asian buyers, who then themselves went into processing.

Tanganyika

Tanganyika was formally speaking not a colony, but a territory mandated to Britain by the League of Nations (forerunner of the UN) as an outcome of the defeat of Germany at the end of the First World War. Open segregation or discrimination against Indians was more difficult in this context, and, as in the case of Uganda, there was no sizeable European settler class here either. In Tanganyika, then, Asians were not prohibited from buying land, and along with Germans and other Europeans a few Asians carved out a niche for themselves as owners of sisal plantations. In Tanganyika, as in Uganda, however, the main concern of the colonial government was to establish African peasant production of crops which would satisfy the insatiable need of industrial enterprises in Britain for cheap raw materials. As in Uganda, the Asians became the commercial intermediaries in this process (Shivji, 1976).

Some general points

Although there were differences in the role and position of the Asians in each territory, it is possible to generalize about Asians in East Africa simply because their socio-economic position was so distinctively different from that of either Europeans or Africans. Commercial activity was dominated by them in every territory, and roughly half of all Asians were to be found in this occupation by the 1960s (Ghai, 1965, p.92). The rest of the Asians were mainly employees in the middle reaches of the economy, especially in the public service (conversely, such positions were almost wholly occupied by Asians), but a minority had moved from commerce into processing agricultural products, the repair of machinery or small-scale manufacturing. The vast majority were urban dwellers.

This intermediate position in the economy was reflected in income levels. Within the East African Civil Service there were, until 1955, discriminatory salary scales for the same grade of jobs, with Europeans on a higher scale than Asians, and Asians on a higher scale than Africans. Apart from this, Asians enjoyed a higher standard of living than Africans simply by virtue of the positions they occupied in the occupational hierarchy. Average annual earnings for all three territories in the early 1960s show up the glaring inequality (Ghai, 1965, p.99): Europeans £1560; Asians £564; Africans £75.

18.2 Race, ethnicity or class as explanatory frameworks

There is no doubt that most people in East Africa viewed colonial society through a lens of racial differentiation. First and foremost, people were perceived as belonging to a particular category defined in racial terms, and their worth evaluated accordingly. This was not merely individual prejudice; its 'truth' was confirmed in everyday social experience. European shop assistants were unknown, while people expected to find, and did find, Indian shopkeepers and bank clerks, African road sweepers and domestic servants. To a large extent Asians accepted their disadvantaged position in this society relative to Europeans, because it gave them certain privileges over the African majority. Yash Ghai goes so far as to assert that Asians had internalized the racist values of colonialism: 'the Asians had come to believe in the myth of white superiority. [They also] began to believe that the Africans were inferior to themselves' (in Ghai, 1965, pp.132–3). There were always exceptional Asians, of course – politically conscious individuals who wrote or spoke against racism, colonial rule, the African land issue or self-determination. Examples include: in the 1920s, in Kenya, supporters of the nationalist leader, Harry Thuku

(Thuku, 1970, p.82); Pio Pinto, assassinated in 1965; or Makan Singh, one of the first trade unionists in East Africa.

I have already illustrated the racist views of Europeans towards the Asians. What of the Africans? Yash Ghai asserts that: 'The Asians are a more hated minority than the Europeans', drawing upon themselves epithets like 'bloodsuckers' and 'exploiters' (in Ghai, 1965, pp.134, 139; for confirmation see Coulson, 1982, p.60). Whether we consider these African views 'racist' or not, Africans had no means, economic or political, to discriminate against Asians in colonial society.

More generally, as Bharati concedes, there was the 'African and European allegation that 'Asians don't mix'' (quoted in Ghai, 1965, p.33), with the implication that they were therefore to blame if they were marginalized in society. It is certainly true that Asians had, and have, their own social lives, that they do not intermarry with Africans and only rarely with Europeans, that they are uninterested on the whole in proselytization (the Ahmadiya Muslims are somewhat exceptional here). The older generation inveighs against 'Western values' as 'loose, latently immoral…ethically degenerate' (Bharati in Ghai, 1965, p.59). We might well concur then with Tandon's characterization of: 'Asian ethnocentrism…isolationist group mentality…closed society of language, religion' (in Ghai, 1965, p.86) were it not that this description does not embrace an 'Asian' ethnicity, for there is no such thing. This category, thrown together in the context of East African colonialism, is in fact more culturally divided among itself than against anyone else. There is no single Asian language or religion; Asians marry almost solely within the confines of their own religious sect or caste or linguistic subgroup (Salvadori, 1983, gives a detailed account of this diversity in Kenya).

If 'ethnicity' offers little in the way of explanation for the position of Asians in colonial society, what other interpretations might we consider? In his account of the Asians in colonial Tanganyika, Shivji (1976, p.40) explicitly rejects the view that 'the relation between Asians and Africans was essentially racial or ethnic.' His justification for this is as follows: 'We are fully aware of ethnic consciousness developed over almost three generations of colonial history. This in itself needs to be explained, rather than be made an independent but decisive variable.'

Shivji proposes a quite different method of analysis, one which focuses on 'the production relations between the two communities within the context of the whole socio-economic structure.' For him, then, the Asians by and large constituted a class in colonial and post-colonial Tanganyika – the 'commercial bourgeoisie', understandably at odds with emergent African class formations. For Mahmoud Mamdani too, writing of Uganda, the Asians were an intermediary class, the 'petty bourgeoisie', playing the vital role of linking the African producer with metropolitan (British) capital. Not only this, but: 'The petty bourgeoisie in the colony had, for political reasons, to be an ethnically alien petty bourgeoisie', because such a category could 'easily be segregated from the mass of the colonized and thus rendered politically safe' (Mamdani, 1976, p.71).

From this perspective, Asians were logically deprived of the option of land ownership in Kenya and Uganda, in order to restrict them to the much more useful role of mediating capitalist relations via commerce, buying agricultural produce from peasant farmers and selling back to them imported manufactures. In addition, they could make a valuable short-term contribution in public service, as 'much cheaper than Europeans' (Lord Lugard, quoted in Mamdani, 1976, p.71), thus freeing Europeans for the most exalted and demanding positions in society. Not only were the Asians politically neutralized by being an alien and subordinate minority, they also acted as a political buffer between African and European. The legitimate anger of Africans against exploitation by those who monopolized the marketing of their crops (Asians) deflected attention from the ultimate beneficiaries – Western capital, and its local agents. In short, the claimed 'ethnic exclusivity' or 'racial' distinctiveness of the Asians, though ideologically prominent, was either spurious or irrelevant – what counted was the economic role they were forced to play and the political consequences of that role.

Viewed from the perspective of colonial class relations, then, it is clear that African antagonism to Asians was anchored in more than mere 'attitudes'. Looking a little more closely at these relations may help us to understand post-colonial developments.

Colonial class relations

Relations between Africans and Asians were at their most hostile where the Asians monopolized the means by which Africans sold their surplus produce. A report on retail and wholesale trade in Tanganyika immediately prior to independence indicated that although profit margins were not as high as Africans no doubt believed, 'cheating, and...short weight was widespread' (Coulson, 1982, p.290). Dharam Ghai points out that these practices 'are the stock in trade of businessmen all over the world, especially in underdeveloped countries. But in East Africa, because of the dominance of Asians in wholesale and retail trade, criticism of such practices is often couched in racial terms' (Ghai, 1965, p.104). While conceding this valid point, it is also necessary to recognize that when Africans expressed hostility towards Asians there was an underlying material basis which lent legitimacy to their perceptions.

In colonial economies dominated by peasant production (territories which were harnessed to producing cheap raw materials for metropolitan industry), the role of the commercial intermediary was pivotal. Not only were Asians encouraged to play this role by the British (and by the Germans), and deterred from engaging directly in agricultural production by prohibitions on their ownership of land, Africans were also actively discouraged from becoming involved in commerce. In colonial Tanganyika, the Credit to Natives (Restriction) Ordinance of 1931 required that specific government permission be given before an African could take out a loan (Coulson, 1982, p.61). Similar restrictions existed in colonial Uganda, as well as the device of setting the cost of trading licences so high that it would 'safeguard the genuine [i.e. Asian] trader' and 'remove the small pettifogging [native] traders' (1901, quoted in Mamdani, 1976, p.72).

Africans thus had two legitimate grievances against Asians in these territories. On the one hand, as peasant producers they resented the low prices offered by Asian traders and the cheating endemic in such relationships. On the other hand, when they themselves wished to trade, they found the officially favoured Asians in their way. In Tanganyika this led Africans into setting up co-operatives in order to circumvent Asian buyers. In Uganda there was cut-throat competition between African and Asian traders at the lowest levels, with Africans often finding themselves as mere subagents for Asian wholesalers or retailers. As one African wrote to a newspaper at the time: 'We Africans are anxious to start trading among our own people but are at a loss to know how to buy on a basis that will allow of a margin of profit...Those with sufficient capital to start trading are entirely at the mercy of Indian traders' (1946, quoted in Mamdani, 1976, p.167).

Kenya, being a settler economy, did not allow Asians to play such a central commercial role – the purchasing of European produced commodities was largely in European hands, although there was still space for the Asian middleman to operate in African areas, and retailing in all the major towns was predominantly in the hands of Asians. This accounts for the larger proportion of Asians who were in wage or salary employment here (70%, compared with only 45% in Uganda). Their stranglehold on middle-level positions in the occupational hierarchy created another grievance for educated Africans, who were thereby unable to rise. In all three countries, the education system was racially segregated, with considerably more spent on European education than on Asian, and minimal expenditure on African education. In 1925, the Tanganyikan Director of Education noted a 'growing race-consciousness among the Africans and a growing feeling of resentment that the Asiatics get so many of the plums', but in response to suggestions that separate schools for Africans and Asians be abolished, he expressed the fear that this might 'eventually lead to [them] making common cause for political ends' (quoted in Coulson, 1982, p.84).

In all three territories there were Asians who, after accumulating capital in trade and commerce,

had gone into manufacturing and industry; and exceptions were later made both in Kenya and Uganda to allow Asians to establish sugar plantations. In such enterprises Asian capitalists began to confront Africans as wage workers, with themselves again as the beneficiaries.

To sum up the colonial period then: relations between Africans and Asians were not only anchored in material differences, they were expressed in class relations of exploitation central to the overall functioning of the colonial economy. Following the independence of these three territories in the early 1960s, however, the world of the Asians was turned upside down.

18.3 From Amin to Nyerere: Asians in East Africa after independence

If the 'Asians' in East Africa were forced to play a class role in the colonial economy which suited the colonial power, but which also gave them certain privileges over the African majority, it is understandable that they would view the coming of African government with considerable unease and ambivalence. I want to quote the following passage from Yash Ghai at length, for it captures the general mood, and the degree to which Asians themselves were infected by the racist idiom of colonial society:

> "If the African got less wages than the Asian, if he had to live with the whole family in the one small room, if he had to walk miles or lift heavy weights, the Asian conscience was untroubled because the African was different, he was inferior; he was used to these things; he did not want and certainly would not know what to do with modern conveniences and gadgets. It was because now the same African, who they think has low intelligence and no experience, is in control of government, that the Asian has tremendous problems of reconciling himself to the new order. It is one thing to accept the rule of a superior race, indeed, one even tries to

imitate them, but how humiliating to be bossed around by members of an inferior race!"

(Yash Ghai in Ghai, 1965, p.133)

This expresses frankly and vividly what was going on at the level of consciousness. But if we fail to move beyond this perspective, then the only solution would be for Asians to be 're-educated', to change their attitudes to Africans. What I want to argue is that it was not so much the attitudes of Asians towards Africans (or vice versa) that were the problem: it was the social relations in which Asians and Africans were placed, and which entailed not just inequality but exploitation. Without the transformation of such exploitative social relations, there could be no real improvement in Asians' views of Africans, only hypocritical lip-service to the new political order.

After independence, each country attempted to transform those social and economic relations which were perceived as embodying racial privilege. What is striking is the different ways they went about this project. Broadly speaking:

- In Uganda the Asian population was expelled *en masse* on an overtly racial and racist basis, following the military coup of Idi Amin in 1971. For Mamdani and others, this was the 'Fascist' solution to the Asian problem.

- In Kenya, the government pursued a policy of what we might call 'positive action' in favour of the excluded – a policy which was locally denoted 'Africanization', and which operated within the parameters of developing a petty capitalist economy.

- Tanzania (the name which Tanganyika adopted after independence when it united with the island of Zanzibar) took an avowedly 'socialist' path, which appeared to side-step the issue of race altogether, and concentrated on reducing class privilege and expanding the role of the state in ways which 'just happened' to hit the Asians harder than any other social category.

One immediate issue, that of citizenship, was dealt with fairly uniformly. All those Asians who had either been born in East Africa, or one or both of whose parents had been born there, automatically became citizens. The rest were given a grace period of two years to decide whether to become citizens or to retain their British citizenship (no dual citizenship was allowed).

Kenya

The policy of 'Africanization' adopted by Kenya after its independence in 1963 was one tried initially in all three territories. 'Africanization' meant giving the fruits of independence to the African majority, redressing the racial inequities of the past. Not only did this mean the return of the 'White Highlands' to African ownership; it also meant the rapid promotion of Africans in the Civil Service, and the removal of informal colour bars still operating in job recruitment to public service and in private firms. Pressure was also brought to bear on foreign firms to Africanize their local management and make their shares available 'to Africans who wish to buy them'. In addition: 'Africans would be established in private enterprise by all possible means, such as loans and extension services' – these last two points being written into an important government document published a couple of years after independence (Sessional Paper No.10, paraphrased by Leys, 1975, p.222).

What is clear from this formulation is that Asians were to be elbowed out from the comfortable niche they had occupied in the colonial economy. In relation to trade and commerce the government passed a Trade Licensing Act which made it illegal for non-citizens to operate in rural and outlying urban areas, or to trade in certain specified goods. As many of the Asians had failed to take out citizenship because of their anxieties about the future, they were effectively the losers here. The list of goods which they were forbidden to sell also grew to include most basic mainstays of trade. Another ploy was simply to force them to sell their businesses to 'certain well-connected Africans' (Anon, 1982, p.43). Some Asians were able to get round this by trading through African 'fronts', but many others were forced out of business (Figure 18.2). The government also organized loans and credit on a large scale to allow African traders to replace the Asians, hence operating a policy of 'positive discrimination' against Asians.

Figure 18.2   Kenyan Asians had a comfortable niche as traders during the colonial period, but many have been forced out of business since independence.

'Private enterprise' has remained the keynote of Kenyan government policy, although it is enterprise sponsored and encouraged by the state, and its beneficiaries are generally those who can exercise political influence, rather than 'Africans in general'. The response of Asians has been mixed. Many have left the country voluntarily and made new lives in Britain, in Canada and in America. Those who have stayed on have adjusted to the new scheme of things, finding ways round the new restrictions, new fields of endeavour where their skills and accumulated capital can be put to profitable uses. One of these is the tourist trade and the travel industry; conversely, by taking on African partners or paying off politicians they have managed to retain a visible presence in the retail sector in Nairobi's main streets. Hence it is still the case that, as Tandon (in Ghai, 1965, p.79) says of the colonial period 'the Asian shops are the first to be looted' when there is trouble, for example in the attempted coup of 1981.

Tanzania

Tanganyika became independent in 1961, and Tanzania was created in 1964. Initially, here too, 'Africanization' policies were devised to transform the colonial economy, and many Asians who felt insecure left the country, though not as many as in Kenya. In 1967, however, Tanzania changed political direction and adopted the banner of 'socialism'. The famous 'Arusha Declaration' was a very different document from Kenya's Sessional Paper No.10. It pledged the state 'to prevent the exploitation of one person by another, or one group by another, and…to prevent the accumulation of wealth to an extent which is inconsistent with the existence of a classless society' (quoted in Coulson, 1982, p.176). It was followed up by an extensive programme of nationalizations of key sectors of the economy.

At the same period, President Nyerere took a stand against racial discrimination: 'Both as a matter of principle and as a matter of common sense, discrimination against certain Tanganyikan citizens, on grounds of origin, must go' (1964, *Tanganyikan Standard*, quoted in Ghai, 1965, p.154). And Tanzania was the only East African state in which an Asian held government office for many years (Amir Jamal, Minister of Finance). Reporting Parliamentary debates of the 1960s, however, Hartmann notes that hostility towards Asians was frequently voiced: 'an hostility close to racism, although it was usually clothed in the ideological fabric of socialist rhetoric' (1990, p.240). And there is no denying that although the main target of the Arusha policies was foreign capital, the major local losers in the nationalization process were Asians. The sisal plantations were largely nationalized, as were several Asian-owned grain-milling factories. Later the government decreed that wholesale trade was to be taken over and run by municipal authorities, and then that all butcheries (mostly run by Asian Muslims and by Arabs) were to become municipal enterprises. More drastic still, all rent-earning property over a certain value was taken out of private ownership and reallocated as housing or as state commercial property. This particularly hit Asians, as many had invested commercial profits in real estate as an insurance against old age. Although compensation was paid for nationalized property, it was often felt to be meagre compared with the investment income. Asian-owned manufacturing companies were, however, not taken into state ownership.

This policy was carried out in the name of socialism, and with the avowed goal of reducing class privilege and restructuring the class composition of society. This can be contrasted with the Kenyan policy of preserving the structure intact, but simply changing the colour of those who occupied the higher positions within it. In both cases, however, it is the Asians, as a social category, who have been decimated. From Tanzania too there has been an exodus of Asians, to the UK and to Canada. Those who remain are often the first object of suspicion when the state uncovers 'econ-omic sabotage' – illegal currency dealing, hoarding or smuggling (Maliyamkono & Bagachwa, 1990).

Uganda

In Uganda the problem of making space for African advancement was solved by ejecting the whole Asian population, citizens and non-citizens alike. The political history of Uganda is a complex and tangled one and I shall not go into the details of it here. For a decade after independence in 1961 the Asians seemed to be adapting to the new situation, which, as in Kenya, revolved around the 'Africanization' of government personnel and the provision of credit to allow for the expansion of African entrepreneurship. Some Asians left, but others prospered, given the limited development of African entrepreneurship. At this stage, as in Tanzania, there was an attempt by the state to substitute for private capital by way of nationalizations and economic intervention. But Uganda's 'Move to the Left' failed to gain popular support, and there was a coup led by the military. It was under the ensuing dictatorship of Idi Amin, as Uganda lurched from one economic and political crisis to the next, that the expulsion of the Asians was seen as a solution. An 'Economic War' was declared against the Asians in August 1972 (Figure 18.3).

This was not the first time that an 'alien' category had been scapegoated in Uganda. Just prior to the Move to the Left, in 1969, there was considerable unrest among workers in response to unemployment and rising prices, marked by increased union militancy and strikes. It happened that a fair proportion of Uganda's workforce were Kenyans, largely Kikuyu thrown out of Nairobi during the Mau Mau period (Box 18.2). In a bid to gain popular support, the President of the time (Obote) expelled all the Kenyan workers.

Figure 18.3   The smashed windows of an Asian-owned shop in Kampala.

Box 18.2   The Mau Mau movement

Kikuyu workers were scapegoated by the Ugandan state at a time of economic crisis. Their existence in Uganda at all was a legacy of another crisis which, at the time, had been presented by many as explicable only in ethnic terms. In the early years of colonial rule in Kenya large swathes of land were alienated from Africans and allocated to European settlers, with the aim of recouping the heavy costs of building the railway. This was fertile land in the central area of the country, accessible to the railway line and to the capital – land which had previously been largely occupied by the Kikuyu people. The indigenous population was restricted to 'reserves' in this area, which quickly became overcrowded. In order to survive, Kikuyu men and women found themselves having to work as casual labourers for European settler farmers who were backed by the colonial state. In the 1950s, Kikuyu grievances about land and about lack of political rights exploded in violence. A movement, which came to be known as Mau Mau, waged protracted guerilla warfare against European settlers and the colonial government. In its bid to suppress this rebellion the colonial government mounted a massive military and psychological campaign. Eleven and a half thousand Africans were killed (compared with less than a hundred Europeans), thousands of Kikuyu were expelled from urban areas, 'rehabilitation' camps were set up to contain and to remould suspected terrorists, the land of known guerillas was confiscated, and other Kikuyu resettled in stockaded villages.

For the colonial government and for most Europeans in Kenya, Mau Mau was essentially a Kikuyu phenomenon: it was viewed as 'an atavistic tribal rising aimed against western civilization and technology' (Corfield Report, 1960, quoted in Rosberg & Nottingham, 1966, p.378). Kikuyu people did indeed form the core of Mau Mau, and their grievances were often expressed through quite traditional cultural forms: oathing ceremonies, for example, which drew on potent myths and symbols of Kikuyu ethnic identity. The symbols and the language were traditional, but the purposes to which they were now put were quite dramatically new. As the Kikuyu saw it, Mau Mau was not a struggle to assert traditional 'tribal' values, but to regain their land.

Apart from this, there are other considerations which make an understanding of Mau Mau as an 'ethnic' uprising less than convincing. The Mau Mau fighters were feared by some, but supported by many other Africans in Kenya who saw in their struggle the general anti-colonial issue of 'Africa for the Africans'. Secondly, the Kikuyu people were divided by Mau Mau as much as they were united: the assassination of Kikuyu 'collaborators' (government chiefs and other functionaries, and those whom the government viewed as 'loyalists') was a marked feature of the struggle. In the end some Kikuyu benefited at the expense of others: freedom fighters lost their land to loyalists, Kikuyu political leaders detained throughout the Mau Mau Emergency later returned as martyrs and seized the political initiative which took Kenya to independence and themselves to power as 'national' rather than Kikuyu leaders. Meanwhile, the forest fighters were politically repudiated and the memory of their struggle was buried by the new state.

Although Mau Mau was by no means a 'tribal' rebellion – a simple assertion of cultural identity – Kenyan politics has, ever since, worn the guise of 'tribalism'. Because the Kikuyu area was the most economically 'developed' during the colonial period, with the establishment of commercial farming and infrastructure to support it, and because it was the site of heavy missionary activity, the Kikuyu were the most exploited but also became the most educated and the most politicized of peoples in Kenya. Kikuyu personnel therefore tended to dominate the new government, while those with political and economic aspirations in other areas organized defensively against what was seen as 'Kikuyu' domination (though perhaps you can now see that it was only a minority of Kikuyu who gained such politico-economic benefit). Their own political organizations played on these fears and mobilized support around alternative 'ethnic' loyalties.

Since political leaders are able to dispense rewards to their followers in the form of jobs and opportunities, it is also in the interests of ordinary people struggling to survive to mobilize ethnicity in their own interests. But people may be conscious of having several such identities which may be drawn on for different ends – for example, for people from the Kenya coast, it may pay on some occasions to make claims on the basis of a particular ethnic loyalty (e.g. as a Bajuni as opposed to a Giriama), while on others the loyalty claimed encompasses all those coastal people who see themselves as Swahili-speaking. The same people, in other situations, may play on loyalty to birthplace (the people of Lamu, compared with those of Mombasa) or yet again an instrumental ethnic identity may be forged out of common religious adherence to Islam. In the struggle to survive and to succeed, the language of negotiation, claim and appeal often takes an 'ethnic' form, but the ethnicity it describes is fluid and constantly shifting.

On what grounds did Amin resort to a similar tactic in the year following his coup? The rationale (delivered over Uganda Television) went like this: 'Asians came to Uganda to build the railway. The railway is finished. They must leave now... Asians have discriminated against Africans... Asians have stayed aloof. They have never mixed with Africans... the Asians are a British responsibility' (quoted in Mamdani, 1973, p.15). At a meeting with students at Makerere University later, he added another justification: that the 'Economic War' would create 'black millionaires' (Mamdani, 1983, p.39).

At one level there was a certain rough justice in this. Asians had profited by their key position within the colonial economy, and this privilege did not disappear with the end of colonial rule. It had become institutionalized, built into the very structures of the economy. In that position the Asians blocked the way for Africans to advance. On the other hand, the Asians had not been the prime architects of this situation – they were simply, as Fanon (1967) puts it, beneficiaries 'at a discount' of British economic and political interests. Mangat (1969) writes of the colonial period that there was 'the tendency to use the Indians as a scapegoat', meaning that they provided a diversion in times of economic or political upheaval. In Uganda after Amin, the Asians once again came to play this role. Falling world prices for Uganda's cotton and coffee, severe balance of payments deficits, internal political unrest or the withdrawal of foreign investment were not of their making. But the Asians did represent a highly visible concentration of wealth and privilege in a country where life was becoming increasingly hard for the mass of the people in city and countryside alike.

Since the Asians were forbidden to transfer any of their wealth out of the country on their departure, the capital assets they had accumulated were now available for redistribution. Among ordinary people throughout Africa, Amin was initially hailed as one who had the courage to 'pay back' Asians and Europeans for their racist treatment of Africans. Within Uganda there was popular support for the move at first, but we need to dissect this a little more carefully. As Mamdani tells us: 'The first signs of jubilation and support for Amin's decree came from the African trading centres around Kampala' (1973, p.46) – in other words those Africans who hoped to take the place of Asian retailers. Backing also came from workers on Asian-owned sugar plantations. They struck in support of Amin's decree, but – and here is where the true character of the move becomes apparent – the Army was sent in to force the workers back to work. University students openly resisted Amin, displaying placards saying 'We oppose all forms of racism' when Amin went to speak to them. Again, the Students Union was banned and paratroopers were sent into the University. Finally, 'the property appropriated from Asian capitalists and small proprietors was distributed to big business and military circles' (Mamdani 1983, p.39). In other words the political allies of the President were now rewarded for their loyalty.

Thus, when we speak of 'Africans' benefiting by Amin's action, it is necessary to underline that only a few Africans gained. The position of the massive majority remained unchanged. What is more, although the army under Amin became better equipped than almost any other in Africa, the economy continued to stagnate, political chaos reigned and terror increased (not Asian scapegoats now, but African opponents of Amin brutally murdered, tortured, detained). It was the Tanzanians who finally invaded to overthrow Amin with the help of many exiles who had fled to Tanzania in the course of Amin's regime.

The other end of this story is what happened to the Asians who were deported *en masse* to Britain. First there was the rude awakening they had to the racist reality of British society (Mamdani, 1973, pp.106–8). Conversely, some Ugandan Asians did find ways of transferring assets to Britain, so that now they form a relatively privileged stratum among Asians in Britain, setting up small and large businesses there. A few even went back to Uganda after Amin's downfall.

18.4 Concluding remarks

Wallerstein has argued that: 'Racism and underdevelopment...are...constitutive of the capitalist world economy as an historical system...It is impossible to conceptualize a capitalist world economy which did not have them...and in the long run the system cannot operate with them' (1989, pp. 9–10). For Wallerstein 'race' and 'ethnicity' are infinitely malleable social categories defined and redefined by capital in its search for exploitable labour power. They are social fictions, whose power derives from their apparent foundation on ascriptive criteria: genetics, or unchanging cultural forms.

The creation of a category of 'Asians' in East Africa, spurious in a cultural sense, but whose social reality was confirmed by the economic role it was set up to play, seems to support Wallerstein's view. Certainly the Asians were subject to exclusionist practices both before and after independence which we may describe as racist. But this very exclusion served to cement a new identity in which new cultural styles and an idiom of 'race' was also present. It should be noted that the culture was one of a privileged group with much to defend; it was also one whose interests were tied up with capital and the extension of capitalism, rather than a category whose existence served merely the labour needs of capital. However, in one key sense the Asians fit Wallerstein's conception perfectly: as a social category they were dispensable in the phase of neo-colonialism. Their economic power relative to Africans in the colonial period (and their new-found social identity) did not extend to political leverage, so they were unable to protect themselves from expropriation in the new political order after independence.

It is useful to see this case as an issue of development strategy for newly independent African governments. In order to promote economic development, it was seemingly necessary for them to act against the accumulated privileges of the Asian minority. However, given that a section of the Asian community could have constituted the core of an expanding industrial bourgeoisie, were the exclusionary policies which prevented this outcome ill-advised, and 'racist' rather than 'developmentalist'? Much depends here on whether you define 'development' very narrowly in terms of production, and more particularly capitalist production, or whether it is defined so as to include social justice among its goals. If the latter, then redressing the racially defined inequalities of the past could be seen as a legitimate objective of development policies. Tanzania's approach, whereby the question of the Asians was dealt with as a class rather than a race issue, was a different way of reaching these same goals. Was it thereby less 'racist'?

Cultural differences are real, and the identification of individuals with them goes deep. Labelled as 'ethnicity', such differences are lent additional force by the language of 'blood' and birth. But cultural differentiation is socially constructed, not a product of biology. Its myriad and changing forms reflect power relations and material inequalities, oppression as well as resistance, privilege as well as exploitation. When situations are described as ethnic or racial conflict, you might pause to consider: who is making this claim, who is naming the categories, and to what extent do the category boundaries coincide with other forms of inequality?

Further reading on race and ethnicity

Richardson, J. & Lambert, J. (1985) *The Sociology of Race*, Causeway Press, Ormskirk (a helpful, clear introduction to the major debates in the field).

Rex, J. & Mason, D. (eds) (1986) *Theories of Race and Ethnic Relations*, Cambridge University Press, Cambridge (a more advanced collection of articles written by theorists from a range of social science disciplines).

19

THE POLITICS OF CULTURAL EXPRESSION: AFRICAN MUSICS AND THE WORLD MARKET

RICHARD MIDDLETON

Figure 19.1   Celebrating the setting up of an ANC branch for West Johannesburg.

In recent years, the increasing globalization of culture has had important consequences. It is often assumed (and you may hold this view) that one of the effects of the electronic communications media, especially in the hands of organizations with an international or even world spread, is the standardization of cultural forms (the 'cola-ization' of world culture). Cheap cassette players, radios and TV sets, not to mention satellites, imply the possibility of a more or less universal access to the same messages. Indeed the same dance records as are popular in London and New York can turn up in Sahel villages and Indonesian taxis. This process threatens indigenous forms, usually replacing them with imported, often 'international' styles or with 'modernized' hybrids. The 1970s saw the emergence of some variant of 'pop-rock' (generally local techniques crossed with Western pop rhythms and instrumental sound) in almost every Third World country. The flow is overwhelmingly from rich to poor countries – hence the appearance of the theoretical label 'cultural imperialism' to describe it. As a Sri Lankan writer puts it:

> "My fear is that in another 10 or 15 years' time, what with all the cassettes that find their way into the remotest village, and with none of their own music available, people will get conditioned to this cheap kind of music. Then they will lose their own culture…If this disappears, then the whole world culture will lose one little aspect. However small a nation we are, we still have our own way of singing, accompanying, intonating, making movements and so on…But we could lose it due to the lack of organization and finance."
>
> (quoted in Wallis & Malm, 1984, p.269)

Western publishers, music corporations and broadcasting organizations dominate the world cultural system. 'Language imperialism' is acutely felt by many writers. Kenyan author, Ngugi wa Thiong'o, has described his despair at the thought that 'The very people about whom I was writing were never going to read the novel' (*A Grain of Wheat*) and his 'sense of belonging' when in 1983 he turned from using English to writing in Gikuyu (*The Guardian,* 21 October 1989).

But this is not the simple either/or issue that you may imagine. People in Third World countries often feel that the local language and the traditional culture are restrictive, preventing access to the 'modern' world. They may be regarded as a brake on possibilities for change, or worse, they may be cynically exploited by political élites in order to keep peasants and workers in unquestioning subservience. There is a constant tension between indigenous ways regarded as a source of cultural strength and as reactionary anachronisms; just as there is between the global culture seen as an engine of progress and modernization, and as a force for exploitation and homogeneity.

We see this clearly in the phenomenon of 'world music' – which became a distinct trend (and fashionable label) in the 1980s. Western musicians such as Peter Gabriel, Paul Simon, Sting and David Byrne collaborated with Third World musicians and were influenced by them. Reggae was

widely influential and Cuban salsa, Trinidadian *soca,* Zaïrean *soukous* and South African *mbaqanga* achieved significant visibility. African performers like Youssou N'dour, Ladysmith Black Mambazo, Salif Keita, Sunny Ade and many more toured Western countries and their records attracted a minority but substantial and committed audience. The search for 'roots' was everywhere.

But the trend was (and is) surrounded by controversy. Does world music originate in a simple desire that people previously without a musical voice on the world stage should be heard? Or in the Western music industry's relentless need for new sources of exploitable 'product'? Are Western fans motivated by tolerance and curiosity? Or by revulsion at the sterility of commodified Western pop music and a romantic idealization of the 'simple', 'authentic', 'genuinely creative' music cultures of the Third World? Some might argue that both industry and fans are practising a form of colonialism. The industry wants 'raw materials' which, while 'different' enough to attract consumer attention, can be turned into a product that will sell in a global market. The fans want 'roots', and if the musicians modernize their styles, or use the latest technology, they are often disowned. According to Salif Keita, 'Africa will never become anything other than what the West wants it to be' (*The Guardian*, 1 September 1989). I hope that, as you read this chapter, you will formulate your own views on these questions, which might be encapsulated in the following general question:

> **Q** What roles do Third World musics play in defining cultural identity and how is this affected by their relationship with the wider world, particularly the global music market?

In point of fact, these issues are not new. From the nineteenth century on, 'folk' music cultures *within* the West were at once attacked, and drawn on, by the rising culture industries of the metropolitan centres. From the early years of this century, music styles associated with, derived from or influenced by the 'African diaspora' have gradually arrived at a predominant position in the formation of world popular music. African and African–American musics have crossed and

Figure 19.2 '*In a world in which authoritarian power is maintained by means of superior technology, and the superior technology is supposed to indicate a monopoly of intellect, it is necessary...to understand why a madrigal by Gesualdo or a Bach Passion, a Sitar melody from India or a song from Africa...a Balinese gamelan or a Cantonese opera...may be profoundly necessary for human survival, quite apart from any merit they may have as examples of creativity and technical progress' (Blacking, 1976, p.116). A resident makes her voice heard as council workers, guarded by riot police, bulldoze shacks in a Cape Town squatter camp during April 1990.*

recrossed the Atlantic. In particular, US Afro-American music, in its position of 'internal colonialism', has been so crucial that its history can be regarded as a model for the situation of many Third World musics now. There is the same tension between the music's revitalizing role and its manipulation and exploitation by the (white) music industry, between 'roots' and 'modernity', music as commodity and music as authentic self-expression. On this level of analysis, such ambiguities seem to be part of the very essence of the late capitalist world cultural system.

There are thus sound reasons for choosing examples in this chapter mainly if not exclusively in African music – particularly southern Africa where Afro-American music has played an important role. (Against this, there are parts of the world where the Afro-American influence has been much less, especially in the Middle and Far East. The most recorded voice in the world belongs apparently not to Michael Jackson or the Beatles but to Indian film music singer Lata Mangeshkar, and one might suppose that Egyptian star Umm Kulthum is not far behind.) Of course, the peculiarities of southern Africa must be borne in mind. At the same time, an extended colonial history, more than a century of rapid industrialization and urbanization, a lengthy tradition of political struggle, and widespread dispossession and proletarianization of peasants, together with a crucial position within the world capitalist economy, have combined to produce a situation which, in cultural as well as other aspects, presents many typical Third World trends in extreme form. Music in this region seems at present to be on the brink, teetering between incorporation in the world system and continuing to speak in unique ways to and for its own people (Figure 19.2).

Various theoretical approaches can be taken to the processes outlined above. The view adopted here recognizes the importance of the Western music industry hegemony but insists on the individuality of specific Third World music cultures, the uneven and differentiated nature of change and the contradictions and ambiguities to be found throughout the system.

19.1 Tradition, modernization and cultural difference

'Although we will freely allow innovation in our own music, we somehow think that African music must be traditional' (Kauffman, 1979–80, p.42). The comment applies beyond the specific case of Africa. For many of us in the West, music here has a richly documented history of change, whereas *there* its origins are lost in the mists of myth, and people have always beaten the same age-old rhythm as they do now. Often this is part of why the music appeals to us but at the same time it lays the ground for assumptions, explicit or implicit, about its 'primitivism'. What do you understand by the word 'tradition'? The concept is vague and rarely disinterested. Generally it carries the implication that the music concerned is (1) changeless, (2) simple and (3) artless (that is, created spontaneously rather than through craft, discipline and knowledge). For most if not all Third World musics, these assumptions can easily be shown to be false. Confining ourselves to southern Africa, we now have many detailed studies (e.g. Blacking, 1976; Berliner, 1978) making clear the complexity of the musical structures, the demanding nature of the performance techniques and learning processes, and important changes in style, instruments and repertory which have occurred in the past. *Mbira* instruments, among others, are often wonderful demonstrations of the craftsman's and decorator's skills, and they have a documented history of 400 years (Figure 19.3).

Looking more widely, there is evidence of contacts between sub-Saharan musics and a range of external music cultures stretching back to Ancient Greece and including the Middle East, India, China and Indonesia. The musics of European colonialists had a big impact, from at least the seventeenth century. For instance, there were European-style military bands in West Africa then, and by 1750 they included local players. The penetration of the mass media began remarkably early. According to Manuel (1988), there was an Egyptian record market by 1890, and many other countries were not too far behind. Radio generally followed in the 1920s.

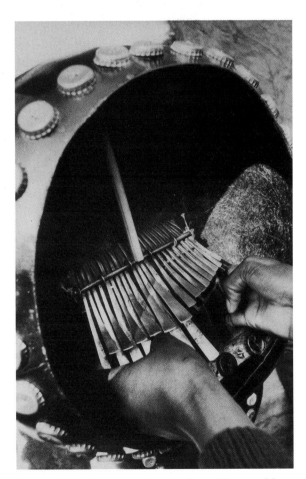

Figure 19.3 A modern Shona mbira. The metal keys are plucked with the fingers and thumbs, and the notes resonate in the steel drum. Older mbira use a calabash to amplify the sound.

Change, then, has been normal and in this century is increasing; acculturation (between different non-Western musics and between non-Western and Western musics) has been commonplace. You should not take this to mean that there are *no* differences in rates of change or musical complexity. In some cases (e.g. the Vedic chant of India) there seems to have been little change for thousands of years. Similarly, many societies make distinctions of complexity and seriousness between different types of music within their own cultures. But such differences do not correlate with any developed/developing world typology. All societies, including ours, have traditions, just as 'all societies are developing. Each of them is, in different degrees, at the same time backward and modern' (Eisenstadt, 1973, p.107). Societies in which traditions are relatively few, weak or marginal may *invent* or *fake* 'traditions'. Nineteenth and twentieth-century Europeans have done this not only for their own cultures (Scottish 'tartanry'; 'Constable's England') but also – starting with the 'Noble Savage' – for their colonies (Hobsbawm & Ranger, 1983). It is for this reason that we should be wary of describing any Third World musics as 'traditional', for the term cannot now help but mythologize. As musician and anthropologist Johnny Clegg explains, 'The musicians see the Western roots attitude as a form of slavery. African musicians want to experiment with Western music…And, if not experiment, then to be able to use the techniques and the technology available to them. But when they start to use drum boxes, they get branded as cultural sell-outs. There's a feeling among roots people that African music has to be frozen into a timewarp' (quoted in Stapleton & May, 1989, p.211).

This is not meant to suggest that there are no important distinctions between musics, or that all musics are becoming the same. Many long-established styles, extremely diverse in their characteristics, are still practised, and, because they are still vigorous, as well as because of their frequent influence on more recently developed styles, it is important to be aware of them. They help to confirm the astonishing diversity of the world's musics, offering an 'otherness' that must be respected. The nature of this diversity does not permit any theory of 'evolutionism'. African musics, for example, are not 'survivals' from a stage the West has outgrown, nor are they 'on the way' to practices developed in Europe; they are responses to different contexts, needs, functions, problems and tastes, and, if and when they are 'modernized', this will take forms specific to their own cultures and circumstances.

Thus, choral music has been a common feature in various Western traditions for centuries, but in southern Africa, this medium – central to the repertory – is approached in ways that are unfamiliar to Westerners.

Key characteristics are:

- short phrases, much repeated, sometimes with variation, sometimes not;

- plentiful use of pitch effects outside what we think of as standard scales;

- structures based on antiphony (call and response) between voice-parts; often call and response overlap, resulting in intricate contrapuntal textures marked by staggered voice entries;

- voice-parts moving in parallel;

- structures based on a repeating sequence of two or three 'harmonies'; often the melodic content of successive phrases is based on 'harmonic roots' a tone apart;

- interplay of different rhythms in the different voice parts, within, however, an underlying strong regular beat.

Conceptually similar in some ways (though quite different in sound) is the *mbira* music of the Shona people of Zimbabwe. And again, while solo and small group instrumental music, and solo singing with such instruments, is familiar enough to Westerners, *mbira* music sounds and functions like nothing we are used to. In the most highly developed style, pieces are open-ended elaborations of basic frameworks consisting of cyclical melodic patterns. The pitches of each phrase (of which there are usually four to a cycle) are distributed between three registers (high, middle and low) so that, as the notes are prolonged by the gourd resonator and overlap, the effect is of an intricately shimmering texture made up of interlocking polyrhythmic parts, their relationships ever changing as the performance proceeds. Often a second or third *mbira* varies or supplements this material, and often too a vocal part is superimposed. This is deeply serious, intellectually demanding music closely linked in the culture with religious revelation and the historical identity of Shona society.

These two styles represent still living music cultures – though, like many long-established styles, they are often most associated nowadays with older generations, while younger people may prefer more 'modern' kinds of music. Sometimes, too, there is a functional differentiation, with old styles reserved for special occasions (initiations or other ceremonies, for example), perhaps respected rather than loved, while 'everyday' music takes on a more 'up-to-date' character. Thus Berliner (1978) describes how for Shona youngsters the *mbira* came to seem old-fashioned compared with the guitar, which symbolized the possibilities of town life, Westernization and social change (Figure 19.4). Even though there has been a considerable *mbira* revival since the 1970s, the cultivation of the music is now at least partly more deliberate and self-conscious than before, through broadcasts,

the activities of special societies and college teaching programmes. For many of the young, the *mbira* now has a kind of 'classical' status.

Old styles can also be used (or misused), under the guise of 'tradition', for definite political ends. The Zimbabwean freedom-fighters, who invoked the 'spirit mediums' closely associated with the fight against Cecil Rhodes 80 years earlier, also benefited from the revival of *mbira* music, which became linked with the rebirth of black national culture. At the same time, the white regime's policy was to co-opt the chiefs, working through the 'traditional' tribal system, just as they too tried to make use of the music in their propaganda. Similarly, the South African government has made cultural conservatism and separatism a cornerstone of its *apartheid* policy. Radio Bantu, for instance, is divided into seven 'ethnic' channels, each centring its content of plays, documentaries and music on the appropriate 'homelands'. Music is old-fashioned and lyrics are censored, not only for political themes but also urban slang and anything smacking of modern life. The 1974 musical show *Ipi Tombi* was the first of several to promote 'tribal exoticism', and the Cape Town 'Coon Carnivals', which from the late nineteenth century have played an important role in the evolution of various acculturated musical styles,

Figure 19.4 The guitar symbolized the possibilities of town life: but foreign instruments and musical forms may quickly become indigenized.

Figure 19.5   Traditional music may be promoted for a political purpose: Amandla, the ANC cultural ensemble

have also become in part a tourist attraction showcasing the traditions of 'happy coloureds'.

The phenomenon of 'fake folk music' is widespread wherever tourists have penetrated the Third World. At the same time, it is no less common for post-independence governments to promote traditional music as an instrument of policy, setting up national troupes of musicians and dancers, subsidizing tours and recordings, encouraging local competitions and defining genres and their status. Often, the concepts of 'folk' and 'tradition' are elided with that of 'nation' as part of a strategy to provide historical legitimacy for the modern state. At other times, the motivation is more localist. The Indian governments of Gujarat and Punjab, in an attempt to contest the hegemony of the commercial cinema of Madras and Bombay and its music, subsidize the production of local films containing folk-derived

dance and music. Or, different again, cultural nationalism can be tied to the needs of religion. The xylophone music of Chopi migrant workers from Mozambique is being used in black South African churches, just as *mbira* music and locally styled singing has been taken up by black Zimbabwean Christians. Everywhere, then, 'tradition' has become a tool of politics, but in contradictory ways (Figure 19.5). In a still potent analysis, Fanon (1967) describes the defects in both the nationalist revival of tradition and the Westernization against which it reacts, arguing that the materials and lineages of indigenous culture must be brought into alignment with contemporary popular struggles. 'We must work and fight with the same rhythm as the people to construct the future. A national culture is not a folklore, nor an abstract populism that believes it can discover the people's true nature' (Fanon, 1967, p.187).

19.2 Music and politics

How, in clear terms, is music to do what Fanon suggests? Some music has an explicit political role, namely songs with lyrics containing overt political or social content. Many established genres fall into this category, including those of the *griots* of West and Central Africa (these are musicians paid to praise, condemn or satirize prominent individuals in song). Political themes can be found in many urban popular genres as well (rumba, calypso and the various types of Latin American *nueva canción*, for instance), even if, often, social comment is now mixed in varied and ambiguous ways with entertainment and escapist functions, as in samba, salsa, Tex–Mex conjunto and South African *mbaqanga*. Despite the pressures of the commercial market and, sometimes, of censorship, similar types of lyric content appear in some contemporary styles lying closer to the international mainstream: for instance, the Afro-rock of Nigerian Fela Kuti and a good deal of reggae, notably Bob Marley's. In the 1980s, reggae was widely popular in Africa, probably for political as much as musical reasons; Marley starred at the concert held to celebrate Zimbabwe's independence in 1980.

Politicized lyrics can take 'official' form. Almost all nationalist and liberation movements and post-independence governments and political parties promote such songs for agitation and mobilization purposes. There is an irony here, however. Presumably because political élites have almost always received a Western education and absorbed Western cultural norms, their propaganda, party and state songs have typically drawn on European models for their musical style. This is not so true of South Africa, where the relative informality and grassroots character of much political struggle has resulted in a rich tradition of protest songs, chants and slogans with musical links to indigenous styles, dating back to the late nineteenth century and including such famous songs as '*Azikwelwa*' (We Refuse to Ride), thrown up by the Johannesburg bus boycott of 1956, '*Thinantsha*' (We are the Youth), and '*Thina Sizwe*' (We the Blacks).

Already we can see that musical style can be as important as lyric content. Indeed, it is a mistake to think that music's political role relies solely on verbal messages. As several writers have argued (Coplan, 1979, 1982; Blacking, 1980; Erlman, 1985), in the right context feelings of solidarity, group identity, collective pride and even outright resistance can be articulated through musical style; at the extreme, musical practice may condition participants' consciousness in such a way that prepares the way for social decisions and action. Even in 'message songs', musical style often plays a role. While much of the *chimurenga* (struggle) music of the Zimbabwean liberation war was relatively Europeanized (especially that broadcast by the liberation movement choirs on the Voice of Zimbabwe), much also (notably that produced by musicians inside the country, like Thomas Mapfumo) was grounded in the *mbira* revival, using old-established tunes, rhythms and musical patterns, either on the *mbira* or in adapted forms on electric guitars. The most serious and highly valued type of *mbira* music has deep philosophical, religious and ritualistic significance and undoubtedly these associations enabled the apparently ambiguous and highly coded lyrics of Mapfumo's songs to take on clear national–political meanings for sympathetic listeners.

Often, style is the main or only factor. The waves of cultural nationalism that have swept through many Third World countries have affected music deeply. Existing relatively Europeanized urban genres such as West African highlife and ju-ju were often 'Africanized'; in Zaïre, the style which had emerged in the 1950s under the strong influence of imported Cuban rumba was transformed by an infusion of indigenous elements in association with President Mobutu's '*authenticité*' campaign; the resulting *soukous* became perhaps the dominant modern style of sub-Saharan Africa. (At the same time, musicians were often caught between becoming political tools of the government, and censorship or worse for songs that did not conform.) Afro-American soul music, with its 'black power' connotations, has played a similar role in Africa. In South Africa, where Afro-Americans have been important role models for blacks since the nineteenth century, its influence has been particularly strong. Interestingly, while black South Africans have responded positively to successive Afro-American styles, from minstrel songs and ragtime through jazz to funk and disco, rock 'n' roll seems to have been an exception. Only white rock 'n' roll performers were marketed in the country, so blacks could not identify with what they perceived to be a white style. More recently, several South African bands (Malombo, Harari, Malopoets, Savuka) under the influence of 'black consciousness' philosophies, have pursued a deliberate 'back-to-the-roots' policy. Jazz musician Dollar Brand, whose later music shows similar aims, changed his name to Abdullah Ibrahim to mark his conversion to Islam and his embrace of 'black consciousness'.

Of course, 'roots' can be used for reactionary purposes, as the *apartheid* regime (not to mention various black dictatorships further north) has proved. Also, it can be argued that the more mass-produced, international pop styles, such as disco, can act as an opiate, escapist entertainment diluting political energy. Some critics have suggested that the large disco systems deployed in the townships for marketing and market-research purposes are regarded by the authorities as instruments for pacifying (passive-ifying?) the people. *Mbaqanga* guitarist Marks Mankwane dismisses 1980s disco-influenced township pop as 'nothing but love, love, love' (quoted in Prince, 1989, p.105), and jazz trumpeter Hugh Masekela says 'Disco is a social tranquillizer' (Mutloatse, 1987, p.86).

The more formalized South African political movements, led by a middle-class élite, have generally favoured a relatively Europeanized style for their songs, as is the case with 'Nkosi Sikelel iAfrika' (God Bless Africa), anthem of the African National Congress. Even here, however, things are not so simple. John Knox Bokwe's influential hymns (published 1876–84), adaptations of compositions by the early Xhosa Christian convert, Ntsikana Gaba, display an admixture of European harmony and stolid rhythm with the call-and-response structure and parallel voice movements typical of indigenous Xhosa music. The political significance of these songs, and of their style – which became the foundation of *makwaya* ('choir') music, itself the basis of subsequent

middle-class political song – lies wholly in the associations of the music; the verbal texts are religious. This musical style was taken over (along with many others) by working-class organizations too, and, mediated by later developments such as the *ingom'ebusuku* choral music of the 1920s and 30s, it formed an important ingredient of a trade union and workers' choir repertory which is still vigorously alive.

I hope you can see, then, that the interrelations of words and music are complex. But what we can say is that, for oppressed people, simply to own a music which is felt to be one's own is already a gesture of self-assertion. To create an *alternative* to what is given is a political act, an impulse by people 'to gain control of their national culture and to use it to regain control of their individual and national lives' (Coplan, 1985, p.1). This is certainly clear in the history of South African township music, as we shall see in the next section.

19.3 Urbanization: the example of South African township music

In recent decades, throughout the world, vast migrations to explosively expanding cities has forced on people new ways of life, marked by unfamiliar economic relationships, a novel social fluidity and an increase in cultural choices. What are the consequences for music? The answer, you might assume, is obvious. Surely one would expect a shift from 'traditional' to 'modern', and from indigenous to Westernized practices. Moreover, as the mass media disseminate the new practices everywhere, an urban cultural hegemony is established. I think, though, that these assumptions are over-simple. Certainly, all music forms change under the impact of urbanization. And, looking at South Africa as an example, Rycroft's assertion that 'musical taste in town is closely correlated with economic, social and educational standing' (1957, p.35) undoubtedly identifies a general tendency. But these changes do not take place automatically, homogeneously or in a generalizable manner; nor do cultural distinctions necessarily correlate in a neat way with other categories (rural/urban, traditional/modern, indigenous/ Westernized). For example, older cultural values may act as a 'brake' on the impact of new patterns. Or adaptations (of old to new and vice versa) may work by helping to *manage* new conditions. In such situations, musicians may function as 'cultural brokers', 're-establishing bases of social communication and order' and supplying 'metaphors of cultural identity and change' (Coplan, 1982, pp.120, 121). These processes are mediated by differences in *social class*, in *social conditions* (in South Africa, between, for instance, temporary migrant and settled townsman, between Christian and pagan), in *race* and *ethnic group*, and in *political outlook*. Larger scale processes thus take variable forms in specific contexts.

Because of South Africa's relatively long history of urbanization, and because white political policies have ensured that contacts between city and country, 'modern world' and 'tribal homeland', have remained close, excellent examples of the kind of processes I am talking about can be found there – tendencies that are widespread in the Third World are here writ large. Nineteenth-century urban settlements in the Cape saw a multitude of acculturated styles. Carried by itinerant Cape Coloured and Xhosa musicians, these were fed into an even richer mix in the Kimberley and Johannesburg areas when, in the 1870s and 80s, gold and diamond mining began, revolutionizing the South African economy. Zulu, Sotho, Tswana, Chopi and Venda rhythms, tunes and polyphonic structures were adapted to cheap concertinas and homemade guitars and fiddles for the entertainment of migrant workers, and were crossed with European hymn and folk tune, minstrel songs learned from visiting Afro-Americans, and syncopating brass bands. Many migrants hung on to 'neo-traditional' variants of this acculturated repertory, and, in essentials, these could still be heard decades later, especially from street musicians (Rycroft, 1959, 1977). It was often these styles that were taken back to the rural areas, mediating urban tendencies and preserving 'old-fashioned' approaches after more progressive urban taste had abandoned them. Thus the Zulu *isishameni* dance style and its music, which developed in rural Natal in the 1930s and 40s, evolved from older dances through the influence of the hybrid styles which local men had encountered on their periodic migrant labour visits to Durban and Johannesburg. Similarly, the constant interaction of urban and rural influences in the evolution of the working-class town style of choir singing, *ingom'ebusuku* (associated above all, as it was, with Zulu migrants), demonstrates that 'music and performance are part of a complex network of production and reproduction that spans town and countryside' (Erlman, 1990, p.201) (Figure 19.6).

Between 1921 and 1936, the black urban population doubled, to 1 252 000 (Dubb, 1974, p.442), and many of the working-class inhabitants of the townships and slums now regarded themselves as permanent city dwellers (even if the white authorities did not agree). The music and dance culture, which developed for their entertainment and to ease their otherwise dreadful lot, was called *marabi* and was associated with *shebeens* (illegal drinking houses) and *stokfel* meetings (*stokfels* are credit clubs closely related to mutual assistance principles common in the rural African economy). The music drew on a rich, dynamically evolving and variable mixture of sources: existing acculturated hybrids; rural black song; American ragtime, jazz and swing, freely available in South Africa on imported records; brass band music; and the already popular choral style fusing *makwaya* and ragtime, associated above all with the middle-class composer, educationalist and conductor, Reuben T. Caluza. *Marabi* was basically a keyboard music, but bands including brass, guitars and percussion were also characteristic. *Shebeen* culture was steamy, erotic, often violent; gangsters were prominent. The role of lumpen groups, *abaqhafi* ('careless livers'; urban drifters with allegiance neither to traditional African nor Westernized culture), in the *shebeens* and in the supply of musicians was important. The same pattern occurred in the development of jazz and blues, and in similar situations elsewhere in the world (Argentina and the development of tango, for example). Such people, moulded by the city environment, are culturally mobile, free to develop new forms, offering models to less uninhibited groups.

No recordings were made of *marabi*, though 'higher-class' variants by more up-market jazz bands exist, and pianist Dollar Brand's later

Figure 19.6   Musical forms have been diverse in southern Africa: a township scene from the 1940s.

evocation of the style, *Mannenburg,* gives us some idea. The taste of many middle-class blacks ran to Western-style concert and choral music, but Caluza's 'ragtime' was popular with many of them, and by the 1930s jazz bands were too. *Marabi* was looked down upon: 'The 'marabi' dances and concerts and the terrible 'jazz' music banged and wailed out of the doors of the foul-smelling so-called halls are far from representing real African taste', as one critic insisted (quoted in Coplan, 1985, p.109).

Despite these differences, however, both working-class and middle-class blacks responded to types of 'ragtime' and 'jazz'. For both groups they represented city as opposed to country life and the 'modern' strategy of a pan-tribal culture. Similarly, working-class *ingom'ebusuku* absorbed the influence of Caluza, just as *marabi* percolated into the music of jazz and dance bands like the Jazz Maniacs. There were no clear, separate 'class cultures'. Rather (as is usually the case), a range of elements, some held in common, were selectively moulded into a variety of *formations* related to the specific circumstances of different groups.

Here, class was notably mediated by ethnicity. The middle-class strategy of 'advance' through 'Westernization', closed off by white intransigence, was diverted into the quest for a progressive (i.e. non-tribal, 'modern') African National Culture, which was bound to draw on many of the same elements as workers' music. At the same time, working people, looking to locate themselves as urbanites in contrast to 'backward' rural folk, found many tendencies in middle-class music suitable for adaptation to their needs.

As the slums were cleared, the focus of black culture shifted to the fast-growing western areas of Johannesburg, notably Sophiatown. Here in the 1940s and 50s, there was something of a cultural 'golden age' – albeit marked, still, by old tensions between classes, between city slickness and the older fashions of rural migrants, between the desire to create a modern indigenous culture and the appeal of European and Afro-American models. Sophiatown jazz acquired an international reputation, climaxing in the emergence of such stars as saxophonist Kippie Moeketsi and trumpeter Hugh Masekela, both members of the celebrated Jazz Epistles (along with pianist Dollar Brand from Cape Town). American-styled dance music and black variety and musical theatre had broad appeal, and also threw up an international success in the musical *King Kong* (1959). Professionalism, individual ambition and an international as against local perspective were inevitable developments. So, too, was the growing involvement of white entertainment impresarios and of the white-owned local record industry; exploitation of black performers and composers was rife.

Working-class people, to some degree left outside this glittering scene, created a new dance form to follow *marabi*, called *tsaba-tsaba*. But this too produced an international hit: August Musururgwa's 'Skokiaan' (1947, US hit as 'Happy Africa', 1954). Similarly, the continuing *ingom'ebusuku* style, often known as 'bombing' during the 1940s (in description of the characteristic shouts), was also christened *mbube* after the success of Solomon Linda's song of that title with his Original Evening Birds (*c.*1939) (Figure 19.7); this was later a success for Pete Seeger in the West under the title 'Wimoweh'. *Kwela*, originally a street music played on penny whistles and homemade guitars, basses and percussion, also achieved international success in the 1950s (notably through

Elias Lerole, Spokes Mashiyane and Lemmy 'Special' Mabaso) with its blend of jazz and local dance rhythm and improvisatory, quasi-vocal melody. In all these styles we can hear the characteristic elements of what has become a recognizable twentieth-century hybrid musical practice:

- moderately complex rhythmic patterns;

- simple chord sequences endlessly repeated (often a 'merciless three-chord vamp', as Rycroft has put it);

- short-phrased, repetitive but highly ornamented melody;

- call-and-response structures;

- staggered voice entries in multi-voice textures.

All these can be traced back to rural roots.

As Coplan notes (1985, p.139), 'Performers who wished to appeal to the large African working-class audience had to draw upon indigenous performance culture. Nevertheless, the internationalism of African jazz became part of a struggle against cultural isolation and segregation and expressed the aspirations of the majority of urban Africans.' Many jazz and dance bands and vocal groups concentrated on covering American

Figure 19.7 Solomon Linda (left) and the Evening Birds, in 1941.

material and styles; some, though (such as the Shantytown Sextet and the Manhattan Brothers, featuring the young singer Miriam Makeba), absorbed aspects of *tsaba-tsaba* and *kwela* into their music. The most 'homely' and accessible of such music was named *mbaqanga* ('steamed mealie bread', i.e. everyday music) or *msakazo* ('broadcast'). It was studio assembled for the mass record market and radio programming. After the destruction of Sophiatown (late 1950s) and the Sharpeville massacre (1960), and with the tightening of *apartheid,* the jazz scene was crippled. *Mbaqanga*, along with imported and assimilated soul, disco and funk, became the dominant styles catering for a more restrictive cultural situation. The collision between mass musical production and the pull of international trends, on the one hand, and the force of local traditions, on the other, set an agenda which is now duplicated worldwide.

19.4   The world music industry

In the 1960s and 70s, the record industry underwent a prodigious expansion. In many countries, in both First and Third Worlds, record and tape sales increased three or fourfold or even more. For example, in Mexico, 1967 sales were 14.8 million, 1980 sales 82.6 million; in Brazil 6.9 million in 1965 went up to 56 million in 1980; in South Africa a 1965 figure of 5.5 million increased to 13.3 million in 1981 (IFPI, 1981; Gronow, 1983). The industry was now a vital part of musical life virtually everywhere. In the 1980s, under the impact of recession and economic crisis, the increase flattened out, but activity remains high. During the same period, a process of takeover and conglomeration resulted in the appearance of five multinational corporations, based in the USA, Japan and Western Europe (Sony-CBS, EMI, WEA, BMG and Polygram), which between them dominate the world industry. Through their operations, and those of their subsidiaries and licensees, successive waves of Western music (rock, soul, disco) have been marketed with considerable success throughout the world. New technology made production and consumption cheaper and

more flexible, and the emerging world market made huge profits from 'mega-sellers' a possibility. But *local* Third World industries grew too. Cassettes – much easier and cheaper to produce and distribute than records – took on a central role in poor countries. Strategies have been evolved to control the importation of foreign music. Pirate cassette producers (with between 50% and 90% of the tape market in many Third World countries; IFPI, 1989, pp.58–60) can undercut major companies. While hurting musicians as well as companies, this can also have the effect of freeing up markets, preventing big-industry dominance, and subverting the influence of 'official' cultural taste (e.g. that of state broadcasting companies). Under these conditions, local 'grassroots' music styles, dependent on the mass media but derived from indigenous rather than (or as well as) Western roots, continued to develop: West Indian reggae, salsa and *soca*, Chilean *nueva canción*, many modernized forms of South Asian folk music, and Indonesian *jaipongan* are examples of 'modern' music which challenge the hegemony of the emerging global mainstream style.

The 'cultural imperialism' thesis is too simple for this situation. While statistics are incomplete and often untrustworthy, it seems likely that the multinationals' world market share is actually rather less than 50% (Laing, 1986, p.334). Their presence is variable. In some places (West Indies, Indonesia, Sri Lanka) it is small or even non-existent, while in others it is overwhelming: for instance, EMI dominates East Africa and CBS dominates Chile; in some cases, a multinational shares the market with a dominant local company (e.g. CBS and Gallo in South Africa). Music industry growth can lead not to a multinational but to a local hegemony – as is the case with Indian film music, whose tentacles stretch to South-east Asia, the Middle East, Africa and even Eastern Europe. Large firms often produce local music if the market justifies it, and musicians sometimes prefer them because, however rapacious, they are usually bureaucratically efficient and legalistically honest, while small outfits may flout regulations, avoid paperwork and cheat them with impunity. Conversely, much of the pirate repertory is actually international, since existing hits carry less marketing risk. In any case, it is by no means universally accepted that a local music policy is progressive: one Kenyan listener in 1980 who stated that 'we should promote even the meagre remnants of our shattered culture as much as possible' was accused by another of 'having fallen prey to Ayatollah Khomeini's sermon that Western music is like opium which was to be banished. To some of us, Western music is the only form of entertainment' (Wallis & Malm, 1984, p.259).

It is a moot point (a) whether *mbaqanga* star Mahlathini (Figure 19.8) is best described as a guardian of authentic black culture and its rural roots—his visual image, dance style and 'groaning' deep bass voice, the call-and-response patterns of the music and the lyrics' traditional references suggest so; or (b) whether he is a product of a manipulative assembly-line system designed to satisfy easily identified needs and fit stereotypes of African tribalism: 'It was pure exploitation. There'd be hundreds of groups churning the stuff out. The studios would do eighteen groups in eight hours,' according to one musician (in Stapleton & May, 1989, p.190; see also Prince, 1989, pp.102–4). Similarly, top *mbube* group Ladysmith Black Mambazo, with their short-phrase, often parallel chording, soulful glides and quasi-improvisatory rhythmic flexibility, rouse associations going deep into black music history and appeal especially to semi-urbanized workers. But they have been criticized for the often idyllic, religious or rural themes of their lyrics and their readiness to accept radio censorship – not to mention the wealth their prolific output has brought them and, latterly, their pitch for success in the West, which risks cutting them off from the audience at home.

To some extent, differing views of these performers correlate with differences in class and generation. Working-class listeners, especially the middle-aged and elderly, and especially if unschooled, only semi-urbanized, migrant or rural, are more likely to respond positively to Mahlathini and Ladysmith Black Mambazo than middle-class people and the city-wise, politically aware youth who came to the fore after the Soweto riots of 1976. To the latter groups, rock, soul, disco and reggae are 'hip', 'Michael Jackson is more of a role model than Mahlathini' (Prince, 1989, p.100), and local imitators such as Brenda Fassie and Steve Kekana are the new stars. But there is something of a 'roots revival' too, and the conflicts must be regarded as ideological and political, rather than purely sociological. The large music industry corporations, whose economic interest lies precisely in being all things to all men (or at least all consumers), readily exploit such conflicts, implicitly or explicitly selling different images to different groups.

Both Mahlathini and Ladysmith Black Mambazo were produced by the local (but large) South African company Gallo. Commercial imperatives, for them, are little different from those affecting the multinationals. And, for all the qualifications mentioned above, there is no doubt that these have deleterious effects:

- There is a constant pressure to maximize markets: financial support goes preponderantly to 'international' rather than local repertory, or to local artists with 'crossover' potential (in South Africa, Brenda Fassie, Steve Kekana and the like), and this results in a tendency towards stylistic homogenization.

- The large companies, especially the multinationals, do seek to dominate.

- Musicians are exploited: royalties have been avoided where possible, with small, one-off fees paid for studio sessions; recordings can be bought up cheaply, then released on the world market without the musicians knowing; and there is a long history of Third World compositions being copyrighted by Western companies, musicians and producers, or of fees not finding their way back to Third World copyright holders.

- Huge imbalances in music technology, recording facilities and market potential often drive musicians to London, Paris or the USA, either to record or to work permanently; obviously cultural roots are at risk. This affects political exiles too. Some would say that the world success of Miriam Makeba, for example, came with an easily accessible, rather bland 'international folk' style, and that even South African material such as the famous 'Click Song' is given inappropriately glossy and inflated recording production.

Figure 19.8   Mahlathini performing with the Mahotella Queens.

International success, then, tends to come on the industry's terms. The Zimbabwean Bhundu Boys tried a more grassroots strategy. Settling in Britain in 1986, they built a reputation on the basis of an exhausting programme of local gigs together with the support of cult disc jockeys like John Peel and Andy Kershaw, rather than a large promotional budget. This took their *Shabini* album to the top of the 'independent' charts in 1987. Their *jiti* style (a young people's dance music in Zimbabwe, fusing local rhythms and tunes with the interactive guitar duetting of Zaïrean *soukous*) would seem to have been tailor-made for the dance music boom then sweeping Britain. Yet they never broke through to a mass audience, and despite a record contract with WEA, subsequent albums moved towards a more mainstream style which achieved little commercial success and risked losing their Zimbabwean listeners. According to fellow Zimbabwean musician, John Chibadura, 'Two years ago there was magic in these guys' music, but their style has changed now and it is not good to me. Back home the Zimbabwean people do not buy the Bhundu Boys' music' (quoted in *The Guardian*, 7 June 1990).

Most of the issues raised in this section are given a sharp focus by the *Graceland* affair. Paul Simon's phenomenally successful 1986 album (on WEA) and the subsequent world tour he organized drew together musicians and musical materials from the USA and from Africa (preponderantly South Africa). 'Collaboration' was a key factor in the appeal. Simon used compositions, performers and recordings with a variety of origins, which were then mixed together in New York; some lyrics even mix English and Zulu, and many tracks fuse aspects of different genres and styles (Figure 19.9). However, Simon clearly dominates, both musically and commercially (he takes the lion's share of royalties), and despite the 'liberal' image of the project, there is no overt political content in the songs.

Controversy surrounded *Graceland* from the start. Many white South Africans, including the government, approved of it because it could be represented as doing an ambassadorial job for the country and the existing social relations. Many black South Africans approved too, liking the black/white collaboration, the international links and the world exposure for their music. But others objected, because the record broke the United Nations cultural boycott of South Africa or because they saw Simon as stealing their culture. Some listeners – black and white – liked the way Simon had 'cleaned up' his 'ethnic' sources, feeling this brought the music into the modern world; others regarded the refined production as an example of the exploitation, corruption and commodification of authentic musical roots. The reputations – and incomes – of some of the performers involved (notably Ladysmith Black Mambazo) benefited, as record contracts and foreign tours were arranged; but some criticized 'the perpetuation of popular stereotypes, e.g. that Africa provides the rhythm section, the body of the pop music world, while Europe provides the melody, the head…that it is the Black man's job to help the White man do his thing…Is African music only good for backings, not frontings?' (Tracey, 1987). The most radical view was probably that the entire project was designed to boost a flagging career (Simon's) and fits into a history in which he had already exploited Afro-American, Peruvian and Cajun music and reggae (Hamm, 1989). Subsequently (1990), Simon applied the *Graceland* approach again, on his *Rhythms of the Saints* album, using musicians from Bahia in Brazil.

Figure 19.9   Ladysmith Black Mambazo performing a song from the Graceland album together with Paul Simon.

Meintjes (1990) argues that the politics of *Graceland* are not secondary but are embedded in the music, and that the varied responses to this exceptionally multivalent text correlate with listeners' existing attitudes to cultural identity and socio-political relationships. 'The political is simultaneously articulated with the musical in the international market to create a transnational flow of meaning...*Graceland* illustrates that the meanings of transcultural musical styles are located at the conjunction between the multilevelled global economic and political system and the local lived experience of specific creators and interpreters' (Meintjes, 1990, p.69).

Summary and conclusions

Your own view of *Graceland* might well provide a useful indication of your attitude to 'music development' issues. In a field of such complexity and ambiguities, definite conclusions about the current situation are hard to come by. We can say that Third World musics are affected by many of the same factors as those in the First World: rapid change, threats to existing styles, the industrialization, commodification and internationalization of cultural forms, and Americanization – but these appear unevenly and in specific and varied forms. Beyond this are factors – peculiar by their nature or by degree – to developing countries, principally:

1   the dominance of the rich countries, particularly the multinational music corporations, in the world cultural economy;

2   the effects of rapid urbanization, and, conversely, of the overlaying of the electronic media on societies with a large rural sector;

3   the significance of cultural identity ('roots'), both *within* developing societies and *to* Western consumers, and, conversely, the attractions of 'Westernization';

4   the closeness of music and politics, for example, in relation to class structure and conflict, to nationalist ideologies and movements, to censorship and state repression.

Perhaps, finally, the overriding issue is that of control. In the words of South African musician Ray Phiri:

"Who's fooling a who?

who's fooling a me?

are you fooling a me or am I fooling a you?

who's using a who?

who's using a me?

are you using a me, baby, or am I using a you?"

(from the album, *Look, Listen and Decide* by Stimela, quoted in Meintjes, 1990, p.69)

20

PROSPECTS AND DILEMMAS FOR INDUSTRIALIZING NATIONS

TIM ALLEN

Figure 20.1   Petro-chemical plant on the South Korean coast.

This book has been ambitious: its aim has been to serve as an introduction to development in the 1990s. I hope that reading it has proved thought provoking and enlightening. But it would be foolish to suggest in the final chapter that this book has achieved an all-encompassing overview. Many important issues have been left out or barely touched upon.

20.1　What has been left out?

- Two thirds of the globe's surface is covered with water, but this book has had nothing to say about the increasing exploitation of the oceans.

- The 1990s are the United Nations Decade for Natural Disaster Reduction, but the authors have made no assessment of efforts to mitigate the terrible effects of earthquakes, floods and cyclones.

- There have been comments on the roles of the World Bank, the IMF and some of the UN institutions, but almost nothing has been said about country-to-country aid, nor about the work of NGOs like Oxfam and Save the Children Fund.

- Some chapters have made references to the crucial issues of war, education, human rights, religion, urbanization and nationalism, but these topics would each merit a chapter to themselves.

It is likely that many readers will be surprised by such omissions, and it is almost certain that someone looking into this book early in the next century will smile knowingly at what the authors have chosen to emphasize, and remark on their lack of foresight. Doubtless many things this book has ignored will seem overwhelmingly important in the years to come.

Of course the choice of what to include has been constrained by space. This is already a long book. Several of the omissions on my list are dealt with in the other volumes of this series, and a halt had to be called somewhere. In any case, it is my belief that the underlying messages of these chapters are widely applicable. Even if it becomes clear that an author has been wrong about a particular topic, the approach should remain valuable, and should help you take part in virtually any discussion about development. What has been argued here about poverty, colonialism, the environment, population growth, employment, gender and commoditization will enable you to put into

context the simplistic views regularly propagated by charities, politicians and the media. It will be apparent to you that there are no easy solutions to development problems, and that an understanding of what is happening in a particular place requires a balancing of different kinds of information and a combination of various modes of analysis. Studying development necessitates a multi-faceted and interdisciplinary perspective. It is not adequate to keep, say, an economic or technical analysis at the centre of our thinking, and just add a bit of anthropology, ecology, political science or history. We have to question our frames of reference, and re-examine our assumptions.

It is because of the desire to discuss a variety of approaches to development, and to highlight the interconnections between them, that this book has underemphasized something else. Many analysts would regard industrialization as *the* fundamental issue of development studies. It has frequently been mentioned and attempts at promoting it discussed (such as the state socialist model described in Chapter 12), but the debates about it have not been tackled directly. To do so in any depth would have required a review of theories that have tended to be based on the 'economistic' presuppositions critiqued in Chapter 17. To have adequately engaged with these arguments here would have narrowed the focus of the book, and it was decided that the topic needed a volume to itself (*Hewitt et al., 1992*). However, it is implicit in much of what has been written in this book that rethinking industrialization is likely to be the greatest challenge for both development analysts and practitioners in the 1990s. It therefore seems appropriate to end by posing the following question.

Q Is industrialization essential for development?

In the next section, I present the case for an affirmative answer drawn from the work of Gavin Kitching. Then in Section 20.3, I comment on his views, taking into account various insights and arguments which have been put forward by the authors of this book, and asking what the options are for the 1990s.

20.2 The necessity to industrialize

Not long ago, to have questioned the need to industrialize would have been considered strange. Indeed, 'modernization', 'development' and 'industrialization' have sometimes been treated as synonymous concepts. This is not to suggest that the benefits of heavy industry have been universally applauded. There has in fact been a long tradition of trying to improve living standards by other means, such as by improved agricultural production and 'alternative technology'. Recently these efforts have been encouraged by the decline of heavy industry in some countries such as Britain, and by an increased recognition of the appalling environmental effects of rapid industrialization in, for example, the USSR and India. Nevertheless, the view that development is impossible without industrialization remains a potent one. It is an argument that links together Marxist and classical economic theory with most neo-classical and development economics. In the terms of Chapter 6, it links neo-liberal, structuralist and most interventionist views. It has been put succinctly and compellingly by Gavin Kitching in *Development and Underdevelopment in Historical Perspective* (1982; 1989). Much of the following draws from his exposition.

The conventional wisdom

Kitching attacks those who he believes look at the world through rose-tinted spectacles. For him economic development is a long-term process of structural change, which poses awesome moral and political dilemmas. To confront these dilemmas adequately requires hard and informed thinking and, for the policy makers themselves, considerable courage and self-discipline. He starts by explaining that there is an established orthodoxy that 'if you want to develop you must industrialize', and then proceeds to defend this contention by highlighting inadequacies in alternative approaches, particularly those put forward by populists (see Chapter 6). An important part of his argument is that people who suggest that development can be made more palatable by means of rural orientated, grassroots strategies are not

facing facts, and that there is simply no viable alternative to what he calls 'the old orthodoxy'. There is no room here for a full review of his detailed critique of populism, so I will mainly overview the arguments he presents for why industrialization is necessary (rather than why other strategies do not work).

Kitching accepts that, just because things have tended to happen in a certain way does not preclude the possibility that a viable alternative may emerge, but points out that if there is a lesson for poor countries to learn from the rich ones, it would seem to be that industrialization is the key to affluence. In countries where per capita incomes have risen, the importance of industry in their economies has usually increased, while the importance of agriculture has diminished. This seems to be clearly demonstrated by the figures he presents (Table 20.1 reproduces Table 1 from his revised edition of 1989), and he allows these data to stand for themselves without additional comment (I will have something to say about them in Section 20.3).

Kitching then turns his attention to the influential theoretical argument which suggests that there is a limit to the levels of prosperity forthcoming from agricultural production alone. According to Kitching, this requires the construction of an ideal model of a *closed* economy (that is, not open to foreign trade or investment, etc.), made up of small-scale farmers who do not export to or import from any other economy. Once this is done, certain evolutionary assumptions might be made. At first the farmers will produce mainly for their own consumption, and will therefore grow a wide variety of food crops. Then they will begin to specialize in growing crops suited to their particular area. As their skills improve, they will be able to produce ever more crops from the same total land area, possibly even with a decreasing input of labour. They will be able to exchange more and more food with their neighbours, which means that their individual incomes and their total income will grow along with their total output. But the process has a definite limit. Thus the requirements for food will no longer grow as fast as the capacity to produce and exchange it. Once this point has been

Table 20.1 Share of agriculture and industry in the economies of selected developed and underdeveloped countries c.1800–1985 and their per capita GNP in 1985 (from Kitching, 1989, Table 1, p.7)

Country	Share of agriculture (%)	Share of industry (%)	GNP per capita 1985 ($)
UK			
1801	32	23	
1901	6	40	8460
France			
1835	50	25	
1962	9	52	9540
Germany			
1860	32	24	
1959	7	52	10 940
USA			
1869	20	33	
1963	4	43	16 690
Japan			
1878	63	16	
1962	26	49	11 300
USSR			
1928	49	28	
1958	22	58	4550[a]
Bangladesh			
1960	61	8	
1985	57	14	150
Kenya			
1960	38	18	
1985	41	20	290
Thailand			
1960	40	19	
1985	27	30	800
Bolivia			
1960	26	25	
1985	17	30	470
Côte d'Ivoire			
1960	43	14	
1985	21	26	660
Turkey			
1960	41	21	
1985	27	35	1080

Source: Kitching, G. (1989) *Development and Underdevelopment in Historical Perspective*, Routledge, London, using data from Kuznets, S. (1966) *Modern Economic Growth: rate, structure and spread*, Yale University Press, New Haven, Table 3.1, pp.88–92; and World Bank (1987) *World Development Report 1987*, Oxford University Press, Oxford, Tables 1 and 3, pp.202–7.

[a]1980 figure from World Bank (1982) *World Development Report 1982*, Oxford University Press, Oxford, Table 1, p.111.

reached, the farmers will want to exchange their surplus food for something else. They will want to obtain clothing, footwear, better housing and other manufactured goods. But this will only be possible if there is some sort of industry making these things.

A logical progression of this kind may also be interpreted as indicating that an agricultural surplus is a precondition for the emergence of producers of non-agricultural commodities.

> "For the emergence of such a surplus makes it possible for some people to give up subsistence agriculture entirely and trade non-agricultural products for food, and at the same time it enables an 'effective demand' for these goods to emerge, i.e. it creates a 'surplus' of food which can be exchanged for such goods (and indeed for non-material services such as are provided by priests or government officials)."

> (Kitching, 1982;1989, p.8)

The next step is to assess how far this logical model corresponds to the real world.

Refining the arguments

There are several ways in which so crudely sketched an exposition of the conventional wisdom fails to correspond with what actually happens. But Kitching maintains that the model can be made more subtle in order to incorporate such apparent weaknesses. Here I highlight some of the queries that can be raised, and summarize Kitching's response to each of them in turn.

- Agricultural production is not restricted to food

This is obviously true. Vegetable fibres, oil seeds, wool and cotton are agricultural products as well as being industrial raw materials. When non-agricultural production begins, it may produce an increased demand for such products, and this may replace the slowing demand for extra food. But these are not grounds for adopting non-industrial development strategies, because demand for agricultural raw materials implies the presence of industry (or at least of small-scale non-agricultural production).

- Agricultural production can itself be industrialized, and this makes it problematic to separate the concepts of 'agriculture' and 'industry'

It is certainly the case that agriculture can be industrialized. This involves the enlargement of the scale of production on estates and plantations, the increased use of machinery, and the employment of wage labour by profit-maximizing enterprises. Nevertheless, it makes sense to maintain the distinction between 'industry' and 'agriculture', for reasons which have been emphasized by many agrarian populists.

There are two major differences between agricultural and non-agricultural production.

> "First, the environment of agricultural production tends to be less controllable because of the variability of the weather and (to a lesser extent) of soils and pests. Second, a number of crucial operations in agriculture (particularly harvesting and weeding) tend to be technically difficult to mechanize effectively. As a result it is difficult to obtain the degree of 'capital intensity' of production in agriculture which can be obtained in a lot of industry."

> (ibid., p.13)

A consequence of these factors is that small-scale farming can sometimes compete effectively with industrialized agriculture, particularly when people are prepared to work very long hours for very low remuneration. The production of coffee, tea and pyrethrum by Kenyan peasants is a good contemporary example. Thus populists are sometimes on firmer ground when arguing about the viability of smallholder agricultural production than they are when assessing the viability of small-scale non-agricultural production or 'cottage industries'. This needs to be recognized, but it does not add up to a general argument against industrialization. After all, the markets which allow peasant production to survive are to a large extent urban centred. Moreover, it is always possible that an innovation may break through the technical barrier, making peasant production uncompetitive. This happened in the late nineteenth century, when the introduction of the combine harvester in North America, along with other innovations in transportation, made a great deal of peasant wheat production in Europe non-viable.

• Economies are not closed

The theoretical argument about the limits to agricultural production assumed that the economy made up of small-scale farmers was closed. But of course in the real world there are a multitude of economies (mainly nation states) with trade links between them.

An implication of this is that an agricultural economy may grow by means of the export of food and/or agricultural raw materials to other economies. If the exports are raw materials for industry then there may be continually rising output and incomes in the agricultural economy as a result of continually rising demand in the industrialized or industrializing economies. Well-known examples are the economic development of Denmark, New Zealand and Australia in the late nineteenth century. In these countries economic growth was initially based on the export of meat and dairy products to the rapidly growing industrial economies of Europe. More recently, a similar situation has evolved in Africa and South and South-east Asia, where development has mainly been based on the export of food and agricultural raw materials to the industrialized economies of the West. Thus, continuously rising output and income may occur in an agricultural economy provided that there are industrial (or at least non-agricultural) economies *somewhere* to provide continually expanding markets.

Refining the theoretical argument along these lines takes on board the possibility that development strategies might be oriented towards agriculture in some countries. But in the long run this rarely proves to be the case (even when smallholder agriculture remains an economically viable proposition), because there is an 'economo-political' logic which prompts the rulers of agricultural economies to adopt programmes of national industrialization. Global competition between states causes a conceptual linking of an industrial base with national independence. National economic policies therefore aim at decreasing dependence on foreign suppliers of industrial goods

by encouraging manufacture within the country (this is one of the basic principles of import substituting industrialization, discussed in Chapter 11).

• Industry is something distinct from the small-scale production of non-agricultural commodities

The conventional theoretical argument about industrialization undoubtedly uses the term 'industry' in a misleadingly loose way. In effect, it is being used as a residual category, to refer to all kinds of non-agricultural production. Populists have argued that this side-steps an important issue: the difference between 'industry' and 'small-scale non-agricultural production' or 'cottage industries'. Nowadays the distinction is commonly ignored, but it was important for nineteenth-century thinkers, and remains useful because it concentrates attention on the ways in which small-scale production becomes large scale and capital intensive.

When Marx and others wrote about small-scale non-agricultural production, what they usually had in mind was a form of manufacturing with the machinery owned by its operators, and the labour being provided by the family. In contrast, the machinery and raw materials for industry were owned by a capitalist, work took place in a factory rather than the homes of producers, and the labourers depended on wages paid by the capitalist for their livelihood. The change from small-scale non-agricultural production to industry therefore involved fundamental shifts in social relations, as well as a change of scale. Industrial forms of non-agricultural commodity production required fixed capital (buildings and machinery) and the employment of much labour. It also meant the concentration of wealth in the hands of industrial capitalists, and the destruction of much of the earlier household production which was unable to compete. There was also a tendency for capital-intensive industries to become spatially concentrated in towns, and for peasants and former artisans to migrate to such towns to seek work as propertyless proletarians.

The debate between those who hold with the 'old orthodoxy' on industrialization and the populists is not about the need for small-scale non-agricultural production, but about industry. Nineteenth-century populism was no simplistic defence of agricultural development. It was a critique of the social and spatial concentration of production, income and power associated with industrialization. Against it was juxtaposed an ideal of a society of small-scale agricultural and non-agricultural producers living in villages or, at most, in small towns.

Similarly, modern populists (neo-populists) tend to bemoan the effects of industrial growth, while promoting the development of non-agricultural production in conjunction with family farming. The issue is how to respond to the demand for non-agricultural commodities. On the one hand, those who think that industrialization is necessary for development point to 'economies of scale' (discussed below), and deny that small-scale production can transform income levels. On the other, populists question the benefits of industrial development. They argue for policies aimed at reversing or avoiding the industrialization process, and promoting social and economic equality. It is easy to feel sympathetic to their position, but the trouble is that they cannot provide good examples of where their schemes have worked. Up to now, success has been possible only for relatively short periods, in very particular locations, and usually with considerable amounts of outside support.

- Bigger is not always better

Accepting this does not necessarily lead to the populist view that 'small is beautiful'. It simply recognizes that increasing the size and concentration of production may not always lead to 'economies of scale'. 'Economies of scale' is the basic idea behind the development of mass production. The term is used to indicate a situation where the long-run average costs of production fall, because the expansion of production causes total production costs to increase less than proportionally with output. This is crucial when an industrial process requires large amounts of fixed capital (plant and machinery).

"Up to a certain size of operation the volume of output from that fixed capital (we might think of an example such as the production of strip steel) grows proportionately with the size of the investment. But beyond a certain point, which varies with the technology being employed, the volume of output grows more than proportionately to the capital investment required to produce it – hence the cost in terms of fixed capital of each unit of output falls. Since highly 'capital-intensive' industrial technologies tend also to reduce the amount of labour employed per unit of output as the scale of production grows, then, all other things being equal, the enterprise using such techniques stands to reduce all costs per unit of output and thus to gain more profit per unit of output as output rises, and hence more profit overall."

(Kitching, 1982; 1989, p.12)

The likelihood of 'economies of scale' occurring varies from one place to another, and from one product to another. It has already been noted that in some countries industrialized agriculture is less efficient than labour-intensive, peasant farming. Similarly, for some non-agricultural products small production units may be more efficient.

"If, for example, the industrial process involved is a very complex one in which it is technically difficult to design or utilize machinery to replace human labour, and if, in addition, there is for some reason an abundance of labour seeking employment and wages are low, then it may be more profitable for an enterprise to continue to operate in small units and to increase production by multiplying the units rather than by enlarging the scale of production in big plants. This has been the case in the production of electronic and optical equipment in Japan and other parts of South-east Asia."

(ibid., pp.12–13)

But it cannot be concluded from this that the populist ideal of rural-based, small-scale, non-agricultural production is feasible. Across a wide range of industries (and particularly heavy 'producer goods' industries like iron and steel, chemicals and cement manufacture) the world-wide tendency from the nineteenth century onwards has been towards a larger and larger scale of production. This has usually been the cheapest means of producing the commodities required for material affluence. At the same time there has been a tendency towards urbanization, with factories cutting costs by sharing essential public facilities (roads, sewage, water and energy), and taking advantage of the large markets afforded by dense populations.

It is widely recognized that life is grim for poor factory workers and their families who are caught up in these processes, living in the urban slums of industrializing countries. They are among the 'wretched of the earth' – those people who, if the world has progressed, have certainly been the victims of that progress. Nevertheless, it is hardly likely that 'grassroots', rural development strategies are going to do anything to help them, nor to do much to discourage migration out of the countryside. Where economic development occurs, there have always been many losers. Sadly there are no panaceas.

Kitching's view

Thus, Kitching's defence of the 'old orthodoxy' on industrialization does not simply dismiss populism out of hand, but keeps coming back to the issue of what has actually happened where economic affluence has been achieved. In essence his contention is that, while industrialization is a grim business, there is no real alternative. Populist strategies are flawed by wishful thinking. His arguments may be summarized as follows:

- Empirical evidence indicates that high living standards are linked to industrialization.

- A widely accepted theoretical argument suggests that agricultural development will lead to a demand for manufactured products.

- While it is possible that manufactured products might be purchased from abroad by means of the export of agricultural produce (including non-food produce), this is only possible for some countries (with surplus farm land), and even in these cases various political and economic processes will eventually encourage the development of national industries.

- Although agriculture can be industrialized, it is difficult to obtain the degree of 'capital intensity' which can be obtained in a lot of industry. This means that small-scale farming can sometimes compete with industrialized agriculture.

- It is useful to make a distinction between 'industry' and small-scale non-agricultural production. This helps clarify the historical processes involved in industrialization: a shift in ownership of the means of production to capitalists, a concentration of capital-intensive production, the organization of labour in factories, and a socially transforming rise in output.

- Sometimes industries can be run viably in small units, but more usually there is a tendency towards an ever larger scale of production (due to 'economies of scale'), as well as towards the urban concentration of production.

Kitching's attack on populism, together with his emphasis on the increasing scale and concentration of industry, has given some readers of his book the impression that he favours a Stalinist 'heavy industry' model of development (see Chapter 12). But in a postscript to the second edition, he states categorically that this is not so. He explains that his view is that industrialization strategies should be adapted to the particular context in which they are applied. What is appropriate depends on a host of factors:

"including the demographic and geographical size of the economy involved, its resource endowment (including its 'human capital' endowment) and its role in the world economy at the point at which industrialization is attempted. For small, resource-poor Third World countries a development sequence which begins with the expansion of primary product exports, moves to the manufacture of simple inputs and basic consumption goods for the primary producers (usually, though not always, peasant producers) and from there to the manufacture of labour-intensive consumer and producer goods for export and domestic consumption, is a particularly appropriate strategy. Certainly it is a more appropriate industrialization strategy for small peripheral economies than either 'crash' heavy industrialization under state auspices or luxury 'import substitution' industrialization undertaken under the auspices of multinational corporations. Such a strategy is particularly desirable in that it can accommodate forms of rural 'agro-industry' which can act as a counter-balance to over-rapid urbanization."

(ibid., p.192)

Nonetheless, even the most appropriate of industrialization strategies is going to involve difficult decisions. Kitching concludes:

"development is an awful process. It varies only, and importantly, in its awfulness. And that is why my most indulgent judgements are reserved for those, whether they be Marxist-Leninists, Korean generals, or IMF officials, who, whatever else they may do, recognize this and are prepared to accept its moral implications. My most critical reflections are reserved for those, whether they be western liberal-radicals or African bureaucratic élites, who do not, and therefore avoid or evade such implications and with them their own responsibilities.

(ibid., p.195)

20.3   A world of industrialized nations

I have examined the arguments for industrialization made by Kitching at some length because his is a strong case, and because its ramifications are enormous. Nevertheless, there are grounds for questioning parts of it, and for asking if it is in fact going to be much help as an overarching, guiding prognosis for development in the 1990s. In this section I am going to raise a further set of problems with the conventional wisdom on industrialization, ones which Kitching does not attempt to resolve or even compounds. I will then end by commenting upon where all of this debate leaves us. I will not be directly tackling Kitching's critique of populism, although it will be apparent that I have rather more sympathy with some forms of it than he does, and rather less sympathy

with ruthless 'Marxist-Leninists, Korean generals, or IMF officials'.

Some problems with the old orthodoxy on industrialization

What follows are reservations about the proposition that industrialization is the only means of achieving development. I do not attempt to replace this old orthodoxy with a fully worked-out alternative, but point to some flaws and limitations in the kinds of arguments that Kitching and others have put forward.

- Development is more than material affluence

Kitching is aware that there are various meanings ascribed to the word 'development'. He notes that the eighteenth and nineteenth century political economists (like Smith, Ricardo and Marx) who established the old orthodoxy thought that economic progress was occurring if the volume and value of output or production was continuously rising. In contrast, populists have been more concerned with distribution, and reject the idea that the coexistence of great wealth with mass poverty can be regarded as progress in any sense. Nevertheless, like most supporters of the need-for-industrialization thesis, Kitching himself uses an extremely 'economic' definition of development. He defines it as 'high and continuous growth of output and incomes'. Is this adequate? In Chapter 2 it was explained that although Cuba is a poor country, it has health statistics comparable to rich countries. Has a kind of development occurred here or not?

- The empirical evidence on the link between material affluence and industrialization is not so straightforward

As Table 20.1 shows, Kitching used per capita GNP as a development indicator, and associates this with a shift in the labour force out of agriculture and into industry. Leaving aside the ambiguities involved in contrasting 'agriculture' and 'industry' in this way, the use of GNP data can be very misleading (a point commented on in Chapter 6). In this particular context it needs to be borne in mind that GNP may reflect increased per capita income from industrial production more accurately than per capita income from agriculture, because non-agricultural commodities are more likely to be marketed through 'official' channels. Also, in recent years, it has become more obvious that the 'real' GNP figures for Second World countries are far lower than has been estimated. In the second edition of his book (1989), Kitching recognizes this for China, but if anything the figures from the USSR are even more open to doubt. Such factors do not in themselves refute Kitching's assertions, but they do suggest that the empirical evidence needs to be examined more closely. It may be that the importance of industrialization appears overwhelming, partly because of the categories used for assessing it, and partly because its economic benefits have sometimes been exaggerated for political purposes. Furthermore, although a few developing countries certainly have increased material affluence through industrialization (e.g. Hong Kong, South Korea), there is considerable evidence that such cases are exceptional, and that their achievements will not be repeated elsewhere (an issue taken up in Book 2 of this series, *Hewitt et al.*, 1992) (Figure 20.2).

- **Development planning cannot simply be based on 'economies of scale'**

Although Kitching has defined his own position more fully in the postscript to the second edition of his book, it is not altogether clear if he is arguing that development ultimately means achieving and maintaining 'economies of scale'. If so, then countries like Britain, the USA and the USSR, where capital-intensive manufacturing industry has been in relative decline, could be seen as 'not developing'. A danger of taking such an approach is that economic development becomes a pre-ordained historical process, something which has to be harnessed or tapped into. Large-scale industrialization is an inevitable and necessary step to development. The implication is that development policy and political struggle mean no more than facilitating a situation in which history can speed up.

Such a grimly determinist view could lead to a position that there is little point in doing anything to help the poor or change society. Although life will be hard at first for workers, eventually industrialization will improve the lot of everyone, either

"Join us. It's only a step."

Figure 20.2

due to the distribution of benefits through the market, or by a socialist revolution once the organization of labour has been established. It can also lead to an end-justifies-the means argument with respect to dictators, like Stalin, who might be seen as historical catalysts. For many concerned people this is simply unacceptable – whatever the strength of the logic that lies behind it. Many socialists have argued that it is possible to mobilize a peasantry and to change social relations without an industrial proletariat. Moreover, leaving aside grand theories, there is surely a value in trying to improve the lot of some of the people living in atrocious circumstances. It may be a drop in the ocean, and the prospects for establishing rural utopias in developing countries are certainly slight, but if promoting development does not involve immediate assistance to those most in need, one is left wondering if there is much point in it.

- **If development is equated with nations industrializing, that is tantamount to saying it is impossible, given resource constraints**

Although Kitching recognizes that industrialization need not always be attempted within national boundaries, he argues that an 'economo-political' logic ensures that, in the long run, this is almost invariably what happens. When he outlines what he sees as the appropriate forms of industrialization for particular contexts, his particular contexts are all countries. Ideally each country should have its own thriving industrial sector and, it seems to be suggested, each should aspire towards the kinds of lifestyles associated with the West. But if this were possible, could the world afford it? Many years ago Gandhi observed:

> "God forbid that India should ever take to industrialization after the manner of the West. The economic imperialism of a single tiny island kingdom is today keeping the world in chains. If an entire nation of 300 million took to similar economic exploitation, it would strip the world bare like locusts."
>
> (Gandhi, 1928, p.422)

Chapter 5 pointed out that sweeping statements about the global environment need to be treated with caution. But there is no doubt that if every country in the world used up resources as fast as the 'consumer societies' of Japan and the USA, there would quickly be a major crisis. Thus, even if

it is true that the material affluence of part of the human race has only been possible because of the capital-intensive industrialization of nation states, an alternative is going to have to be found if there is to be an improvement in the lot of the rest.

Where does this leave us?

This has been a pessimistic discussion with which to end this book (Figure 20.3). Some reservations have been expressed about the old orthodoxy that development necessitates industrialization with all its grim implications, but the basic contention remains: that is how it has been done in the past. Even if we adopt a less 'economistic' definition of the term 'development', it cannot be denied that industrial products are essential for a better life. How can these be provided without repeating the mistakes of the past?

If there are answers to this question, it is likely that they are going to be found in new industrial technology, and in global agreements about the use of the world's resources. The 'economo-political' logic which causes countries to industrialize has to be reconsidered. While not wishing to indulge in the kind of wishful thinking Kitching castigates, I believe that such developments may

Look Bud, If God had intended us to share the World's resources he wouldn't have created Nation States!

Figure 20.3

be more possible in the 1990s than they were in the mid-1980s.

Since the industrial revolutions of Europe, industrial technology has tended to evolve in such a way as to make production ever more capital intensive and concentrated. Kitching would maintain that this was inevitable because it made production more efficient. But it can also be argued that the technology of mass production developed as a form of social control, linked to the ownership of factories by capitalists (or by the state). For some products (particularly electronic products), technology has more recently emerged which makes industrial production in smaller units at least as efficient as production in large factories (e.g. in Italy and Japan). It may well be that in years to come this will be the way forward. More products might be assembled from parts made in small, resource-efficient units of production, either collaborating locally or perhaps located in several countries.

At the same time, the collapse of the state socialist experiment in eastern Europe and the cutting back of state interference in national economies have had the effect of making economic nationalism seem less important than hitherto. National integrity is no longer seen as something that necessarily has to be underpinned by economic autonomy. Globalism and regionalism are on the increase (see Chapter 13). Is it far-fetched to expect a federated Europe to come to some kind of fair arrangement with former European colonies which involves a sharing of wealth and resources, and for the USA to come to similar agreement with Latin American countries? Perhaps it is, but it is something to hope for and work towards, and it is certainly a key debating issue for those who want to see world development.

Figure 20.4   Soviet miners.

References

Agarwal, B. (1985a) 'Women and technological change in agriculture: the Asian and African experience', in Ahmed, I. (ed.) *Technology and Rural Women*, Allen and Unwin, London.

Agarwal, B. (1985b) 'Rural women and high yielding variety rice technology', in International Rice Research Institute (eds) *Women in Rice Farming*, Gower Press, Aldershot.

Algar, H. (1972) 'The oppositional role of the Ulama in twentieth-century Iran', in Keddie, N. (ed.), *Scholars, Saints and Sufis*, University of California Press, Berkeley.

Amin, S. (1976; 1978) *The Arab Nation*, Zed Press, London.

Amin, S. (1990a) *Maldevelopment*, Zed Press, London.

Amin, S. (1990b) *Delinking,* Zed Press, London.

Anderson, M. B. (1985) 'Technology transfer: implication for women', in Overholt, C., Anderson, M. B., Cloud, K. & Austin, J. E. (eds) *Gender Roles in Development Projects: a casebook*, pp.60–8, Kumarian Press, Connecticut.

Anon (1982) *Independent Kenya*, Zed Press, London.

Antrobus, P. (1988) 'Consequences and responses to social and economic deterioration: the experience of the English-speaking Caribbean', Workshop on Economic Crisis, Household Strategies and Women's Work, Cornell University, Ithaca, NY.

Arnold, D. (1986) *Police Power and Colonial Rule: Madras, 1859–1947*, Oxford University Press, Delhi.

Arrighi, G. (1990) 'Marxist century, American century: the making and remaking of the world labour movement', *New Left Review*, 179, pp.29–63.

Asia Labour Monitor (1988) *Min-Ju No-Jo: South Korea's new trade unions*, Asia Monitor Resource Centre, Hong Kong.

Atkinson, L. (1989) 'Two decades of contraceptive research: prospects for the 21st century', in UN (eds) *World Population Prospects 1988*, United Nations, New York.

Azarya, V. (1988) 'Reordering State-Society Relations: incorporation and disengagement', in Rothchild, D. & Chazan, N. (eds) *The Precarious Balance: state and society in Africa*, pp. 3–21, Westview Press, Boulder.

Bagchi, A. K. (1972) *Private Investment in India, 1900–1939*, Cambridge University Press, Cambridge.

Baker, J. (1990) Speech, Washington DC, October 22 1990, US Information Agency, Washington DC.

Barbier, E. (1989) *Economics, Natural Resource Scarcity and Development: conventional and alternative views*, Earthscan, London.

Bardhan, P. (1974) 'On life and death questions', *Economic and Political Weekly*, 9 August, pp.1293–1304.

Barratt Brown, M. (1963) *After Imperialism*, Heinemann, London.

Bates, R. H. (1981) *Markets and States in Tropical Africa*, University of California Press, Berkeley.

Bayly, C. A. (1983) *Rulers, Townsmen and Bazaars,* Cambridge University Press, Cambridge.

Bayly, C. A. (1989) *Imperial Meridian: the British Empire and the World, 1780–1830*, Longman, London.

Beecham, D. & Eidenham, A. (1987) 'Beyond the mass strike: class, party and trade union struggle in Brazil', *International Socialism*, Series 2, 36, pp.3–48.

Beneria, L. (1982) *Women and Development: the sexual division of labor in rural societies*, Praeger, New York.

Berliner, P. (1978) *The Soul of Mbira*, University of California Press, Berkeley.

Bernier, F., trans. Constable, A. & Smith, V. (1916) *Travels in the Mogul Empire 1656–1668*, Oxford University Press, Oxford.

Bernstein, H. (1983) 'Development', in Thomas, A. & Bernstein, H. (eds) *The 'Third World' and 'Development'*, Block 1 of the Open University course U204 Third World Studies, The Open University, Milton Keynes.

Bernstein, H. (1988) 'Production and producers', in Crow, B. & Thorpe, M. (eds) *Survival and Change in the Third World*, Polity Press, Oxford.

Blacking, J. (1976) *How Musical is Man?*, Faber, London.

Blacking, J. (1980) 'Political and musical freedom in the music of some black South African churches', in Holy, L. & Stuchlik, M. (eds) *The Structure of Folk Models*, pp.35–62, Academic Press, London.

Boyd, R. (1988) *A Global View of Women after the UN Decade*, Centre for Developing Area Studies, McGill University, Montreal, Working Paper No.56, October 1988.

Brandt, W. *et al.* (1980) *North–South: a programme for survival, Report of the Independent Commission on International Development Issues*, (The Brandt Report), Pan Books, London.

Brown, S. (1974) *New Forces in World Politics*, Little Brown, Boston.

Buchanan, A. (1982) *Food, Poverty and Power*, Spokesman, Nottingham.

Bundy, C. (1977) 'The Transkei Peasantry, c.1890–1914: Passing through a time of stress', in Palmer, R. & Parsons, N. (eds) *The Roots of Rural Poverty*, Heinemann, London.

Bundy, C. (1979) *The Rise and Fall of the South African Peasantry*, Heinemann, London.

Burridge, K. (1969) *New Heaven, New Earth*, Blackwell, Oxford.

Burton, J. (1972) *World Society*, Cambridge University Press, Cambridge.

Buvinic, M. (1986) 'Projects for women in the Third World: explaining their misbehaviour', *World Development*, 14(5).

Callinicos, A. (1987) 'Imperialism, capitalism and the state today – a review of Nigel Harris's *The End of the Third World*', *International Socialism*, Series 2, 35, pp.71–115.

Cardoso, F. H. & Faletto, E. (1979) *Dependency and Underdevelopment in Latin America*, University of California Press, Berkeley (first published 1971).

Carney, J. (1988) 'Struggles over crop rights and labour within contract farming households in a Gambian irrigated rice project', *Journal of Peasant Studies*, 15(3) (April).

Carson, R. (1962) *Silent Spring*, Houghton Mifflin, Boston Mass., and Penguin, Harmondsworth.

Cernea, M. (1985) *Putting People First: sociological variables in rural development*, Oxford University Press, New York.

Chambers, R. (1983) *Rural Development: putting the last first*, Longman, London.

Chambers, R. (1988) 'Sustainable rural livelihoods', in *The Greening of Aid: sustainable livelihoods in practice*, Earthscan, London.

Chambers, R. (1989) *The State and Rural Development: ideologies and an agenda for the 1990s*, IDS Discussion Paper No.269, Institute of Development Studies, Brighton.

Chandra, Bipan *et al.* (1989) *India's Struggle for Independence*, Penguin, Harmondsworth.

Chen, L. C., Huq, E. & D'Souza, S. (1980) *A Study of Sex-Biased Behaviour in the Intra-Family Allocation of Food and the Utilization of Health Care Services in Rural Bangladesh,* International Centre for Diarrhoeal Disease Research, Bangladesh, and Department of Population Sciences, Harvard School of Public Health.

Chen, R. S. (ed.) (1990) *The Hunger Report*, The Alan Shawn Feinstein World Hunger Program, Brown University, Providence, Rhode Island.

Choi Hyung Sup (1988) 'Guidelines for Future Development of Steel Technology', *Journal of Materials Education*, 10(3), pp. 286–99.

Cliffe, L. & Cunningham, G. L. (1973) 'Ideology, organization and the settlement experience in Tanzania', in Cliffe, L. & Saul, J. S. (eds) *Socialism in Tanzania*, vol.2, East African Publishing House, Nairobi.

Colclough, C. & Manor, J. (1991) *States and Markets? Neo-liberalism and the Development Policy Debate*, Oxford University Press, Oxford.

Connolly, B. & Anderson, R. (1988) *First Contact: New Guinea's highlanders encounter the outside world*, Penguin Books, Harmondsworth.

Coplan, D. (1979) 'The African musician and the Johannesburg entertainment industry, 1900–1960', *Journal of Southern African Studies*, 6(2) (April), pp.135–64.

Coplan, D. (1982) 'The urbanization of African music: some theoretical observations', *Popular Music*, 2, pp.113–29.

Coplan, D. (1985) *In Township Tonight! South Africa's Black City Music and Theatre*, Longman, London.

Cornia, G. A. (1987) 'Adjustment policies 1980–85: effects on child welfare', in Cornia, G. A., Jolly, R. & Stewart, F. (eds), *Adjustment with a Human Face: protecting the vulnerable and promoting growth, a study by UNICEF*, Clarendon Press, Oxford.

Coulson, A. (1982) *Tanzania: a political economy*, Clarendon Press, Oxford.

Crowder, M. (1968) *West Africa under Colonial Rule*, Hutchinson, London.

Crowder, M. (1987) 'Whose dream was it anyway? Twenty-five years of African independence', *African Affairs*, 86(342) (January), pp.7–24.

CSCE Joint Declaration (1990) Joint Declaration on CSCE, 19 November 1990, US Information Service, Washington DC.

Cummings, B. (1989) 'The abortive abertura: South Korea in the light of Latin American experience, *New Left Review*, 173 (January–February), pp.5–34.

Curtin, P. D. (1969) The Atlantic Slave Trade, University of Wisconsin Press, Wisconsin.

Curtin, P. D. (1984) *Cross-Cultural Trade in World History*, University Press, London.

Das Gupta, J. (1989) 'India: democratic becoming and combined development', in Diamond, L., Linz, J. & Lipset, S. (eds), *Democracy in Developing Countries: Asia*, pp.53–104, Adamantine Press, London.

Davey, B. (1975) *The Economic Development of India*, Spokesman University Paperback 20, Nottingham.

Davidson, B. (1978) *Africa in Modern History*, Penguin, Harmondsworth.

Day, R. (1977) 'Trotsky and Preobrazhensky: the troubled unity of the Left opposition', *Studies in*

Comparative Communism, 10, pp.77–91.

De Lancey, M., Elliot, S., Green, D., Menkhaus, K., Moqtar, M. & Schraeder, P. (1988) *World Bibliographical Series: Somalia*, Clio Press, Oxford.

Denitch, B. (1990) *The End of the Cold War*, Verso, London.

Dessalegn Rahmato (1987) *Famine and Survival Strategies: a case study from Northeast Ethiopia*, Addis Ababa University, Institute of Development Studies, Addis Ababa. Food and Famine Monograph Series No 1.

Dey, J. (1982) 'Development planning in The Gambia: the gap between planners' and farmers' perceptions, expectations and objectives', *World Development*, 10(5).

Dhaouadi, M. (1988) 'An operational analysis of the phenomenon of the other underdevelopment in the Arab world and in the third world', *International Sociology,* 3(3) (September), pp.219–34.

Diamond, L., Linz, J. & Lipset, S. (eds) (1988) *Democracy in Developing Countries: Africa*, Adamantine Press, London.

Diamond, L., Linz, J. & Lipset, S. (eds) (1989) *Democracy in Developing Countries: Asia*, Adamantine Press, London.

Diaz, B. (1963) The *Conquest of New Spain*, Penguin, Harmondsworth.

Dicken, P. (1986) *Global Shift,* Harper and Row, New York.

Dirks, N. B. (1988) *The Hollow Crown: ethnohistory of an Indian kingdom*, Cambridge University Press, Cambridge.

Drèze, J. & Sen, A. (1989) *Hunger and Public Action*, Clarendon Press, Oxford.

Dubb, A. (1974) 'The impact of the city', in Hammond-Tooke, W. D. (ed.) *The Bantu Speaking Peoples of Southern Africa*, 2nd edn, pp.441–72, Routledge & Kegan Paul, London.

Dwyer, D. & Bruce, J. (eds) (1988) *A Home Divided: women and income in the Third World*, Stanford University Press, Stanford, Cal.

Dyson, T. & Moore, M. P. (1983) 'On kinship structure, female autonomy and demographic behaviour in India', *Population and Development Review*, 9, pp.35–60.

Ebrahim, G. J. & Ranken, J. P. (1988) *Primary Health Care: reorienting organisational support,* Macmillan, London.

Eckstein, A. (1977) *China's Economic Revolution*, Cambridge University Press, Cambridge.

Edwards, M. (1989) 'The irrelevance of development studies', *Third World Quarterly,* 11(1) (January), pp.116–35.

Eisenstadt, S. (1966) *Modernization: protest and change*, Prentice-Hall, Englewood Cliffs, N.J.

Eisenstadt, S. N. (1973) *Tradition, Change and Modernity*, Wiley-Interscience, London.

Ellman, M. 'Did the agricultural surplus provide the resources for the increase in investment in the USSR during the first Five Year Plan?', *Economic Journal*, 85, pp.844–63.

Elson, D. (ed.) (1991) *Male Bias in the Development Process*, Manchester University Press, Manchester.

Enloe, C. (1989) *Bananas, Beaches and Bases*, Pandora, London.

Erlich, A. (1960) *The Soviet Industrialization Debate 1924–8*, Harvard University Press, Cambridge, Mass.

Erlich, P. (1968) *The Population Bomb*, Ballantine Books, New York.

Erlmann, V. (1985) 'Black political song in South Africa – some research perspectives', in Horn, D. (ed.) *Popular Music Perspectives*, vol.2, pp.187–209, International Association for the Study of Popular Music, Exeter.

Erlmann, V. (1990) 'Migration and performance: Zula migrant workers' Isicathamiya performance in South Africa, 1890–1950', *Ethnomusicology*, (Spring/Summer), pp.199–220.

Fallers, L. (ed.) (1964) *The King's Men*, Oxford University Press, Oxford.

Fanon, F. (1961; 1963; 1967; 1969) *The Wretched of the Earth*, trans. C. Farrington, Penguin, Harmondsworth.

Farooq, G. M. & DeGraff, D. S. (1990) 'Fertility and development: an introduction to theory, empirical research and policy issues', *Training in Population, Human Resources and Development Planning*, 7, International Labour Office, Geneva.

Feldman, R. (1989) 'Women for a change: the impact of structural adjustment on women in Zambia, Tanzania and Mozambique', War on Want, London.

Frank, R. & Chasin, B. (1981) 'Peasants, peanuts, profits and pastoralists', *Ecologist*, 11, pp.156–68.

Freire, P. & Shor, I. (1987) *A Pedagogy for Liberation*, Macmillan, London.

Friedman, S. (1987) *Building Tomorrow Today: African workers in trade unions* 1970–84, Ravan Press, Johannesburg.

Fukuyama, F. (1989) 'The End of History?' *The Public Interest*, Summer, p.63.

Fukuyama, F. (1990) 'Forget Iraq – history IS dead', *The Guardian*, 7 September.

Galbraith, J. K. (1972) *The New Industrial State*, Andre Deutsch, London (2nd edn) (Penguin, Harmondsworth 1974).

Galbraith, J. K. (1990) 'The price of world peace', paper prepared for 1990 Oslo conference sponsored by the Norwegian Nobel Peace Committee and the Elie Wiesel Foundation for Humanity on the continuing issues of hate and conflict as the Cold War ends, reprinted in *The Guardian*, 8 September 1990.

Gandhi, M. K. (1928) *Young India*, Viking Press/Ganesan, London.

Gann, L. H. & Duignon, P. (1967), *Burden of Empire: an appraisal of Western Colonialism in Africa South of the Sahara*, Hoover Institution Press, Stanford.

Gastil, R. D. (1988) *Freedom in the World: political rights and civil liberties, 1987–88*, University Press of America, Lanham, MD.

Geertz, C. (1984) 'Culture and social change: the Indonesian case', *Man*, 19, pp.511–32.

George, S. (1976) *How the Other Half Dies*, Penguin, Harmondsworth.

Ghai, D. (ed.) (1965) *Portrait of a Minority: Asians in East Africa*, Oxford University Press, Oxford.

Gheerbrant, A. (ed.) (1961) *The Incas*, Orion Press, New York (Avon Books, 1966).

Ginwala, F., Mackintosh, M. & Massey, D. (1991) *Gender and Economic Policy in a Democratic South Africa*, DPP Working Paper No. 21, The Open University, Milton Keynes.

Goldsworthy, D. (1988) 'Thinking politically about development', *Development and Change*, 19(3)(July), pp.505–30.

Gordon, D. (1988) 'The global economy: new edifice or crumbling foundations? *New Left Review*, 168 (March), pp.24–65.

Gordon, S. (1985) 'Ladies in limbo', Commonwealth Secretariat.

Goulet, D. (1971) *The Cruel Choice: a new concept in the theory of development*, Atheneum, New York.

Gramsci, A. (1988) 'Some Theoretical and Practical Aspects of 'Economism' ', in Forgacs, D. (ed.), *A Gramsci Reader*, Lawrence and Wishart, London.

Gronow, P. (1983) 'The record industry: the growth of a mass medium', *Popular Music*, 3, pp.53–75.

Guha, R. (1989) *The Unquiet Woods: ecological change and peasant resistance in the Himalaya*, Oxford University Press, Delhi.

Guia del Tercer Mundo (1981) Periodistas del Tercer Mundo, Mexico.

Gwynne, R. B. N. (1990) *New Horizons? Third World Industrialisation in an International Framework*, Longman Scientific and Technical, London.

Habib, I. (1969) 'Potentialities of capitalistic development in the economy of Mughal India', *Journal of Economic History*, 29, pp.32–78.

Halliday, F. & Molyneux, M. (1981) *The Ethiopian Revolution*, Verso, London.

Halliday, F. (1978) 'Revolution in Afghanistan', *New Left Review*, 112, pp.3–44.

Halliday, F. (1979) *Iran: dictatorship and development*, Penguin, Harmondsworth.

Halliday, F. (1980) 'War and Revolution in Afghanistan', *New Left Review*, 119, pp.20–41.

Halliday, F. (1981) 'The North Korean enigma', *New Left Review*, 127, pp.18–52.

Halliday, F. (1987) *The Making of the Second Cold War*, Verso, London.

Halliday, F. (1989) *Cold War, Third World*, Abacus, London.

Halliday, F. (1990a) 'The End of Cold War', *New Left Review*, 180, pp.5–23.

Halliday, F. (1990b) *Cold War, Third World*, Hutchinson, London.

Hamm, C. (1989) '*Graceland* revisited', *Popular Music*, 8(3), pp.299–304.

Han Sung-joo (1989) 'South Korea: politics in transition', in Diamond, L., Linz, J. & Lipset, S. (eds), *Democracy in Developing Countries: Asia*, pp.267–303, Adamantine Press, London.

Harris, M. (1969) *The Rise of Anthropological Theory*, Routledge and Kegan Paul, London.

Harris, N. (1986) *The End of the Third World*, I. B. Tauris, London.

Harris, N. (1987) *The End of the Third World*, Penguin, London.

Harriss, B. (1989) 'Differential female mortality and health behaviour in India', *Ld'A–QEH Development Studies Working Papers*, Oxford.

Hartmann, B. & Standing, H. (1989) *The Poverty of Population Control: family planning and health policy in Bangladesh*, Bangladesh International Action Group (BIAG), London.

Hartmann, B. (1987) *Reproductive Rights and Wrongs: the global politics of population control and contraceptive choice*, Harper & Row, New York.

Hartmann, J. (1990) 'The rise and rise of private capital', in O'Neill, N. & Mustafa, K. (eds) *Capitalism, Socialism and the Development Crisis in Tanzania*, Avebury, Aldershot, pp.233–54.

Hayter, T. (1989) *Exploited Earth*, Earthscan, London.

Headrick, D. P. (1988) *The Tentacles of Progress: technology transfer in the Age of Imperialism, 1850–1940*, Oxford University Press, New York.

Heering, L. (1990) *Health Inputs and Impacts: a case study on the mediating role of mother's characteristics and practices on health of children in Indonesia*, Institute of Social Studies Working Paper No. 81, PO Box 90733, 2509 LS, The Hague, Netherlands.

Held, D. (1987) *Models of Democracy*, Polity Press, Cambridge.

Henderson, J. (1989) *The Globalisation of High Technology Production*, Routledge, London.

Hewitt, T., Johnson, H. & Wield, D. (eds) (1992) *Industrialization and Development*, Oxford University Press/The Open University, Oxford (Book 2 of this series).

Hinton, W. (1972) *Fanshen*, Penguin, Harmondsworth.

Hobsbawm, E. & Ranger, T. (1983) *The Invention of Tradition*, Cambridge University Press, Cambridge.

Hobsbawm, E. (1968) *Industry and Empire*, Penguin, Harmondsworth.

Hübi K. (1986) 'The nomadic livestock production system of Somalia', in Conze, P. & Labahn, T. (eds) *Somalia: agriculture in the winds of change*, Epi Verlag, Hamburg.

IFPI (1981) *1981 World Sales*, International Federation of Phonogram and Videogram Producers, London.

IFPI (1989) *The I.F.P.I. Review 1988/89*, International Federation of Phonogram and Videogram Producers, London.

Iliffe, J. (1979) *A Modern History of Tanganyika*, Cambridge University Press, Cambridge.

ILO (1976) *International Recommendations on Labour Statistics*, International Labour Organization, Geneva.

ILO (1989–90) *Labour Statistics*, International Labour Organization, Geneva.

ILR (1988) 'Interview with Crispin Beltran', *International Labour Reports*, 27/28, Summer, pp.18–19.

IWHC (1986) *The Contraceptive Development Process and Quality of Care in Reproductive Health Services*, International Women and Health Coalition (with The Population Council), Meeting, 8–9 October, New York.

Jackson, K. (1989) 'The Philippines: the search for a suitable democratic solution, (1946–1986)', in Diamond, L., Linz, J. & Lipset, S. (eds), *Democracy in Developing Countries: Asia*, pp.231–265, Adamantine Press, London.

Johnson, R. W. (1972) 'French imperialism in Guinea', in Owen, E. R. J. & Sutcliffe, R. B. (eds) *Studies in the Theory of Imperialism*, Longman, London.

Johnson, H. & Bernstein, H. (eds) (1982) *Third World Lives of Struggle*, Heinemann, London.

Johnson, Samuel (1810) *Dictionary of the English Language*, F. & C. Rivington, London.

Jordan, B. (1985) *The State: authority and autonomy*, Blackwell, Oxford.

Joss, S. (1990) 'Your husband is your God', GADU Pack No.12, OXFAM, Oxford.

Kaplinsky, R. (1990) *The economies of small: appropriate technology in a changing world*, Intermediate Technology Publications, London.

Karl, T. (1990) 'Dilemmas of democratization in Latin America', *Comparative Politics*, 23(1) (October), pp.1–21.

Kauffman, R. (1979–80) 'Tradition and innovation in the urban music of Zimbabwe', *African Urban Studies*, 6, pp.41–8.

Kegley, C. W. & Wittkopf, E. R. (1989) *World Politics*, Macmillan, London.

Khan, A. (1990) 'The impact of international labour migration on the rural 'Barani' areas of modern Pakistan', PhD thesis, University of Sussex, Brighton.

King, M. (1990) 'Health is a sustainable state', *The Lancet*, 336, 15 September.

Kitching, G. (1980) *Class and Economic Change in Kenya*, Yale University Press, New Haven.

Kitching, G. (1982) *Development and Underdevelopment in Historical Perspective: populism, nationalism and industrialization*, Methuen, London.

Kitching, G. (1989) *Development and Underdevelopment in Historical Perspective*, Routledge, London (2nd revised edition).

Kloosterboer, W. (1960) *Involuntary Labour Since the Abolition of Slavery*, E. J. Brill, Leiden.

Kluckhohn, C. & Kroeber, A. (1952) *Culture*, Peabody Museum Papers, Harvard University, Cambridge, Mass., 67(1).

Krasner, S. (1985) *Structural Conflict*, University of California Press, Berkeley.

Krikler, J. (1987) 'Reflections on the transition to socialism in South African agriculture', *Africa Perspective*, New Series, 5&6, pp.95–120.

Laing, D. (1986) 'The music industry and the 'cultural imperialism' thesis', *Media, Culture and Society*, 8(3), pp.331–41.

Lappé, F. M. & Schurman, R. (1988) *Taking Population Seriously*, Earthscan Publications, London.

Lawand, T. A., Hvelplund, F., Alward, R., & Voss, J. (1976) 'Brace Research Institute's *Handbook of Appropriate Technology* (1975)', in Jequier, N. (ed.) *Appropriate Technology: problems and promises*, pp.124–36, OECD, Paris.

Leach, E. (1954) *The Political Systems of Highland Burma*, Cambridge University Press, Cambridge.

Lenin, V. I. (1939) *Imperialism: the highest stage of capitalism. A popular outline*, English version, International Publishers, New York (first published in Moscow 1916).

Lewis, I. (1965) 'The northern pastoral Somali of the Horn', in Gibbs, J. (ed.) *Peoples of Africa*, Holt, Rinehart and Winston, New York.

Lewis, I. (1975) 'The dynamics of nomadism: prospect for sedentarization and social change', in Monod, T. (ed.) *Pastoralism in Tropical Africa*, Oxford University Press, London.

Lewis, I. (1985) *Social Anthropology in Perspective*, Cambridge University Press, London.

Leys, C. (1975) *Underdevelopment in Kenya*, Heinemann, London.

Lipset, D. (1989) 'Papua New Guinea: the Melanesian ethic and the spirit of capitalism, 1975–1986', in Diamond, L., Linz, J. & Lipset, S. (eds) *Democracy in Developing Countries: Asia*, pp.383–421, Adamantine Press, London.

Low, D. A. (1973) *Lion Rampant: essays on the study of British Imperialism*, Cass, London.

Mackintosh, M. (1981) 'Gender and economics: the sexual division of labour and the subordination of women', in Young, K. *et al.* (eds) *Of Marriage and the Market: women's subordination in international perspective*, CSE Books, London.

Maitan, L. (1976) *Party, Army and Masses in China*, New Left Books, London.

Maliyamkono, T. & Bagachwa, M. (1990) *The Second Economy in Tanzania*, James Curry, London.

Mamdani, M. (1973) *From Citizen to Refugee: Ugandan Asians come to Britain*, Frances Pinter, London.

Mamdani, M. (1976) *Politics and Class Formation in Uganda*, Heinemann, London.

Mamdani, M. (1983) *Imperialism and Fascism in Uganda*, Heinemann, London.

Mangat, J. S. (1969) *A History of the Asians in East Africa*, Clarendon Press, Oxford.

Mann, M. (1984) 'Capitalism and militarism', in Shaw, M. (ed.) *War, State and Society*, Macmillan, London.

Mannoni, D. (trans. Powesland, P.) (1956) *Prospero and Caliban: the psychology of colonization*, Methuen, London.

Manuel, P. (1988) *Popular Musics of the Non-Western World: an introductory survey*, Oxford University Press, New York.

Markowitz, I. L. (1977) *Power and Class in Africa*, New Jersey, Prentice-Hall.

Martorell, R. (1988) 'Nutrition, infection and growth', *Clinical Nutrition*, 7(4), pp.156–67.

Martorell, R. (1989) 'Body size, adaptation and function', *Human Organisation*, 48(1), pp.15–20.

Mass, B. (1976) *Population Target: the political economy of population control in Latin America*, Charters Publishing, Ontario.

Mayall, J. (1990) *Nationalism and International Society*, Cambridge University Press, Cambridge.

McFarlane, B. (1984) 'Political economy of class struggle and economic growth in China 1950–1982', in Maxwell, N. & McFarlane, B. (eds) *China's Changed Road to Development*, Pergamon, Oxford.

McKeown, T. (1979) *The Role of Medicine*, Basil Blackwell, Oxford.

McLelland, D. (1963) 'The achievement motive in economic growth', in Hoselitz, B. F. & Moore, W. E. (eds) *Industrialization and Society*, UNESCO and Mouton, The Hague.

McNeill, W. H. & Waldman, M. R. (1973) *The Islamic World*, Oxford University Press, New York.

Meadows, D. H., Meadows, D. L., Randers, J., & Behrens, W. III (1972) *The Limits to Growth*, Earth Island Ltd, London.

Meintjes, L. (1990) 'Paul Simon's *Graceland*, South Africa, and the mediation of musical meaning', *Ethnomusicology*, 34(1) (Winter), pp.37–73.

Michaelson, K. L. (1981) *And the Poor Get Children: radical perspectives on population dynamics*, Monthly Review Press, New York.

Michalak, S. J. (1983) *UNCTAD*, Heritage Foundation, Washington DC.

Misra, B. B. (1976) *The Indian Political Parties: an historical analysis of political behaviour up to 1947*, Oxford University Press, Delhi.

Mister, R. (1988) 'Refugees and Nomads: successes and failures in Somalia', in Poulton, R. & Harris, M. (eds) *Putting People First*, Macmillan, London.

Moghadam, V. (1991) 'Women's employment in the Middle East and North Africa: gender, class and state policies', Mimeo, UNU/World Institute of Development Economics Research, Helsinki.

Momsen, J. (1991) *Women and Development in the Third World*, Routledge, London.

Moore, B. (1966), *Social origins of Dictatorship and Democracy: lord and peasant in the making of the modern world*, Beacon Press, Boston.

Moore, B. (1967) *Social Origins of Dictatorship and Democracy*, Allen Lane, London (Penguin, 1973).

Moore, B. (1969) 'Tolerance and scientific outlook', in Wolff, R., Moore, B. & Marcuse, H. (eds), *A Critique of Pure Tolerance*, pp.65–91, Cape, London.

Morrison, D. (1982) 'A critical examination of A. A. Barsov's empirical work on the balance of value exchanges between the town and the country', *Soviet Studies*, 34, pp.570–84.

Mueller, R. (1983) 'Measuring women's poverty in developing countries', in Buvinic, M. *et al.* (eds) *Women and Poverty in the Third World*, Johns Hopkins University, Baltimore, MD.

Murphy, Y. & Murphy, R. F. (1974) *Women of the Forest*, Columbia University Press, New York.

Mutloatse, M. (ed.) (1987) *Umhlaba Wethu: a historical indictment*, Skotaville Publishers, Johannesburg.

Nair, S. (1989) *Imperialism and the Control of Women's Fertility*, Campaign Against Long Acting

Contraceptives, Amsterdam.

Nandy, Ashis (1983) *The Intimate Enemy: loss and recovery of self under colonialism*, Oxford University Press, Delhi.

Nathan, A. (1986) *Chinese Democracy*, Tauris, London.

Nolan, P. (1976) 'Collectivization in China: some comparisons with the USSR', *Journal of Peasant Studies*, 3, pp.192–220.

Nolan, P. (1989) *The Political Economy of Collective Farms*, Blackwell/Polity Press, Oxford.

Nzula, A. T., Potekhin, I. I. & Zusmanovich, A. Z. (1979) *Forced Labour in Colonial Africa*, Zed Press, London (first published in Moscow, 1933).

Oliver, R. & Fage, J. D. (1962) *A Short History of Africa*, Penguin, Harmondsworth.

Open University (1989) *Living with Technology: a foundation course, T102; Block 6: Health*, The Open University, Milton Keynes.

Pacey, A. (1990) *Technology in World Civilization*, Basil Blackwell, Oxford.

Payne, P. R. (1990) 'Measuring malnutrition', *IDS Bulletin*, 21(3).

Pearson, R., Whitehead, A. & Young, K. (1984) 'The continuing subordination of women in the development process', in Young, K. *et al.* (eds), *Of Marriage and the Market: women's subordination in international perspective*, 2nd edn, Routledge, London.

Perham, M. & Simmons, J. (1948) *African Discovery*, Penguin, Harmondsworth.

Pielta, H. & Vickers, J. (1991) *Making Women Matter: the role of the United Nations*, Zed Press, London.

Pittin R. (1990) 'Women, work and ideology in the context of economic crisis: a Nigerian case study', *ISS Working Papers*, Sub Series on Women, History and Development, no.11.

Population Reports (1987) 'Norplant implants', *Population Reports, Series K*, 3, March–April, Population Information Program, Johns Hopkins University, Baltimore, MD.

Potter, D. C. (1986) *India's Political Administrators: 1919–1983*, Clarendon Press, Oxford.

Poulton, R. & Harris, M. (eds) (1988) *Putting People First*, Macmillan, London.

Prince, R. (1989) 'South Africa – music in the shadows', in *Rhythms of the World*, pp.98–107, BBC Books, London.

Pye, L. (1985) *Asian Power and Politics: the cultural dimensions of authority*, Harvard University Press, Cambridge, Mass.

Rakovsky, C. (1981) 'The Five Year Plan in crisis', *Critique*, 13, pp.5–54.

Randall, V. (ed.) (1988) *Political Parties in the Third World*, Sage, London.

Ranger, T. (1985) *Peasant Consciousness and Guerilla War in Zimbabwe*, James Currey, London.

Ravenhill, J. (1990) 'The North–South balance of power', *International Affairs*, 66(4), pp.731–48.

Raza, S. H. (undated) Unpublished Papers, India Office Library, MSS, Eur. F.180/29.

Redclift, M. (1987) *Sustainable Development: exploring the contradictions*, Routledge, London.

Reddy, A. K. N. (1979) *Mazingira: the world forum for environment and development*, no.8.

Riskin, C. (1988) *China's Political Economy*, Oxford University Press, Oxford.

Robbins, A. & Freeman, P. (1988) 'Obstacles to developing vaccines for the Third World', *Scientific American,* November, 259(5), pp.90–5.

Robinson, F. C. R. (1971) 'Consultation and control: the United Provinces' government and its allies, 1860–1909', *Modern Asian Studies*, 5(11), pp.313–36.

Robison, R. (1988) 'Authoritarian states, capital-owning classes, and the politics of newly industrializing countries: the case of Indonesia', *World Politics,* 41(1) (October), pp.52–74.

Rogow, D. (1986) 'Quality care in international family planning: a feminist contribution', in IWHC (eds) *The Contraceptive Development Process and Quality of Care in Reproductive Health Services*, International Women and Health Coalition (with The Population Council), Meeting, 8–9 October, New York.

Rosberg, C. G. & Nottingham, J. (1966) *The Myth of 'Mau Mau': nationalism in Kenya*, East African Publishing House, Nairobi.

Rotberg, R. I. (1983) 'Nutrition and history', in Rotberg & Rabb (eds) *Hunger and History: the impact of changing food production and consumption patterns on society*, Cambridge University Press, Cambridge.

Rungta, R S. (1970) *The Rise of Business Corporations in India, 1851–1900,* Cambridge University Press, Cambridge.

Rustow, D. (1970) 'How does a democracy come into existence?', *Comparative Politics*, 2, pp.37–63.

Ruthven, M. (1984) *Islam in the World*, Penguin, Harmondworth.

Rycroft, D. (1957) 'Zulu male traditional singing', *African Music*, 1(4), pp.33–5.

Rycroft, D. (1959) 'African music in Johannesburg: African and non-African features', *Journal of the International Folk Music Council*, 9, pp.25–30.

Rycroft, D. (1977) 'Evidence of stylistic continuity in Zulu 'town' music', in KPW Festschrift Committee (eds), *Essays for a Humanist: an offering to Klaus Wachsmann*, Town House Press, New York.

Sahlins, M. (1976) *Culture and Practical Reason*, Aldine, Chicago.

Salvadori, C. (1983) *Through Open Doors: a view of Asian cultures in Kenya*, Kenway Publications, Nairobi.

Sanders, D. (with Richard Carver) (1985) *The Struggle for Health: medicine and the politics of underdevelopment*, Macmillan, London.

Sandhu, R. & Sandler, J. (1986) *The Tech and Tools Book: a guide to technologies women are using world-wide*, International Women's Tribune Centre, New York, and Intermediate Technology Publications, London.

Schiff, L.M. (1939) *The Present Condition of India: A Study in Social Relationships*, Quality Press, London.

Seers, D. (1963) 'The limitations of the special case', *Bulletin of the Oxford Institute of Economics and Statistics*, 25(2), May.

Seers, D. (1969; 1979) 'The meaning of development', in Lehmann, D. (ed.) *Development Theory: four critical studies*, Frank Cass, London.

Selden, M. (1971) *The Yenan Way in Revolutionary China*, Harvard University Press, Cambridge, Mass.

Selden, M. (1984) 'The logic – and limits – of Chinese socialist development', in Maxwell, N. & McFarlane, B. (eds) *China's Changed Road to Development*, Pergamon, Oxford.

Sen, A. (1975) *Employment, Technology and Development*, Clarenden Press, Oxford.

Sen, A. (1981) *Poverty and famines: an essay on entitlement and deprivation*, Oxford University Press, Oxford.

Sen, A. (1984) *Resources, Values and Development,* Basil Blackwell, Oxford.

Sen, A. K. (1987) 'Gender and cooperative conflicts', Mimeo, World Institute of Development Economics Research, Helsinki.

Sen, G. (1989) 'Fertility decline and women's autonomy: another look', paper presented at the International Economics Association, Athens, 28 August–1 September.

Shaban, M. A. (1971) *Islamic History: a new interpretation*, Cambridge University Press, Cambridge.

Shivji, I. (1976) *Class Struggles in Tanzania*, Heinemann, London.

Silverblatt, I. (1988) ' 'The Universe has turned inside out. There is no justice for us here.' Andean Women under Spanish Rule', in Etienne, M. & Leacock, E. (eds) *Women and Colonization: anthropological perspectives*, Praeger, New York.

Singer, H. W. (1989) *Lessons of Post–War Development Experience: 1945–1988*, Discussion Paper no.260, April, Institute of Development Studies, Brighton.

Sivan, E. (1985) *Radical Islam*, Yale University Press, London.

Smelser, N. J. (1968) 'Toward a theory of modernization', in Smelser, N. J. (ed.) *Essays in Sociological Explanation*, Prentice-Hall, Englewood Cliffs, N.J.

Snow, R. (1972) *Red Star over China*, Penguin, Harmondsworth.

Soto, H. de (1989) *The Other Path: the invisible revolution in the Third World*, Harper & Row, New York.

South Commission (1990) *The Challenge to the South*, Oxford University Press, Oxford.

Standing, G. (1981) *Labour Force Participation and Development*, International Labour Organization, Geneva.

Stapleton, C. & May, C. (1989) *African All-Stars: the pop music of a continent*, Paladin, London.

Starke, L. (1990) *Signs of Hope*, Oxford University Press, Oxford.

Stein, B. (1977) 'Privileged land-holding', in Frykenberg, R. E. (ed.) *Land Tenure and Peasant in South Asia*, Orient Longman, New Delhi.

Stein, B. (1990) *Vijayanagara*, Cambridge University Press, Cambridge.

Stein, S. J. (1972) 'Brazilian slavery re-examined', in Winks, R. W. (ed.) *Slavery: a comparative perspective*, New York University Press, New York.

Thaiss, G. (1972) 'Religious Symbolism and Social Change: the drama of Husain', in Keddie, N. (ed.) *Scholars, Saints and Sufis,* University of California Press, Berkeley.

Thelwell, M. (1982) 'Looking for work in Kingston', in Johnson, H. & Bernstein, H. (eds) *Third World Lives of Struggle*, Heinemann, London.

Therborn, G. (1978) 'The rule of capital and the rise of democracy', *New Left Review*, 103 (May–June), pp.3–41.

Thuku, H. (1970) *An Autobiography*, Oxford University Press, Nairobi.

Tinker, H. (1974) *A New System of Slavery*, Oxford University Press, Oxford.

Tinker, I. (1979) *New Technologies for Food Chain Activities: the imperative of equity for women*, Office of Women in Development, Agency for International Development, Washington DC.

Toye, J. (1987) *Dilemmas of Development: reflections on the counter-revolution in development theory and policy*, Basil Blackwell. Oxford.

Tracey, A. (1987) 'A word from the editor', *African Music*, 6(4), p.3.

Treece, D. (1988) 'Your enemy is our enemy', *International Labour Reports*, 26, March/April, p.13.

Turton, D. (1988) 'Anthropology and Development', in Leeson, P. & Minogue, M. (eds) *Perspectives on*

Development, Manchester University Press, Manchester.

UNDP (1990) *Human Development Report 1990*, United Nations Development Programme, Oxford University Press, Oxford.

UNFPA (1989a) *The State of World Population 1989*, United Nations Population Fund, New York.

UNFPA (1989b) *1988 Report*, United Nations Population Fund, New York.

UNFPA (1990a) *Investing in Women: the focus of the '90s*, United Nations Population Fund, New York.

UNFPA (1990b) *Population Issues: a briefing kit,* United Nations Population Fund, New York.

UNICEF (1988) *Asian and Pacific Atlas of Children in National Development*, UNICEF East Asia and Pakistan Regional Office with the co-operation of the United Nations Economic and Social Commission for Asia and the Pacific (ESCAP).

UNICEF (1990) *State of the World's Children*, Oxford University Press, Oxford.

Visaria, P. (1990) *Concepts and Measurement of Unemployment and Underemployment in ESCAP Countries: a comparative study*, Working Paper No.31, The Gujarat Institute of Area Planning, Ahmedabad.

Walby, S. (1990) *Theorizing Patriarchy*, Blackwell, Oxford.

Wallerstein, I. (1989) 'The Myrdal Legacy: racism and underdevelopment as dilemmas', *Cooperation and Conflict*, 24, pp.1–18.

Wallis, R. & Malm, K. (1984) *Big Sounds from Small Peoples: the music industry in small countries*, Constable, London.

Wallraff, G. (1985) *The Lowest of the Low*, Methuen, London.

Waring, M. (1989) *If Women Counted: a new feminist economics*, Macmillan, London.

Warnock, J. W. (1988) The politics of hunger, Methuen, Toronto.

Warren, B. (1980) *Imperialism: pioneer of capitalism*, Verso, London.

Washbrook, D. A. (1981) 'Law, state and agrarian society in colonial India', *Modern Asian Studies*, 15(3), pp.649–721.

WCED (1987) *Our Common Future* (The Brundtland Report), Oxford University Press, Oxford.

Weiner, M. (1965) 'India: two political cultures', in Pye, L. & Verba, S. (eds) *Political Culture and Political Development*, Princeton University Press, Princeton.

Whitcombe, E. (1972) *Agrarian Conditions in Northern India; vol.1: The United Provinces under British Rule, 1860–1900,* University of California Press, Berkeley.

Whitehead, A. (1985) 'Effects of technological change on rural women: a review on analysis and concepts', in Ahmed, I. (ed.) *Technology and Rural Women*, Allen and Unwin, London.

Whitehead, A. (1990) 'Food crisis and gender conflict in the African countryside', in Bernstein, H., Crow, B., Mackintosh, M. & Martin, C. (eds) *The Food Question: profits versus people*, Earthscan, London.

WHO (1989) *World Health Statistics Annual*, World Health Organization, Geneva.

Willetts, P. (1978) *The Non-aligned Movement: the origins of a Third World alliance*, Francis Pinter, London.

Wink, A. (1986) *Land and Sovereignty in India: agrarian society and politics under the eighteenth century Maratha Svarajya*, Cambridge University Press, Cambridge.

Winner, L. (1977) *Autonomous Technology: technics-out-of-control as a theme in political thought*, Massachusetts Institute of Technology Press, Cambridge, Mass., and London.

Wolf, E. (1982) *Europe and the People without History*, University of California Press, Berkeley.

World Bank (1980) *World Development Report 1980*, Oxford University Press, Oxford.

World Bank (1983) *The Effect of Piped Water on Early Childhood Mortality in Urban Brazil, 1970–76*, Thomas Merrick, World Bank, Washington DC.

World Bank (1984) *World Development Report 1984*, Oxford University Press, Oxford.

World Bank (1985) *Quantitative Studies of Mortality Decline in the Developing World,* by Julie DaVanzo, Jean-Pierre Habicht, Ken Hill & Samuel Preston, World Bank, Washington DC.

World Bank (1986) *Poverty and Hunger*, World Bank, Washington DC.

World Bank (1987) *Sri Lanka and the World Bank: a review of a relationship,* World Bank, Washington DC.

World Bank (1990) *World Development Report 1990*, Oxford University Press, Oxford.

Worsley, P. (1964; 1967) *The Third World*, Weidenfeld and Nicholson, London.

Worsley, P. (1979) 'How many worlds?', *Third World Quarterly*, 1(2), pp.100–8.

Worsley, P. (1984) *The Three Worlds*, Weidenfeld and Nicholson, London.

Wuyts, M., Mackintosh, M. & Hewitt, T. (eds) (1992) *Development Policy and Public Action*, Oxford University Press/The Open University, Oxford (Book 4 of this series).

Young, K. (1989) *Women and Economic Development: local, regional and national planning strategies*, Berg/UNESCO.

Acknowledgements

Grateful acknowledgement is made to the following sources for permission to reproduce material in this book:

Text

Chapter 2, pp.34–5: Ferriman, A., 'The "silent genocide" of millions of children', in *The Observer*, 1 October 1989, copyright *The Observer*, London, 1989; *Chapter 3, pp.65–6:* Extract 'Domitila's day' from Bernstein, H. & Johnson, H. (1988) *Third World Lives of Struggle*, Heinneman; *Box 3.1:* Extract from Tickell, S. 'Domitila – the forgotten activist', *New Internationalist*, October 1989, reproduced by courtesy of *New Internationalist*; *Box 4.1:* Extract from Hartmann, B. (1987) *Reproductive Rights and Wrongs*, copyright © 1987 by Betsy Hartmann, reprinted by permission of HarperCollins Publishers Inc.; *Box 5.2:* Myers, N. 'You can't see the future for dust', *The Guardian*, 8 January 1988; *Box 5.3:* Huband, M. 'Ivorian cocoa, the cash crop nobody wants to buy', *Financial Times*, 22 November 1989, copyright © 1989 *Financial Times*; *Chapters 8 and 9:* Extracts from Crow, B. & Thorpe, M. (eds) (1988) *Survival and Change in the Third World*, Polity Press, reproduced by permission of Basil Blackwell Ltd; *Box 11.3:* UNDP (1990) *Human Development Report 1990*, Oxford University Press; *Box 15.1:* Extracted and adapted by permission from Moghadam, V. (1991) *Women's Employment in the Middle East and North Africa: gender class and state policies*, World Institute for Economic and Development Research of the United Nations University; *Box 15.2:* Extracted from Joss, S. (1990) 'Your husband is your god', GADU Pack no.12, Oxfam; *Box 16.1:* Jenkins, G. 'Cuba: a picture of health', *Financial Times*, 17 February 1989, copyright © Gareth Jenkins (who we were unable to contact); *Box 16.4:* Extract from Sandhu, R. & Sandler, J. (1986) *The Tech and Tools Book: a guide to technologies women are using world-wide*, Intermediate Publications.

Tables

Table 1.1: Sen, A. (1981) *Poverty and Families: an essay on entitlement and desperation*, Clarendon Press, reproduced by permission of Oxford University Press; *Table 13.2:* Hocking, B. & Smith, M. (1990) *World Politics*, Harvester Wheatsheaf; *Table 13.3:* Taylor, P. 'Regionalism: the thought and deed', in Groom, A. J. R. & Taylor, P. (eds) (1990) *Frameworks for International Co-operation*, Pinter Publishers Ltd; *Table 15.3:* adapted from Joss, S. (1990) 'Your husband is your god', GADU Pack no.12, Oxfam; *Table 15.4:* Momsen, J. (1991) *Women and Development in the Third World*, Routledge.

Diagrams

Figure in Box 1.3: from Martorell, R. (1989) 'Body size, adaption and function', in *Human Organization*, 48(1), The Society for Applied Anthropology; *Figure 2.8:* from Ebraham, G. J. & Ranken, J. P. (1988) *Primary Health Care*, Macmillan Education Ltd; *Figure 4.1:* from UNDP (1990) *Human Development Report 1990*, Oxford University Press; *Figure in Box 4.3:* adapted from The World Bank (1980) *World Development Report 1980*, Oxford University Press, *Figure 4.3:* adapted from a figure by Kiss, G. in Crossen, P. K. & Rosenberg, J. 'Strategies for agriculture', *Scientific American*, September 1989; *Figures 4.5 bottom and 8.3:* adapted from Sanders, D. (1985) *The Struggle for Health*, Macmillan Education Ltd; *Figure 5.2:* from ODI (March 1988) *Commodity Prices: investing in decline?* Overseas Development Institute; *Figures 11.5, 11.8 and in Box 11.5:* from The South Commission (1990) *Challenge to the South*, Oxford University Press.

Photographs and cartoons

Title pages: Ann Dalrymple Smith/Oxfam; Malcolm Harper/Oxfam; Mark Edwards/Still Pictures; *Introduction Figure 1:* Tim Allen; *Figure 1.1:* cartoon by R. K. Laxman from *The Times of India*; *Figures 1.2, 1.3, 1.5:* Mark Edwards/Still Pictures; *Figure 1.4:* Mike Goldwater/Network; *Figure 1.6:* reproduced by courtesy of *New Internationalist*; *Figures 2.1, 2.6, 2.7:* John and Penny Hubley; *Figure 2.2:* Julio Etchart; *Figure 2.10:* Oxfam; *Figure 3.1 left:* Roblaw Publishers; *Figure 3.1 right:* Maggie Murray/Format;

Figure 3.2: John and Penny Hubley; *Figure 3.3 top: The Times of India; Figure 3.3 bottom:* Tom Hewitt; *Figure 3.5 left:* Sally and Richard Greenhill; *Figure 3.5 right:* Maggie Murray/ Format; *Figure 3.6 (both):* Tom Hanley; *Figure 3.8:* Ben Crow; *Figure 4.2:* Ines Smyth; *Figure 4.4:* Oxfam; *Figure 4.5 top:* Tom Hewitt/Ines Smyth; *Figure 4.6:* Tom Hewitt; *Figure 4.7 left:* Jenny Matthews; *Figure 4.7 right:* Julio Etchart; *Figure 4.8 left:* International Planned Parenthood Federation; *Figure 4.8 top right:* Maggie Murray/ Format; *Figure 4.8 bottom right:* Sally and Richard Greenhill; *Figure in Box 5.1:* copyright © Caroline Austin/The Environmental Picture Library; *Figure 5.1 left:* Manchester City Art Galleries; *Figure 5.1 right:* Sally and Richard Greenhill; *Figure 5.3:* H. Girardet/The Environmental Picture Library; *Figure in Box 5.3: Financial Times; Figures in Box 5.5:* Nigel Dickinson, Leader Photos; *Figure 5.4 top:* Mark Edwards/Still Pictures; *Figure 5.4 top right:* Hutchison Library; *Figure 5.4 bottom right:* Popperfoto; *Figure 5.5:* cartoon by Bruce Petty, Australia, from Regan, C., Sinclair, S. & Turner, M. (1988) *Thin Black Lines*, Development Education Centre, Birmingham; *Figures 6.1. 6.9:* cartoons by R. K. Laxman, India, from Regan, C., Sinclair, S. & Turner, M. (1988) *Thin Black Lines*, Development Education Centre, Birmingham; *Figure 6.2:* cartoon by Rachid Ait-Kaci, Morocco, from Regan, C., Sinclair, S. & Turner, M. (1988) *Thin Black Lines*, Development Education Centre, Birmingham; *Figure 6.3:* from Regan, C., Sinclair, S. & Turner, M. (1988) *Thin Black Lines*, Development Education Centre, Birmingham; *Figure 6.4:* PANA/Fotomedia; *Figures 6.5, 6.7:* Sharma Studio, New Delhi; *Figure 6.6:* Professor Dr Muhammad Saleem Ahmad; *Figure 6.8:* Tiofoto, Stockholm, photo by Gun Kessle; *Figure 7.1:* reproduced from Roberts, J. (1985) *The Triumph of the West*, BBC Publications, London; *Figure 7.2:* Victoria and Albert Museum, London, photo by Ikon; *Figure 7.3:* copyright © Carlos Pasini, Disappearing World/Alan Hutchison Library; *Figure 7.4:* The Hulton-Deutsch Collection, London; *Figure 7.5:* Popperfoto; *Figures 7.6, 7.7:* Crown Copyright, Victoria and Albert Museum, London; *Figure 7.8:* Mary Evans Picture Library; *Figure 8.2:* The Hulton-Deutsch Collection, London; *Figure 8.4:* National Museum of Denmark, Copenhagen; *Figure 8.5:* reproduced from *Judy*, 28 January 1885; *Figure 8.6 left:* Mary Evans Picture Library; *Figure 8.6 right:* reproduced from *Punch*, 25 July 1900; *Figure 8.7:* cartoon by R. Cobb, from Regan, C., Sinclair, S. & Turner, M. (1988) *Thin Black Lines*, Development Education Centre, Birmingham; *Figure 8.8:* The Hulton-Deutsch Collection, London; *Figures 9.1, 9.2:* Mansell Collection; *Figure 9.3:* cartoon by Murray Ball, New Zealand, from Regan, C., Sinclair, S. & Turner, M. (1988) *Thin Black Lines*, Development Education Centre, Birmingham; *Figure 10.2:* India Office Library, photo by R. B. Fleming; *Figure 10.3:* reproduced from *Punch*, 12 September 1857; *Figure 10.4:* Victoria and Albert Museum; *Figure 10.5:* Mary Evans Picture Library; *Figure 10.6:* Roger-Viollet; *Figure 10.7:* Camera Press; *Figure 11.1:* International Bank for Reconstruction and Development/Wide World Photos; *Figure 11.2:* Alan Hutchison Library; *Figure 11.4:* Oxfam; *Figure 11.7:* cartoon reproduced by courtesy of *New Internationalist*; *Figure 11.9:* cartoon by Wasserman, USA, from Regan, C., Sinclair, S. & Turner, M. (1988) *Thin Black Lines*, Development Education Centre, Birmingham; *Figure 12.1:* Lenin Library, courtesy of John Calmann and King Ltd; *Figure 12.2 left:* Sovfoto; *Figure 12.2 right:* David King Collection; *Figure 12.3:* Sally and Richard Greenhill; *Figure 13.1:* Associated Press Photo; *Figure 13.2:* cartoon by courtesy of Centre for World Development Education; *Figure 13.3 main picture: Financial Times*, photo by Alan Harper; *Figure 13.3 inset: The Independent*, photo by Brian Harris; *Figure 13.4 left:* Mark Edwards/ Still Pictures; *Figure 13.4 right:* Poppperfoto; *Figure 13.5:* cartoon by Simpson, USA, from Regan, C., Sinclair, S. & Turner, M. (1988) *Thin Black Lines*, Development Education Centre, Birmingham; *Figure 14.1:* Popperfoto; *Figure 14.2:* Dean Press Images; *Figure 14.4:* Popperfoto/Reuter; *Figure 14.5: The Hindu; Figure 14.6:* Victoria and Albert Museum, London; *Figure 14.7:* reproduced from *The Graphic*, 19 November 1887; *Figure 15.1:* Tom Hanley; *Figure 15.2 left:* Mark Edwards/Still Pictures; *Figure 15.2 right:* Sue Darlow/Format;

Figure 15.3: Oxfam; *Figure 15.4:* Val Wilmer/ Format; *Figure 15.5:* Ro Cole/Oxfam; *Figure 15.6:* Karen Iles/Oxfam; *Figure 16.1:* copyright © David Beatty/Susan Griggs Agency; *Figure 16.2 (both):* Julio Etchart; *Figures 16.3 top, 16.4:* David Spark; *Figure 16.3 bottom:* CWDE/World Bank/R. Witlin; *Figure 16.5 top:* Fabian Acker, *Electrical Review*, March 1982, p.30; *Figure 16.5 bottom:* Margaret Murray/Format; *Figure 16.6:* G. G. Moreno/Oxfam; *Figure 16.7:* John and Penny Hubley; *Figure 17.1:* Abbas/Magnum; *Figure 17.2:* Popperfoto; *Figure 17.3:* Jeremy Hartley/Oxfam; *Figure 18.1:* Kenya Railways; *Figure 18.2:* Maggie Murray/Format; *Figure 18.3:* Popperfoto; *Figure 19.1:* Maggie Murray/Format; *Figures 19.2, 19.4, 19.5, 19.6:* International Defence and Aid Fund; *Figures 19.3, 19.8:* Jak Kilby; *Figure 19.7:* Erlmann, V. (1990) 'Migration and performance: Zulu migrant workers' Isicathamkiya performance in South Africa, 1890–1950', *Ethnomusicology*, 34(2) Spring/ Summer, pp.199–220; *Figure 19.9:* copyright © SIPA-Press, Rex Features; *Figure 20.1:* Tom Hanley; *Figure 20.2:* cartoon reproduced by courtesy of *South*, the Third World Magazine, London; *Figure 20.3:* cartoon by Murray Bell, reproduced by courtesy of *New Internationalist*; *Figure 20.4:* David King Collection.

Authors' acknowledgements

Andrew Kilmister is grateful to Peter Worsley for comments on earlier drafts of his chapter, to Alex Callinicos and Fred Halliday for their writings, and especially to Chris Corrin, Jeremy Krikler, Matthew Lockwood and Jon Lunn for their discussions.

Richard Middleton thanks the following for advice, information, criticisms and recordings: Christopher Ballantine, David Coplan, Lucy Duran of the National Sound Archive, London, Veit Erlmann, Margaret Ling, Linda Nevill of Kaz Records, David Rycroft, Dennis Walder, and Fred Zindi.

List of acronyms, abbreviations and organizations

ACP	African, Caribbean and Pacific states
AICs	advanced industrial countries
AIDS	auto-immune deficiency syndrome
ANC	African National Congress (South Africa)
ASEAN	Association of South East Asian Nations
BBC	British Broadcasting Corporation
BMG	Business Machines Group (Burroughs Corporation)
CARICOM	Caribbean Common Market
CAT	computerized axial tomography (scanner)
CBS	Colombia Broadcasting System Inc.
CFCs	chlorofluorocarbons
COMECON	economic association of communist countries
CSCE	Commission on Security and Co-operation in Europe
DFI	direct foreign investment
EC	European Community
EACM	East African Common Market
ECLA	Economic Commission for Latin America (United Nations)
EMI	Electrical and Musical Industries
EOI	export oriented industrialization
ESCAP	Economic and Social Commission for Asia and the Pacific (United Nations)
FAO	Food and Agriculture Organization (United Nations)
G7	Group of Seven (leading industrialized countries)
G77	Group of 77 (economic grouping of 'Third World' countries)
GATT	General Agreement on Trade and Tariffs
GDP	gross domestic product
GNP	gross national product
GSP	Generalized System of Preferences
HCFCs	hydrochlorofluorocarbons
HFAs	hydrofluoroalkanes
IBM	International Business Machines
IBRD	International Bank for Reconstruction and Development (now more usually known as the World Bank)
IDS	Institute of Development Studies, Brighton, UK
IFPI	International Federation of Phonogram and Videogram Producers, London
IGOs	intergovernmental organizations
ILO	International Labour Office (United Nations)
IMF	International Monetary Fund
INGOs	international non-governmental organizations
ISI	import substitution industrialization
ITT	International Telephone and Telegraph Corporation
ITTO	International Tropical Timber Organization
IWHC	International Women and Health Coalition
LDCs	less developed countries
LLDCs	least developed countries
MP	Member of Parliament (UK)
NAM	non-aligned movement
NEP	New Economic Policy (USSR, 1920s)
NGOs	non-governmental organizations
NICs	newly industrializing countries

NIDL	new international division of labour
NIEO	New International Economic Order
NIIO	New International Information Order
OAU	Organization of African Unity
OECD	Organization for Economic Co-operation and Development
OPEC	Organization of Petroleum Exporting Countries
PLO	Palestine Liberation Organization
PRI	Partido Revolucionario Institucional (Mexico)
PT	Partido Trabalhista (Brazil)
SALs	Structural Adjustment Loans (World Bank)
TFR	total fertility rate
TNC	transnational corporation
U5MR	under-five mortality rate
UK	United Kingdom
UN	United Nations
UNCTAD	United Nations Conference on Trade and Development
UNDP	United Nations Development Programme
UNEP	United Nations Environmental Programme
UNESCO	United Nations Educational, Scientific and Cultural Organization
UNFPA	United Nations Population Fund
UNICEF	United Nations (International) Children's (Emergency) Fund
UNIDO	United Nations Industrial Development Organization
UNWRA	United Nations Works and Relief Agency (for Palestinian refugees)
USA	United States of America
US-AID	United States Agency for International Development
USSR	Union of Soviet Socialist Republics
WCED	World Commission on Environment and Development (United Nations)
ICS	Indian Civil Service
WEA	Warner-Eddison Associates Inc.
WHO	World Health Organization (United Nations)

Index

J

K

L

Index compiled by Frank Pert